P9-DCR-556

VEGETABLE GROWING HANDBOOK
THIRD EDITION

VEGETABLE GROWING HANDBOOK
Organic and Traditional Methods

THIRD EDITION

Walter E. Splittstoesser, Ph.D.
Professor of Plant Physiology
in Horticulture
University of Illinois
Urbana, Illinois

An avi Book
Published by Van Nostrand Reinhold
New York

An AVI Book
(AVI is an inprint of Van Nostrand Reinhold)

Library of Congress Catalog Card Number 89-28506
ISBN 0-442-23971-8

Printed in the United States of America

Van Nostrand Reinhold
115 Fifth Avenue
New York, New York 10003

Van Nostrand Reinhold International Company Limited
11 New Fetter Lane
London EC4P 4EE, England

Van Nostrand Reinhold
480 La Trobe Street
Melbourne, Victoria 3000, Australia

Nelson Canada
1120 Birchmount Road
Scarborough, Ontario M1K 5G4, Canada

16 15 14 13 12 11 10 9 8 7 6 5 4 3 2 1

Library of Congress Cataloging-in-Publication Data
Splittstoesser, Walter E.
Vegetable growing handbook: Organic and Traditional Methods/Walter E. Splittstoesser.—3rd ed.
p. cm.
Includes bibliographical references.
ISBN 0-442-23971-8
1. Vegetable gardening. I. Title.
SB321.S645 1990
635—dc20 89-28506
CIP

Contents

Preface

The ancient art of vegetable gardening involves judicious timing and many skills. The purpose of The *Vegetable Growing Handbook* is to present the scientific concepts upon which vegetable growing is based and to give practical methods for growing herbs and vegetables. They key to a successful garden is the understanding and utilization of these basic principles. Knowledge and understanding of planning, planting, and producing vegetables will reward the gardener with an abundant supply of quality vegetables. *Vegetable Growing Handbook* has been designed for use in high schools, in vocational and community colleges, and for use at an introductory college level. It will serve as a ready reference for county agents or farm advisors and for the astute and informed gardener. It is easily used by those who have had no special courses in agriculture.

Vegetables can be grown with the use of man-made products or organically. Both techniques are given equal treatment, and a legal definition of organic vegetables is included. The text covers all of the latest biological, chemical, mechanical, and organic methods of cultivation, including integrated pest management. Effective techniques for insect and disease control are presented, together with illustrations of beneficial and harmful insects for easy identification.

There is a chapter on physiological events, such as seedstalk formation, which describes why these events occur and how to prevent them. There is also detailed coverage of the essential plant elements, outlining how these can be obtained from the use of chemical and organic sources and fertilizers. A simple hydroponic system for gardeners has been included.

The *Third Edition* of the book accommodates many of the changes that have occurred in vegetable gardening in recent years. More gardeners are utilizing their gardens for a longer growing season; and a section on plant protectors, plastic and organic mulches, composts, and vegetable covers has been added.

A chapter on the nutritional quality and value of vegetables has been added. The contribution of vegetable fiber, fats, vitamins, minerals, and calories to the diet has been included, as well as the amino acid levels found in vegetables, so that vegetarians can determine the correct vegetable mixtures to insure that the proper ratio of essential amino acids required by humans is eaten. The presence of natural toxicants (some are anticancer agents) and food additives for vegetables are given, as well as the laws which govern their consumption.

A new chapter on special-use vegetables gives detailed information on growing them and their use. For many gardeners, these vegetables are an essential part of the diet. Most of these vegetables are now available in the produce department of supermarkets. The special-use vegetables chapter will also allow consumers to recognize these vegetables and how they are prepared for use.

There has been an increased emphasis placed upon salt-free and caffeine-free foods and beverages. The section on herbs has been greatly expanded to include herbs that are used as herbal teas, and for the seasoning of foods. Many of the common herbs are now readily available as cultivars or varieties with new flavors and fragrances, and the growing and use of these herbs are included. Many herbs are misnamed, so the botanical names are given to insure that the gardener can select the desired herb.

The harvesting and storage of vegetables and herbs are discussed. Detailed information is given on growing over 90 different vegetables, as well as the common herbs and many not-so-common herbs.

This book is an outgrowth of lectures, along with greenhouse, container-grown, indoor gardening, and field procedures, used in the author's courses on "Organic and Traditional Vegetable Gardening" and "Home Vegetable Gardening." Publications from state and provincial agricultural experiment stations and the United States Department of Agriculture have been consulted to ensure that the book can be used in all parts of the USA and Canada.

The author is indebted to agricultural scientists everywhere for providing the research upon which this book is based; to Pamela Splittstoesser and Nancy Slaton for typing; to Shirley, Sheryl, and Riley for critical reviews; and to Eleanor Riemer of Van Nostrand Reinhold for encouragement and assistance in bringing the *Third Edition* of this book into being.

WALTER E. SPLITTSTOESSER

VEGETABLE GROWING HANDBOOK
THIRD EDITION

1

Planning the Garden

Interest in gardening is at its highest level since World War II (Blackwell 1977). The U.S. Department of Agriculture reported that nearly half of the households surveyed either had a garden or intended to have one. People cited three major reasons for having a garden: (1) a desire for fresh vegetables, (2) as a hobby, and (3) a desire to save money and reduce the amount spent for food.

Gardeners and consumers are interested in nutrition and good health, and they wish to avoid any toxic residues that might be within the vegetable. Presently, a major reason for growing our own vegetables is to insure that there is no pesticide residue in the produce.

Many purchased fresh vegetables do not have the taste and flavor of home-grown ones. Asparagus and sweet corn are good examples. Sweet corn loses sugar rapidly upon harvest. Only a home gardener can have the water boiling before harvesting the sweet corn so it can be eaten a few minutes after picking (Gomez 1974).

The average-size vegetable garden in the United States is about 600 sq ft in size and in 1981 brought a net return of $600, tax-free (National Garden Bureau, Inc.). Just how profitable the garden is depends upon (1) the tools and equipment used and their cost; (2) the length of the growing season in your area; (3) the choice of vegetables grown, as some are more space-efficient than others or command a higher price; and (4) your skill in using the area from which vegetables have been harvested to grow additional or different vegetables in succession plantings. In Columbus, Ohio, for example, a garden of 150 sq ft produced enough vegetables to provide a return for labor of $1.08 per hr; and this value was calculated after all expenses, including depreciation on the garden tools, were deducted (Utzinger and Connolly 1978). Few leisure-time activities pay you for doing them.

Some vegetables are relatively space-efficient and return a higher monetary value for the space they occupy than other vegetables. Some vegetables produce little, but the produce represents a high cost per pound. Other vegetables may have a high yield with a low cost per pound. Table 1.1 gives a value rating developed by the National Garden Bureau, which was based upon a survey of many top home garden experts. The vegetables are rated from 1 to 10, and consider total yield per square foot, price per pound harvested, and the time required between seeding and harvest. This rating system is somewhat arbitrary, but if the value of the vegetable is low and you have limited space, you should consider growing a vegetable with a higher numerical rating.

If you are a beginning gardener, be prepared for some problems. Weeds and insects will invade the garden. Weather often does not cooperate and problems with rainfall, drought, early or late frosts, or too much sunlight may occur. Gardening requires work. Do not expect to plant the garden and three months later to return for a bountiful harvest. Certain gardening jobs will need to be done at certain times, which may interfere with one's hobbies. If you spend little time at home, limit the size of your garden.

This book is intended to answer the questions of beginning and experienced gardeners alike. It provides the information required to be a successful gardener. This book gives standard "chemical methods" and gives equal treatment to "organic methods." The methods used are the gardener's decision. However, regardless of the method you decide upon, the specific techniques given herein are those that have been researched and that have proven effective.

ORGANIC GARDENING DEFINED

Definitions of organic gardening vary depending on the people using them. The USDA uses the following definition (Anon. 1980):

> "Organic farming is a production system which avoids or largely excludes the use of synthetically compounded fertilizers, pesticides, growth regulators, and livestock feed additives. To the maximum extent feasible, organic farming systems rely upon crop rotations, crop residues, animal manures, legumes, green manures, off-farm organic wastes, mechanical

TABLE 1.1. The Approximate Economic Value of Growing Individual Vegetables in the home garden.

Vegetable	Value	Vegetable	Value
Tomato	9.0	Beans, lima or pole	6.1
Onion, green	8.2	Radish	6.1
Lettuce, leaf	7.4	Cabbage	6.0
Turnip	7.4	Leek	5.9
Squash, summer	7.2	Collard	5.8
Pea, edible podded	6.9	Okra	5.7
Onion, bulb	6.9	Kale	5.6
Bean, pole	6.8	Cauliflower	5.3
Beet	6.6	Eggplant	5.3
Bean, bush	6.5	Pea, English	5.2
Carrot	6.5	Brussels sprouts	4.3
Pepper, bell	6.4	Celery	4.3
Broccoli	6.3	Corn, sweet	4.1
Kohlrabi	6.3	Squash, winter	3.8
Chard, Swiss	6.3	Muskmelon	3.8
Mustard	6.2	Watermelon	3.8
Spinach	6.2	Pumpkin	1.9

cultivation, mineral-bearing rocks, and aspects of biological pest control to maintain soil productivity and tilth, to supply plant nutrients, and to control insects, weeds, and other pests."

Several states have passed laws regulating the use of the term "organic"; and a formal definition does not resolve the debate. The law passed in (Anon. 1979) California (which is similar to those in Maine and Oregon) will be used as an example of a legal definition.

The California Organic Foods Act suggests that the word organic applies to food which is "naturally grown," "wild," "ecologically grown," or "biologically grown," as well as that which is "organic" or "organically grown."

According to the California law, foods bearing the above labels must meet the following requirements:

(1) "Are produced, harvested, distributed, stored, processed, and packaged without application of synthetically compounded fertilizers, pesticides, or growth regulators.
(2) Additionally, in the case of perennial crops, no synthetically compounded fertilizers, pesticides, or growth regulators shall be applied to the field or area in which the commodity is grown for 36 months prior to the appearance of flower buds and throughout the entire growing and harvest season of the particular commodity.
(3) Additionally, in the case of annual crops and 2-year crops, no synthetically compounded fertilizers, pesticides, or growth regulators shall be applied to the field or area in which the commodity is grown for 12 months prior to seed planting or transplanting and throughout the entire growing and harvest season for the particular commodity."

The law stipulates the following concerning pesticide and fertilizer use:

"Only microorganisms, microbiological products, and materials consisting of, or derived or extracted solely from plant, animal, or mineral-bearing rock substances, may be applied in the production, storing, processing, harvesting, or packaging of raw agricultural commodities, other than seeds for planting, in order to meet the requirements of this subdivision. However, before harvest, the application of Bordeaux mixes and trace elements, soluble kelp, lime, sulfur, gypsum, dormant oils, summer oils, fish emulsion, and soap are permitted, except the application of aromatic petroleum solvents, diesel, and other petroleum fractions, used as weed or carrot oils, are prohibited."

The law further specifies:

That its passage neither denies nor confirms the notion that organic foods are in any way superior to conventionally produced food, and

That foods with pesticide residues "in excess of 10 percent of the level regarded as safe by the Federal Food and Drug Administration" may not be labeled as organically grown.

CULTIVARS AND VARIETIES

Vegetables are unique in that tremendously diverse products and plants are listed under this one general category. Virtually every part of the plant is eaten as some horticultural product. The members of the same botanical genus may be grown for different plant parts. Although some vegetables may look quite different from each other, they may have the same genus and species name and just be a horticultural selection.

Botanical Classification

Garden vegetables are classified botanically, first into the Plant Kingdom and next into the Spermatophyta division. Plants in this division include most of our cultivated crops. Within this division is the class Angiospermae (angiosperms), flowering plants that have ovules and seeds enclosed within an ovary, borne in a flower. The angiosperms are divided into two subclasses, the Monocotyledoneae (monocotyledons) and the Dicotyledoneae (dicotyledons). These two subclasses are divided into orders and then families. Plants in the same family share certain flower, fruit, and often leaf characteristics but differ enough among themselves in some characteristics that they can be recognized and set apart from each other in additional subunits called genera (genus). The genus in turn can be further subdivided into one or more species. The species is the main horticultural group. Plants within a species interbreed freely and resemble each other more than they do other plants. Further subdivisions within a species are sometimes used. The main horticultural subdivision is the variety (var.). Plants within a variety have minor but consistent differences such as those among sweet corn vs. popcorn vs. field corn. Within a variety, such as sweet corn, plant breeders have developed a number of plants that are distinct from each other, such as a white sweet corn ("Silver Queen") or a yellow sweet corn ("Golden Cross Bantam"), which are cultivated varieties or cultivars. Most of the vegetable varieties used are, in fact, cultivars (cv). The term "cultivar" has been adopted by the International Code of Nomenclature for Cultivated Plants and is now used in place of the term "variety." The term cultivar is widely used in Europe and the United States and is used throughout the remaining sections. The identity of a specific type of white sweet corn would be listed as: *Zea mays* var. rugosa cv Silver Queen. When a species or an interspecific hybrid includes many cultivars, those plants which look similar to one another may be arranged in groups. Thus, cultivars of *Brassica oleracea* may be assigned to the Acephala group (collards, kale), the Botrytis group (broccoli, cauliflower), the Gemmifera group (Brussels sprouts), the Capitata group (cabbage), or others. A specific green cabbage cultivar would be identified as: *Brassica oleracea* (Capitata group) cv Stonehead. Table 1.2 lists the botanical classification of many garden vegetables.

TABLE 1.2. Botanical Classification of Some Vegetable Crops.

Family, Genus, Species	Common Name
Monocotyledons	
Amaryllidacea (amaryllis family)	
Allium ampeloprasum, Porrum group	Leek
Allium cepa, Aggregatum group	Shallot, potato onion, multiplier onion
Allium cepa, Cepa group	Onion
Allium cepa, Proliferum group	Egyptian onion, top onion
Allium ampeloprasum, Ampeloprasum Group	Elephant garlic
Allium fistulosum	Welsh onion, Spanish onion, Japanese bunching onion
Allium sativum	Garlic
Allium schoenoprasum	Chive
Amaranthaceae (amaranth family)	
Amaranthus spp.	Amaranth
Araceae (arum family)	
Colocasia esculenta	Taro or dasheen
Gramineae (grass family)	
Zea mays var. praecox	Popcorn
Zea mays var. rugosa	Sweet corn
Liliaceae (lily family)	
Asparagus officinalis	Asparagus
Hemerocalis spp.	Daylily
Dicotyledons	
Cactaceae (cactus family)	
Opuntia spp.	Prickly pear cactus
Nopalea spp.	Prickly pear cactus
Chenopodiaceae (goosefoot family)	
Beta vulgaris, Cicla group	Swiss chard
Beta vulgaris, Crassa group	Beet
Spinacia oleracea	Spinach
Compositae (composite family)	
Arctium lappa	Burdock
Chrysanthemum coronarium	Garland chrysanthemum
Cichorium endivia	Endive
Cichorium intybus	Witloof chicory
Cynara scolymus	Globe artichoke
Helianthus annuus	Sunflower
Helianthus tuberosus	Jerusalem artichoke
Lactuca sativa	Lettuce
Taraxacum officinale	Dandelion
Tragopogon porrifolius	Salsify
Convolvulaceae (morning-glory family)	
Ipomoea batatus	Sweet potato
Cruciferae (mustard family)	
Amoracia rusticana	Horseradish
Barbarea verna	Upland cress
Brassica juncea	Mustard greens
Brassica napus, Napobrassica group	Rutabaga
Brassica oleracea, Acephala group	Kale, collards
Brassica oleracea, Botrytis group	Cauliflower
Brassica oleracea, Capitata group	Cabbage
Brassica oleracea, Gemmifera group	Brussels sprouts
Brassica oleracea, Gongylodes group	Kohlrabi
Brassica oleracea, Italica group	Sprouting broccoli
Brassica perviridis	Mustard spinach
Brassica rapa, Chinensis group	Pak-choi
Brassica rapa, Pekinensis group	Pe-tsai

(continued)

TABLE 1.2. (*Continued*).

Family, Genus, Species	Common Name
Brassica rapa, Rapifera group	Turnip
Lepidium sativum	Garden cress
Nasturitium officinale	Watercress
Raphanus sativus	Radish
Raphanus sativus, Longipinnatus group	Winter radish or daikon
Cucurbitaceae (gourd family)	
Benincasa hispida	Wax gourd or winter melon
Citrullus lanatus	Watermelon
Cucumis anguria	West Indian gherkin
Cucumis melo, Chito group	Vine peach, lemon cucumber, melon apple
Cucumis melo, Flexuosus group	Chinese, Armenian, or snake cucumber
Cucumis melo, Inodorus group	Honeydew or casaba melon
Cucumis melo, Reticulatus group	Muskmelon, Persian melon
Cucumis sativus	Cucumber
Cucurbita maxima	Winter squash, pumpkin, gourds
Cucurbita mixta	Pumpkin
Cucurbita moschata	Winter squash, pumpkin
Cucurbita pepo var. melo pepo	Bush summer squash
Cucurbita pepo var. pepo	Winter squash, pumpkin, gourds
Momordica balsamina	Balsam apple
Momordica charantia	Bitter melon, bitter cucumber
Sechium edule	Chayote
Leguminosae (pea family)	
Arachis hypogaea	Peanut
Glycine max	Soybean
Pachyrhizus erosus	Jicama
Phaseolus coccineus	Scarlet runner bean
Phaseolus limensis	Lima bean
Phaseolus limensis var. limenanus	Bush lima bean
Phaseolus lunatus	Butter bean
Phaseolus vulgaris	Common and kidney bean
Pisum sativum var. macrocarpon	Edible-podded pea
Pisum sativum var. sativum	Garden pea
Vicia faba	Broadbean
Vigna radiata	Mung bean
Vigna unguiculata	Southern pea
Vigna unguiculata, subsp. *sesquipedalis*	Yardlong or asparagus bean
Malvaceae (mallow family)	
Abelmoschus esculentus	Okra
Polygonaceae (buckwheat family)	
Rheum rhabarbarum	Rhubarb
Portulacaceae (purslane family)	
Portulaca oleraceae	Purslane
Solanaceae (nightshade family)	
Capsicum annuum var. annuum,	
Cerasiforme group	Ornamental cherry pepper
Fasiculatum group	Ornamental celestial pepper
Grossum group	Bell, perfection, or pimiento pepper
Longum group	Cayenne, chili pepper
Capsicum frutescens	Tabasco pepper
Lycopersicon lycopersicum	Tomato
Physalis ixocarpa	Tomatillo
Physalis peuviana	Ground cherry or poho berry
Solanum melongena	Eggplant
Solanum tuberosum	Potato
Tetragoniaceae (carpet weed family)	
Tetragonia tetragonioides	New Zealand spinach

(*continued*)

TABLE 1.2. *(Continued)*.

Family, Genus, Species	Common Name
Tropaeolaceae (nasturtium family)	
Trapaeolum majus	Garden nasturtium
Trapaeolum minus	Dwart garden nasturtium
Umbelliferae (parsley family)	
Apium graveolens var. dulce	Celery
Apium graveolens var. rapaceum	Celeriac
Daucus carota var. sativus	Carrot
Foeniculum vulgare var. azoricum	Florence fennel
Pastinaca sativa	Parsnip

Source: Staff, Liberty Hyde Bailey Hortorium (1976).

Classification by Edible Plant Part

Vegetables may also be classified according to the part of the plant from which they come. Plant parts eaten as vegetables include leaves, petioles, bulbs, stems, tubers, roots, flower clusters, fruits, and seeds.

Leaves. Leaves are eaten either raw or cooked. These vegetables include Brussels sprouts, cabbage, chard, Chinese cabbage, collards, cress, dandelion, endive, kale, lettuce, mustard, spinach and beet and turnip greens.

Petioles. A petiole is that part of the plant which supports the leaf blade and is connected to the stem of the plant. Petioles are often called a "stalk." Celery and rhubarb are the vegetables commonly eaten as a petiole.

Bulbs. Bulbs usually grow underground and consist of many fleshy leaves surrounding a very short stem. The swollen base of these leaves is the part usually eaten. These vegetables include garlic, leek, onion, and shallots.

Stems. A stem supports and displays the leaves, flowers, and fruits of a plant. Asparagus and kohlrabi are the two major vegetable stems.

Tubers. Vegetable tubers are modified stems that grow underground. Jerusalem artichoke and potato are vegetable tubers.

Roots. Most roots eaten as vegetables are tap roots that enlarge and grow straight down into the soil. These include beet, carrot, celeriac, chicory, horseradish, parsnip, radish, rutabaga, salsify, and turnip. Sweet potatoes are fibrous roots that have branched, spread sideways underground, and enlarged.

Flower clusters. The most common flower clusters eaten as vegetables are broccoli, cauliflower, and globe artichoke.

Fruits. Fruits are the seeds and the seed enclosures produced by a flowering plant. There is no precise distinction, horticulturally, between the terms "fruit" and "vegetable." Generally a vegetable is eaten with the main meal and a fruit is eaten as a dessert. In 1893, the U.S. Supreme Court legally established the tomato as a vegetable. Fruits used as a vegetable include cucumber, edible podded and sugar snap peas, eggplant, husk tomato, muskmelon, okra, peppers, pumpkin, snapbeans, squash, tomato, watermelon, and yardlong bean.

Seeds. The peanut is a vegetable seed that develops underground, while others develop aboveground. Some seeds are harvested when they are still soft, such as peas and sweet corn, while others, such as lima beans and sunflower, are harvested after they become hard. Vegetables eaten as seeds include broad bean, cowpea, garbanzo bean, kidney bean, lima bean, navy bean, pea, peanut, soybean, sunflower, and sweet corn.

Cultivar Selection

Choosing the best cultivars to grow in the garden is important. Gardeners should obtain a list of cultivars recommended for their specific area from the state agricultural experiment station. (See Table 1.3 for address.) Gardeners too often purchase vegetable seeds without knowing if the cultivar is adapted to local climatic conditions. Most seed companies list their All-America selections and highlight the vegetables which performed best in their trials.

Some cultivars grow best in spring plantings; others grow best in late summer or fall plantings. In the Southwest, for example, some tomato cultivars set fruit in spring and fall but do not set fruit during the summer. At high altitudes, where a short growing season is encountered, the tomato cultivar grown must mature and set fruit rapidly.

All-America Selections

This is a nonprofit organization of seed producers who develop and promote new cultivars of vegetables and flowers. The vegetables are grown in various locations throughout the United States under various soil and climatic conditions. The vegetable cultivar must perform well under all of these conditions to be designated an All-America Selection or Award. Not all new cultivars are tested under this program, however. In addition, a cultivar that performs best in one location may be unsuitable in another location and not be designated an All-America Selection.

TABLE 1.3. State Agricultural Experiment Station Addresses.

State	Address
Alabama	Auburn Univ., Auburn, AL 36849
Alaska	Univ. of Alaska, Fairbanks, AK 99701
Arizona	Univ. of Arizona, Tucson, AZ 85721
Arkansas	Univ. of Arkansas, Fayetteville, AR 72701
California	Univ. of California, 2200 Univ. Ave., Berkeley, CA 94720
Colorado	Colorado State Univ., Fort Collins, CO 80523
Connecticut	Univ. of Connecticut, Storrs, CT 06268
Delaware	Univ. of Delaware, Newark, DE 19717
District of Columbia	Federal City College, 1424 St., N.W., Washington, D.C. 20008
Florida	Univ. of Florida, Gainesville, FL 32611
Georgia	Univ. of Georgia, Athens, GA 30602
Hawaii	Univ. of Hawaii, Honolulu, HI 96822
Idaho	Univ. of Idaho, Moscow, ID 83843
Illinois	Univ. of Illinois, Urbana, IL 61801
Indiana	Purdue Univ., West Lafayette, IN 47907
Iowa	Iowa State Univ., Ames, IA 50011
Kansas	Kansas State Univ., Manhattan, KS 66506
Kentucky	Univ. of Kentucky, Lexington, KY 40506
Louisiana	Louisiana State Univ., Baton Rouge, LA 70893
Maine	Univ. of Maine, Orono, ME 04469
Maryland	Univ. of Maryland, College Park, MD 20742
Massachusetts	Univ. of Massachusetts, Amherst, MA 01003
Michigan	Michigan State Univ., East Lansing, MI 48824
Minnesota	Univ. of Minnesota, St. Paul, MN 55108
Mississippi	Mississippi State Univ., Mississippi State, MS 39762
Missouri	Univ. of Missouri, Columbia, MO 65211
Montana	Montana State Univ., Bozeman, MT 59717
Nebraska	Univ. of Nebraska, Lincoln, NE 68583
Nevada	Univ. of Nevada, Reno, NV 89557
New Hampshire	Univ. of New Hampshire, Durham, NH 03824
New Jersey	Rutgers Univ., New Brunswick, NJ 08903
New Mexico	New Mexico State Univ., Box 3AE, Agriculture Bldg., Las Cruces, NM 88003
New York	Cornell Univ., Ithaca, NY 14853
North Carolina	North Carolina State Univ., Raleigh, NC 27650
North Dakota	North Dakota State Univ., Fargo, ND 58105
Ohio	Ohio State Univ., 2120 Fyffe Rd., Columbus, OH 43210
Oklahoma	Oklahoma State Univ., Stillwater, OK 74078
Oregon	Oregon State Univ., Corvallis, OR 97331
Pennsylvania	Pennsylvania State Univ., University Park, PA 16802
Puerto Rico	Univ. of Puerto Rico, Mayaguez, PR 00708
Rhode Island	Univ. of Rhode Island, Kingston, RI 02881
South Carolina	Clemson Univ., Clemson, SC 29631
South Dakota	South Dakota State Univ., Brookings, SD 57007
Tennessee	Univ. of Tennessee, Box 1071, Knoxville, TN 37901
Texas	Texas A&M Univ., College Station, TX 77843
Utah	Utah State Univ., Logan, UT 84322
Vermont	Univ. of Vermont, Burlington, VT 05405
Virgin Islands	College of the Virgin Islands, St. Croix, 00850
Virginia	Virginia Polytechnic Institute, Blacksburg, VA 24061
Washington	Washington State Univ., Pullman, WA 99164
West Virginia	West Virginia Univ., 294 Coliseum, Morgantown, WV 26506
Wisconsin	Univ. of Wisconsin, 432 N. Lake St., Madison, WI 53706
Wyoming	Univ. of Wyoming, Box 3354, Laramie, WY 82071

Hybrids

Many new cultivars of vegetables are hybrids and are usually superior to the older cultivars. Hybrid cultivars usually have resistance to one or more diseases, grow rapidly, and produce more uniform plants. Hybrids usually cost more; but for most gardeners, the disease resistance alone is worth the extra cost. This is particularly true for gardeners who do not use pesticides.

Hybrids do not reproduce true to type in the second generation, and it is usually not advisable to harvest and save your own seed. It is usually advisable to grow a limited number of new cultivars beside your present cultivar for comparison.

RECORDS

Keep records about your garden. At the end of the season, these records can be reviewed and methods to improve your specific garden can be determined. A loose-leaf or bound notebook is better than individual notes.

Cultivars. Keep a list of cultivars planted and record which ones performed well for you. Record where the seeds or transplants were purchased and how much was purchased and used. By comparing cultivars, poor ones may be eliminated and a different cultivar, which may be an improvement, may be added next year.

Soil Fertility. Record soil test reports and types and amounts of fertilizer used (Marr 1977). It is particularly valuable to record soil pH and how much lime or sulfur was used to adjust the pH. Amounts of organic and natural fertilizer used should be recorded. Most natural deposits release their nutrients slowly and initial rates are high but can be reduced in subsequent years. It is valuable to know which organic fertilizers were used so that different ones can be used in subsequent years. This will help prevent the soil nutrients from becoming out of balance with each other and prevent deficiencies of one nutrient from occurring.

Crop Yields. Record how much was produced by each vegetable and how much was used fresh and how much was processed. Record when a vegetable was harvested and when harvest was complete. Gardeners can then determine if more or less space will be required next year for each cultivar to meet your particular needs. Succession planting may be needed to provide vegetables throughout the season. Alternatively some vegetables may be omitted from the garden, if their time of harvest coincides with the gardener's vacation away from home.

Growing Conditions. Record the time of year transplants were started and when they were planted into the garden. Should they have been planted earlier or later? Record when cool-season, warm season and fall plantings were made so timing can be improved. The last frost in the spring and the first frost in the fall should be noted.

Record insect and disease problems, and what, if any, control measures were taken. If no control measures were used, was the quality of the harvested vegetables satisfactory?

GARDEN LOCATION

You can grow vegetables successfully in full sunlight, on good soil, and away from tree roots. In practice, however, gardening locations are a compromise (Wilson 1977). The area you choose should have good soil, but soil on many homeowner lots is from the basemet or crawl space under the house. This soil can be used if adequate nutrients and organic matter are added, but this requires considerable work and effort over several years. The garden needs at least 6 hr of direct sun each day. Soil receiving direct sun warms up quickly, inducing the seeds and plants to grow more rapidly, reducing insect and disease problems. Gardens on the north side that are within 6 to 8 ft of one-story buildings do not receive enough sunlight (Fig. 1.1).

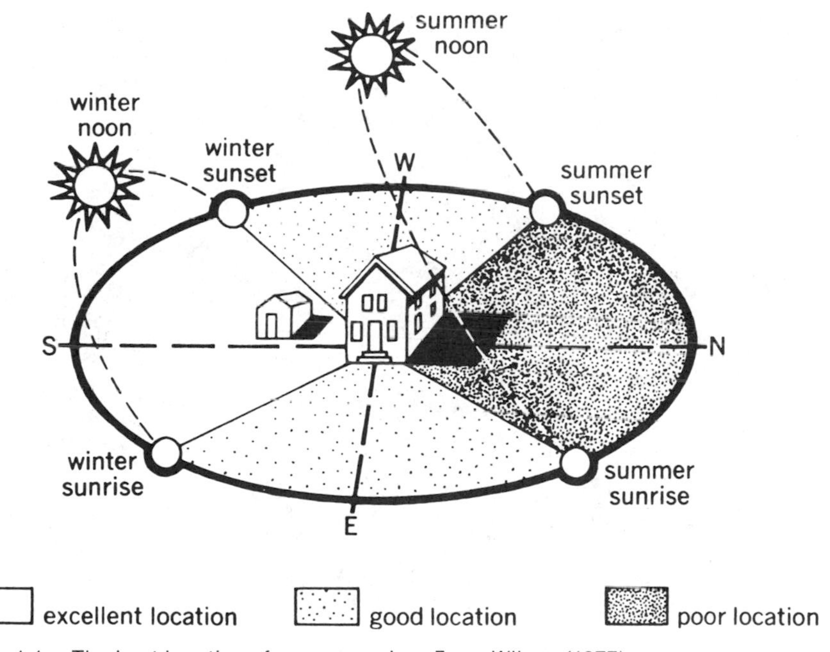

FIG. 1.1. The best locations for your garden. *From Wilson (1977).*

Gardens should not be shaded by trees. Trees and shrubs also compete with vegetables for nutrients and water. Various root barriers generally are of little value. If a garden must be located near trees, use this rule of thumb. The sunniest spot goes to vegetables grown for their fruits or seeds such as corn, tomato, squash, cucumber, eggplant, peppers, beans, and peas.

Plants grown for their leaves or roots like beets, cabbage, lettuce, mustard, chard, spinach, and turnips can be grown in partial shade.

The garden location should be near an abundant water supply, for irrigating your garden during dry spells. Plan your garden near your home if possible. Gardeners are more likely to spend time working in the garden and checking for pests if it is nearby.

If the requirements of good soil, adequate light, and sufficient water are met, gardens can be located almost anywhere (Fig. 1.2): in backyards, in front yards, as minigardens, in a community garden (often sponsored by the local park district), in window boxes, on patios of high-rise apartments, or on the roofs of buildings in large cities.

Backyard and Front-yard Gardens

Most gardens are located in the backyard. However, sometimes the best garden location is in the front yard (Wilson 1977). Front-yard gardens need to be neat and should be mulched or planted to a cover crop during winter.

FIG. 1.2. Cucumbers growing in a small space next to a fence. *Courtesy of W. Atlee Burpee Co.*

Check local ordinances before beginning a garden in the front yard. Some cities have ordinances against permanent fences in the front. A low temporary fence of chicken wire or snow fence may be used to keep out dogs and unwanted visitors. Some front yards contain rapid-growing trees that provide so much shade that even grass will not grow. If the tree is unattractive or not valuable it should be trimmed or removed before beginning a garden. The branches can be ground and used as a mulch.

Front-yard gardens need a flower border with tall annual flowering or foliage plants to screen out the garden and obscure low temporary fences. Lettuce, peas, carrots, beets, potatoes, tomatoes, and many other vegetables can be grown in a front-yard garden. Melons, however, usually disappear if grown in a front yard.

Minigardens

A minigarden requires only a small amount of space. These gardens can easily be made a part of the general landscape design. Any small unused area may be used to grow vegetables. The space may be behind the garage or next to the house (Fig. 1.3).

Some homeowners have erected a greenhouse, attached one to their house (Fig. 1.4), or converted a window into a mini-greenhouse. The greenhouse can be used to start transplants as well as grow plants. During the winter, cool season crops can be grown and ruby lettuce can be grown with greenhouse temperatures set at 50°F. Most locations receive less sunlight during the winter months, and leafy crops are best grown then.

FIG. 1.3. A minigarden along the south side of a house. *From Vandemark et al. (1975).*

FIG. 1.4. Leafy crops can be grown in a cool greenhouse during winter months.

Container Gardens

Gardening can still be done by people with limited space. Apartment dwellers, mobile home residents, and homeowners with yards shaded by mature trees can garden in containers. The containers can be placed on sidewalks, patios, window boxes, porches, or balconies.

Types of Containers. The type of container depends upon the vegetable being grown. As long as there is enough root space, most vegetable plants will thrive. Wooden barrels, decorative boxes, plastic garbage cans, tin cans, plastic laundry baskets, and various pots have been used successfully. To reduce frequency of watering, use containers with a 4-gal capacity or larger. Small containers dry out quickly and may blow over in the wind. Large containers may be placed on low carts and moved easily. Square, rectangular, or circular containers work equally well.

All containers should have holes in the bottom to provide for water drainage. Most vegetables grown in a container need daily watering, and ample drainage is a must. If the container does not have drainage holes, the bottom one-fourth of the container can be filled with rocks or pebbles to hold the excess water until it evaporates or is used. Plastic materials are nonporous and retain more water than clay materials (Vandemark and Splittstoesser 1978). Plants grown in plastic containers dry out less rapidly and can be watered less frequently.

Soil Mix. Most container gardens use a commercial potting mix. These mixes are lightweight, fast-draining, and free of insects, diseases, and weeds. Some commercil mixes are very lightweight. These latter mixes are good for hanging baskets, window boxes, and containers that are moved around. These lightweight mixes are a disadvantage for growing large plants. Sweet corn, staked tomatoes, and eggplant may grow large enough to cause the container to tip over.

Some container gardeners make their own soil mix. These mixes can be made from equal amounts of good garden soil, washed coarse sand and organic material such as peat moss, leaf mold or sawdust (Carbonneau 1969). The mix needs to be free of various pests, which can be accomplished by heating the mix at a low temperature in the oven. For specific times and temperatures your state agriculture experiment station should be contacted (See Table 1.3 for locations.)

Vegetables for Containers. There are a number of cultivars now on the market especially designed for small gardens and container gardens. These cultivars do not grow large but still produce good yields (Arthurs 1977). Many traditional cultivars of vegetables are also easily grown in containers. Larger growing plants need a larger container size. Most herbs, chives, and parsley need a container holding 3 pt of soil (a standard 6 in. pot). Beets, lettuce, onions, and radishes require a container holding about 1 gal. of soil mix. Chard, pepper, and small tomato cultivars grow best with 2 gal. of soil (Fig. 1.5). Large plants such as cucumbers, eggplants, sweet corn, and tomatoes grow best in a 4 gal. container or larger. With proper water and fertilizer most of these plants can be grown in a surprisingly small container.

With the increase in interest in container gardening, most seed companies now list cultivars and varieties suitable for container gardening. Artichoke requires a large one-bushel container. Beans of all types can be grown in containers. Bush beans are best, but pole beans can be grown, if poles or trellises are used for support.

Beets need to be grown in containers about 1 ft deep. Brussels sprouts need cool weather and a large container. Cabbage plants usually grow too large for containers, but early maturing and dwarf cultivars can be used.

Carrots require containers at least 1 ft deep to prevent deformed roots, and should not be grown in soil mixes containing compost, which stimulates the production of root hairs.

Chard requires a container about 2 ft deep. Chinese cabbage grows well in containers for a fall crop. Collards can be grown for a continuous supply of greens. If the outer leaves are routinely harvested, a small container can be used. Cucumbers need a large container and some type of trellis support. There are a number of cultivars available for containers and hanging baskets.

Eggplant needs warm conditions and a large container for best growth, and small plant types are available.

Most herbs can easily be grown in containers, window boxes, or pots. These can be brought into the living area in cold weather and placed near a window for fresh winter herbs.

Endive and kale can be grown in containers. By continuous harvesting of the outer leaves, a small container may be used.

FIG. 1.5. Pixie Hybrid tomatoes growing in a patio container. *Courtesy of W. Atlee Burpee Co.*

Kohlrabi is fast growing and can be grown in small containers. Lettuce is a cool-season plant that can be grown in full sunlight or partial shade. It may be grown indoors in the winter.

Melons do not grow very satisfactorily in containers. They require a large container and their vines require a large amount of space. Several cultivars of small cantaloupes and watermelons are available, but they are best grown in minigardens.

Mustard can easily be grown for greens in a container. Okra is easily grown in a large container. Onions, particularly green onions, can easily be grown in containers.

Peas do not grow well in containers. They need a large container, require staking or trellises, and produce a low yield. The edible-podded types produce a better yield than English peas. Radishes are easily grown in a container. Some cultivars, such as Champion, are adapted to lower light conditions and can be grown indoors in the greenhouse in winter.

Rhubarb can be grown in a large metal container. If taken care of, the plant will grow for many years.

Spinach is a cool-season crop, best grown very early in the spring or as a fall crop. Both regular spinach and New Zealand spinach, which is not a true spinach, can be grown in containers.

Sweet corn is difficult to grow in containers. It requires a large container, sometimes must be hand pollinated, and produces a low yield. However, many container gardens grow good-quality sweet corn. At least four stalks are needed for pollination, and these should be planted near each other. Dwarf cultivars should be considered.

Bush types of summer squash can be grown in containers with 6–8 gallons of soil. Tomatoes can be grown in containers of all types. Standard types need to be staked and grown in large containers. Many dwarf cultivars are available for containers and hanging baskets.

General Care. Plants growing in containers should not be crowded. The number of plants per container must be limited, often to one per container. Root crops and greens should be planted on the basis of the space they need when mature, not as seedlings. The plants in containers should be thinned to allow ample growing space for the plants. Seedlings should be removed by cutting the unwanted seedling's stem at the soil line. If the seedling is pulled out, roots of the remaining seedlings are frequently damaged, particularly when they are growing in soil mixes.

Container-grown plants need frequent watering and fertilizer. Plants receiving too much water develop root rots, and, with too little, wilt and die. Once the edible part (fruit, root, or leaf) is produced, the plants will need more water and fertilizer.

Plants should be watered with cool, not hot, water from a hose, at moderate water pressure. At high pressure, the water will make holes in the soil mix and damage the roots. If plants are watered in the morning, they will be dry by evening and help prevent disease. However, as containers dry out rapidly, plants grown in warm climates may need a second watering in the afternoon. As a general rule, it is better to keep the containers a little dry rather than too wet. This is particularly important for containers without drainage holes.

Containers placed near reflective surfaces warm up rapidly and lose more water than plants placed on black surfaces such as blacktop. Plants growing in containers placed on walks, drives, and concrete patios will require more water.

Many commercial soil mixes contain few plant nutrients, and weekly use of fertilizer is needed. The containers are watered daily, and many added plant nutrients are also removed from the container. Various water-soluble fertilizers may be used. Timed-release fertilizers are often used. These materials release a small amount of fertilizer each time the container is watered and need to be added only once each season. Natural gardeners should make up their own soil mix (see earlier section on Soil Mix). The organic matter can be replaced with fish emulsion, dried blood, or soybean meal to supply nitrogen. The coarse sand in the mix can be replaced with equal parts of greensand and rock phosphate to supply potassium and phosphorus. The containers can routinely receive organic materials as a fertilizer.

Plants growing in containers need adequate sunlight and must be spaced sufficiently apart. If large containers are used, the plants must be spaced further apart than in a full size backyard garden. Table 1.4 gives the number of vegetable plants that a square foot of container space will support. Vegetable plants need at least 6 hr of sunlight each day. If the plants are growing in an area where less than this occurs, reflected light may be used. Aluminum foil, mirrors, chrome reflectors, white gravel, and white buildings reflect light. If this reflected light is of high enough intensity, many vegetable crops can be grown. Lettuce, peppers, and tomatoes have been grown with only 3 hr of direct sunlight and reflected light the remainder of the day (Abraham and Abraham 1977). Containers may also be placed on wheels and moved to take advantage of the available sunlight.

Container-grown vegetables can be spaced so that, with the exception of sweet corn, no two plants of the same cultivar are near each other. This will reduce insect problems. However, insect problems usually develop. Since the containers are watered each day, most insects will be noticed and can be handpicked or other mechanical or cultural methods used (see Chapter 4).

SPRING, SUMMER, AND FALL GARDENS

Temperature and rainfall are the two most important factors that determine when to plant a garden. These two factors also affect crop quality and insect and disease problems.

TABLE 1.4. Spacing of Vegetables Growing in Containers.

Vegetable	Approximate number of plants per square foot	Vegetable	Approximate number of plants per square foot
Beans	3–4	Mustard greens	9
Beets	25[1]	Onions (cooking)	16
Broccoli	3	(hamburger)	9
Brussels sprouts	2	(green bunching)	100[4]
Cabbage	2	Parsley	16
Carrots	100[2]	Parsnips	25
Cauliflower	2	Peanuts	4
Chard, Swiss	9	Peas	25[3]
Corn (dwarf)	4	Peppers	4
Cucumber (standard)	1[3]	Potatoes	1
(dwarf)	2[3]	Sweet potatoes	1
Dandelion	6	Radishes	144[5]
Eggplant	1	Rutabaga	5
Endive	4	Spinach	4
Garlic	36	Summer squash (bush)	1
Kale	4	Winter squash (bush)	1
Kohlrabi	4	Tomato (regular)	1[3]
Leeks	64	(dwarf)	2
Lettuce (head)	4	Husk tomato *(Physalis)*	2
(leaf and semihead)	6	Watermelon (dwarf)	1[3]
Muskmelon	1[3]		

Source: Abraham and Abraham (1977).
[1] Thin at 1 in. diameter for "greens" and let remainder grow.
[2] Thin every other one when "fingerlings" and let others grow.
[3] Train on trellis.
[4] Can thin to eat and let others grow into cooking onions.
[5] Thin small ones to eat and let others grow.

High temperature hastens ripening of fruits (tomato) and reduces quality. These temperatures induce sweet corn to mature so rapidly that the length of the harvest period is reduced and two successive plantings may mature at the same time. Sugar content and color are increased when crops (beets, carrots) mature under sunny, cool, dry conditions. Hot, wet conditions induce the production of fruit that is soft, watery, low in sugar, and of poor storage quality.

Warm, wet periods increase diseases such as scab, bacterial wilts, leaf blights, mildew, and fruit rots. Cool, dry periods help reduce these problems. Hot, dry conditions increase insect problems while cool, moist periods help suppress these pests.

Plant Arrangement Within the Garden. Perennials such as rhubarb and asparagus should be planted in one section of the garden (see Tables 1.5–1.7). They should not be disturbed when the rest of the garden is prepared for annual plants.

The soil composition and drainage of the garden should be considered when planning where to plant individual types of vegetables. Crops such as celery, onions, and late cucumbers can be planted in low moist areas. Crops such as squash, pumpkins, and early season vegetables can be planted in

TABLE 1.5. Spring, Summer, and Fall Gardens Can Use the Same 600 sq ft. Each Bed is About 2.5 ft Wide and 7.5 ft Long. Walkways Between Beds are About 1 ft Wide and About 2 ft Wide between Area A and Area B.

Spring Garden		
Bed number	Area A	Area B
1	Asparagus—3 rows (30 crowns)	Rhubarb—1 row (6 plants)
2	Peas—2 rows (caged or staked)	Cauliflower—2 rows (8 plants)
3	Broccoli—8 plants in 2 staggered rows	Savoy cabbage—1 row (8 plants)
4	Kale—1 row (6 plants)	Early cabbage—1 row (8 plants)
5	Celery or flowers	Kohlrabi—2 rows (24 plants)
6	Mustard greens—2 rows (16 plants)	Carrots—2 rows (40 plants)
7	Parsley, dill, and basil (2 rows—interplanted)	Beets—2 rows (24 plants)
8	Spinach and garlic—2 rows (10 plants—interplanted)	Swiss chard—2 rows (10 plants)
9	Cress/endive/or radish—2 rows (24 plants total)	Bulbing onions—2 rows (28 plants)
10	Leaf lettuce—2 rows (18 plants)	Green onions—3 rows (60 plants)

high areas that are warm and dry. Tall crops such as sweet corn, sunflowers, and pole beans should be planted together. Tall crops should be planted on the north side of the garden so they will not shade low-growing vegetables.

It does not appear to make much difference whether the garden rows are laid out in a north and south direction or in an east and west direction. On gardens planted on a slope, the rows should extend across the slope at right angles to reduce erosion.

A number of seed companies will prepare a computer-generated model for home gardeners. They will use the cultivars that are purchased from them, or provide special types such as bush squash. This will provide a garden arrangement that contains only those vegetables that are going to be grown.

Cool and Warm Season Crops

Vegetables differ in their ability to withstand cool and warm temperatures. Cool season crops can withstand light frosts, and asparagus and rhubarb plants (not the edible part) can withstand winter freezing. Cool season crops are planted early in the spring, but they must have time to mature before temperatures become too warm. They can be planted during hot weather if there is a long period of cool temperature in the fall for them to reach edible maturity. Cultivar differences exist within crops.

Cool and warm season crops differ by more than their susceptibility to cool temperatures. Seeds of cool season crops germinate at a cooler soil temperature. Cool season crops are smaller in size, and their root systems are shallower; therefore, these plants must be irrigated more frequently if rainfall is inadequate. Some cool season crops are biennials, that is, they can produce seedstalks and flowers the second growing season. However, if these biennials such as celery, beet, cabbage, and carrot are exposed to an average temperature of 50°F or lower for several weeks and the plants have enough food reserve, they will produce seedstalks the first year instead of the edible part. The edible portion of cool season crops is not susceptible to chilling injury at temperatures between 32° and 50°F as are some warm season vegetables.

The food value of a cool season crop is usually higher per pound than for a warm season crop (Sims *et al.* 1977). A vegetative part, such as a root, stem, leaf, or flower part, is eaten from cool season crops. The edible plant part of a warm season crop is usually an immature or mature fruit or seed.

TABLE 1.6. A Summer Garden Using the Same Area as a Spring Garden. (See Table 1.5.)

	Summer Garden	
Bed number	Area A	Area B
1	Asparagus—3 rows (30 crowns)	Rhubarb—1 row (6 plants)
2	Midseason sweet corn—2 rows	Early sweet corn/interplant pumpkins—2 rows
3	Same as Bed 2A	Same as Bed 2B
4	Pole lima beans—2 rows (staked)	Pole beans—1 row (staked)
5	Sweet potato or potatoes—2 rows (15 plants)	Caged tomatoes (or okra—1 row (5 plants)
6	Bush winter squash—1 row (5 plants)	Peppers—1 row (7 plants)
7	Bush summer squash—1 row (5 plants)	Eggplant—2 rows (7 plants)
8	New Zealand spinach or cucumbers—2 rows	Melons—1 row (6 plants)
9	Beets—2 rows (24 plants)	Bulbing onions—2 rows (28 plants)
10	Carrots—2 rows (40 plants)	Green onions—3 rows (60 plants)

TABLE 1.7. A Fall Garden Using the Same Area as a Spring Garden. (See Table 1.5.)

	Fall Garden	
Bed number	Area A	Area B
1	Asparagus—3 rows (30 crowns)	Rhubarb—1 row (6 plants)
2	Broccoli—1 row (8 plants)	Collards—1 row (8 plants)
3	Cauliflower—8 plants in 2 staggered rows	Late cabbage—1 row (8 plants)
4	Beets—2 rows (24 plants)	Green or wax beans—2 rows
5	Kohlrabi—2 rows (24 plants)	Chinese cabbage—2 rows (15 plants)
6	Brussels sprouts—1 row (8 plants)	Winter radish—2 rows (16 plants)
7	Turnips—2 rows (24 plants)	Spinach—2 rows (15 plants)
8	Parsnips—2 rows (24 plants)	Mustard greens—2 rows (16 plants)
9	Carrots—2 rows (40 plants)	Green onions—3 rows (60 plants)
10	Peas—2 rows (caged or staked)	Lettuce—2 rows or radish—3 rows

Very Hardy Vegetables. Very hardy vegetable crops can be planted 4 to 6 weeks before the average date of the last frost in the spring. Collards, kale, kohlrabi, lettuce, onions, peas, rutabaga, salsify, spinach, and turnip may be planted from seed. Asparagus, broccoli, Brussels sprouts, cabbage, and onions may be planted as transplants. Onions may also be planted from sets. Horseradish and rhubarb roots and potato tubers may be planted at this time also.

Frost-Tolerant Vegetables. These crops are planted two or three weeks before the average date of the last frost in the spring. These include beets, carrots, chard, mustard, parsnip, and radishes, all planted from seed. Cauliflower and Chinese cabbage may be planted as transplants. Most herbs may be seeded or transplanted on this date.

Tender Vegetables. The tender vegetables should be planted on the average date of the last frost in the spring. These include snap beans, summer squash, sweet corn, and tomatoes as transplants.

Warmth Requiring Vegetables. These crops require the soil to be warm before they will germinate. They should be planted 1 or 2 weeks after the average date of the last frost in the spring. These plants are lima beans, cucumber, cantaloupe or muskmelon, okra, pumpkin, Southern peas, winter squash, and watermelon. Eggplants, peppers, and sweet potatoes can be planted as transplants.

Medium Heat-Tolerant Vegetables. These vegetable crops are good for summer plantings. All beans, including soybean, chard, New Zealand spinach, squash, and sweet corn are included here.

Spring and Summer Gardens

In the Deep South and the low-elevation areas of the Southwest and the West Coast, vegetables may be grown 8–12 months of the year. Three separate and complete gardens on the same plot are possible to achieve. Tables 1.5, 1.6, and 1.7 give some possible vegetables that can be grown. The gardener should select from these plans rather than follow them exactly. For a small garden, vegetables with a high yield per plant space should be chosen, such as bush snap beans, bush lima beans, bush summer squash, Southern peas, leafy greens, tomatoes, and bell peppers. Many gardeners are tempted to plant sweet corn, but a cornstalk will occupy as much space as a tomato plant and produce only one ear of corn. Vining melons, squash, pumpkins, and sweet corn use a lot of garden space for a long time and produce less per plant space.

In areas with a 90–120 day frost-free growing season, spring vegetables will continue to produce through the midsummer. At midsummer, these vegetables can be replaced with ones selected from the fall garden (Table 1.7). Some rows should be left open when the spring garden is planted, so that the warmth-loving plants, chosen from the summer garden(Table 1.6), can be planted after all danger of frost is past. In areas with a short growing season, summer crops should not follow spring crops as these summer vegetables may be killed by a September frost before they have matured.

Most of mid-America has a frost-free growing season of 120 – 140 days. In these areas, separate spring and summer plantings are possible. When the summer vegetables are harvested, they should be removed so that fall vegetables can be planted.

In very hot areas, two successive crops of summer vegetables are planted. The intense heat will be too great for most vegetables, except for Southern peas, okra, butter beans, and sweet potatoes.

Spring and Summer Planting Dates. To determine the specific time for planting each vegetable, the gardener needs to know the average date of the last frost in the spring. Figure 1.6 shows the average dates of the last killing frost for locations in all 50 states. From this figure the planting dates can be determined. Table 1.8 gives planting dates between January 1 and June 30 for spring and early summer crops. This table gives *the earliest time it is safe to*

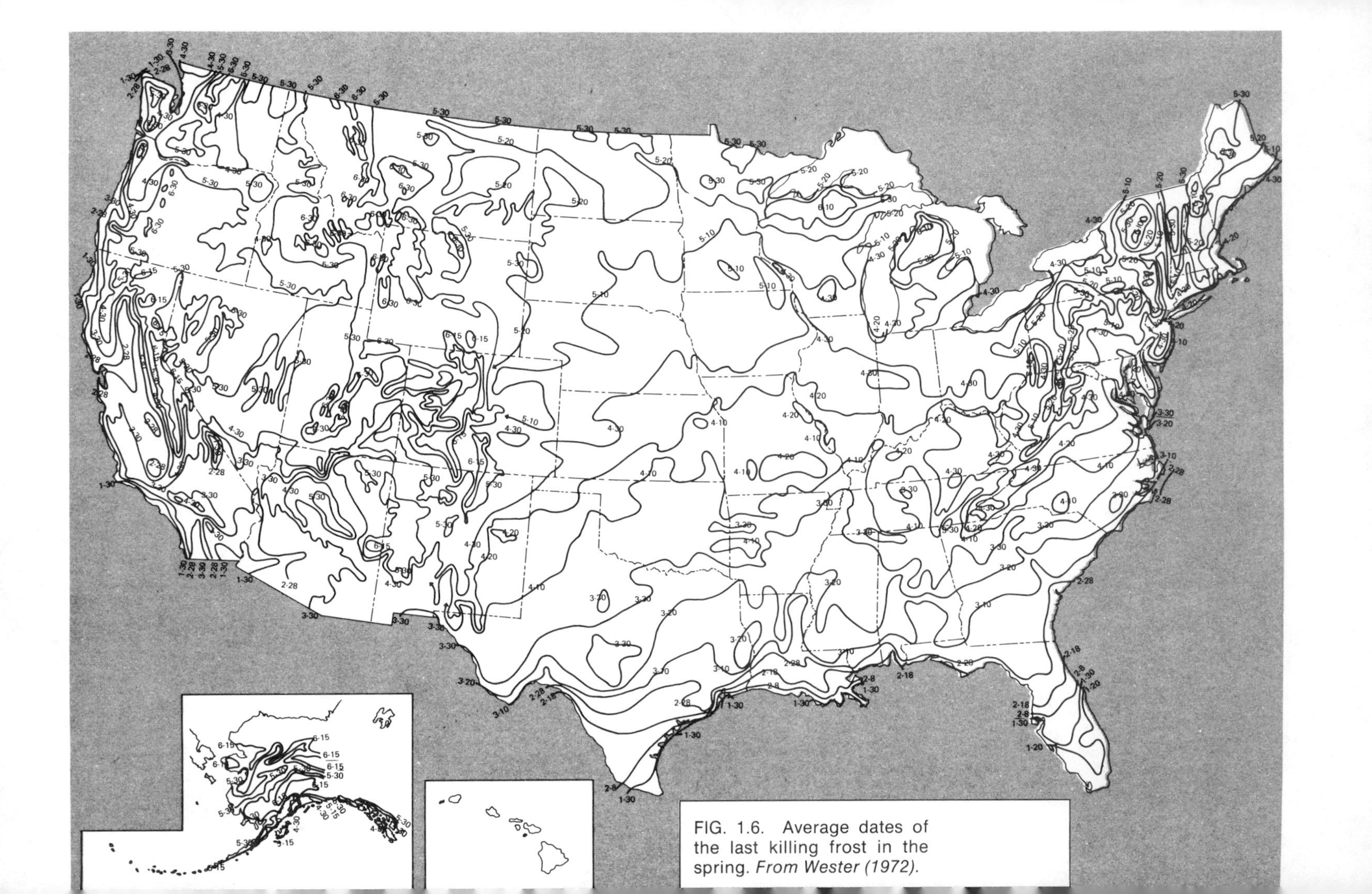

FIG. 1.6. Average dates of the last killing frost in the spring. *From Wester (1972).*

TABLE 1.8. Earliest Dates, and Range of Dates, for Safe Outdoor Planting of Vegetables in the Spring.

Crop	Planting dates for localities in which average date of last freeze is—						
	Jan. 30	Feb. 8	Feb. 18	Feb. 28	Mar. 10	Mar. 20	Mar. 30
Asparagus [1]					Jan. 1–Mar. 1	Feb. 1–Mar. 10	Feb. 15–Mar. 20.
Beans, lima	Feb. 1–Apr. 15	Feb. 10–May 1	Mar. 1–May 1	Mar. 15–June 1	Mar. 20–June 1	Apr. 1–June 15	Apr. 15–June 20.
Beans, snap	Feb. 1–Apr. 1	Feb. 1–May 1	Mar. 1–May 1	Mar. 10–May 15	Mar. 15–May 15	Mar. 15–May 25	Apr. 1–June 1.
Beet	Jan. 1–Mar. 15	Jan. 10–Mar. 15	Jan. 20–Apr. 1	Feb. 1–Apr. 15	Feb. 15–June 1	Feb. 15–May 15	Mar. 1–June 1.
Broccoli, sprouting [1]	Jan. 1–30	Jan. 1–30	Jan. 15–Feb. 15	Feb. 1–Mar. 1	Feb. 15–Mar. 15	Feb. 15–Mar. 15	Mar. 1–20.
Brussels sprouts [1]	Jan. 1–30	Jan. 1–30	Jan. 15–Feb. 15	Feb. 1–Mar. 1	Feb. 15–Mar. 15	Feb. 15–Mar. 15	Mar. 1–20.
Cabbage [1]	Jan. 1–15	Jan. 1–Feb. 10	Jan. 1–Feb. 25	Jan. 15–Feb. 25	Jan. 25–Mar. 1	Feb. 1–Mar. 1	Feb. 15–Mar. 10.
Cabbage, Chinese	(2)	(2)	(2)	(2)	(2)	(2)	(2)
Carrot	Jan. 1–Mar. 1	Jan. 1–Mar. 1	Jan. 15–Mar. 1	Feb. 1–Mar. 1	Feb. 10–Mar. 15	Feb. 15–Mar. 20	Mar. 1–Apr. 10.
Cauliflower [1]	Jan. 1–Feb. 1	Jan. 1–Feb. 1	Jan. 10–Feb. 10	Jan. 20–Feb. 20	Feb. 1–Mar. 1	Feb. 10–Mar. 10	Feb. 20–Mar. 20.
Celery and celeriac	Jan. 1–Feb. 1	Jan. 10–Feb. 10	Jan. 20–Feb. 20	Feb. 1–Mar. 1	Feb. 20–Mar. 20	Mar. 1–Apr. 1	Mar. 15–Apr. 15.
Chard	Jan. 1–Apr. 1	Jan. 10–Apr. 1	Jan. 20–Apr. 15	Feb. 1–May 1	Feb. 15–May 15	Feb. 20–May 15	Mar. 1–May 25.
Chervil and chives	Jan. 1–Feb. 1	Jan. 1–Feb. 1	Jan. 1–Feb. 1	Jan. 15–Feb. 15	Feb. 1–Mar. 1	Feb. 10–Mar. 10	Feb. 15–Mar. 15.
Chicory, witloof					June 1–July 1	June 1–July 1	June 1–July 1.
Collards [1]	Jan. 1–Feb. 15	Jan. 1–Feb. 15	Jan. 1–Mar. 15	Jan. 15–Mar. 15	Feb. 1–Apr. 1	Feb. 15–May 1	Mar. 1–June 1.
Cornsalad	Jan. 1–Feb. 15	Jan. 1–Feb. 15	Jan. 1–Mar. 15	Jan. 1–Mar. 1	Jan. 1–Mar. 15	Jan. 1–Mar. 15	Jan. 15–Mar. 15.
Corn, sweet	Feb. 1–Mar. 15	Feb. 10–Apr. 1	Feb. 20–Apr. 15	Mar. 1–Apr. 15	Mar. 10–Apr. 15	Mar. 15–May 1	Mar. 25–May 15.
Cress, upland	Jan. 1–Feb. 1	Jan. 1–Feb. 15	Jan. 15–Feb. 15	Feb. 1–Mar. 1	Feb. 10–Mar. 15	Feb. 20–Mar. 15	Mar. 1–Apr. 1.
Cucumber	Feb. 15–Mar. 15	Feb. 15–Apr. 1	Feb. 15–Apr. 15	Mar. 1–Apr. 15	Mar. 15–Apr. 15	Apr. 1–May 1	Apr. 10–May 15.
Eggplant [1]	Feb. 1–Mar. 1	Feb. 10–Mar. 15	Feb. 20–Apr. 1	Mar. 10–Apr. 15	Mar. 15–Apr. 15	Apr. 1–May 1	Apr. 15–May 15.
Endive	Jan. 1–Mar. 1	Jan. 1–Mar. 1	Jan. 15–Mar. 1	Feb. 1–Mar. 1	Feb. 15–Mar. 15	Mar. 1–Apr. 1	Mar. 10–Apr. 10.
Fennel, Florence	Jan. 1–Mar. 1	Jan. 1–Mar. 1	Jan. 15–Mar. 1	Feb. 1–Mar. 1	Feb. 15–Mar. 15	Mar. 1–Apr. 1	Mar. 10–Apr. 10.
Garlic	(2)	(2)	(2)	(2)	(2)	Feb. 1–Mar. 1	Feb. 10–Mar. 10.
Horseradish [1]							Mar. 1–Apr. 1.
Kale	Jan. 1–Feb. 1	Jan. 10–Feb. 1	Jan. 20–Feb. 10	Feb. 1–20	Feb. 10–Mar. 1	Feb. 20–Mar. 10	Mar. 1–20.
Kohlrabi	Jan. 1–Feb. 1	Jan. 10–Feb. 1	Jan. 20–Feb. 10	Feb. 1–20	Feb. 10–Mar. 1	Feb. 20–Mar. 10	Mar. 1–Apr. 1.
Leek	Jan. 1–Feb. 1	Jan. 1–Feb. 1	Jan. 1–Feb. 15	Jan. 15–Feb. 15	Jan. 25–Mar. 1	Feb. 1–Mar. 1	Feb. 15–Mar. 15.
Lettuce, head [1]	Jan. 1–Feb. 1	Jan. 1–Feb. 1	Jan. 1–Feb. 1	Jan. 15–Feb. 15	Feb. 1–20	Feb. 15–Mar. 10	Mar. 1–20.
Lettuce, leaf	Jan. 1–Feb. 1	Jan. 1–Feb. 1	Jan. 1–Mar. 15	Jan. 1–Mar. 15	Jan. 15–Apr. 1	Feb. 1–Apr. 1	Feb. 15–Apr. 15.
Muskmelon	Feb. 15–Mar. 15	Feb. 15–Apr. 1	Feb. 15–Apr. 15	Mar. 1–Apr. 15	Mar. 15–Apr. 15	Apr. 1–May 1	Apr. 10–May 15.
Mustard	Jan. 1–Mar. 1	Jan. 1–Mar. 1	Feb. 15–Apr. 15	Feb. 1–Mar. 1	Feb. 10–Mar. 15	Feb. 20–Apr. 1	Mar. 1–Apr. 15.
Okra	Feb. 15–Apr. 1	Feb. 15–Apr. 15	Mar. 1–June 1	Mar. 10–June 1	Mar. 20–June 1	Apr. 1–June 15	Apr. 10–June 15.
Onion [1]	Jan. 1–15	Jan. 1–15	Jan. 1–15	Jan. 1–Feb. 1	Jan. 15–Feb. 15	Feb. 10–Mar. 10	Feb. 15–Mar. 15.
Onion, seed	Jan. 1–15	Jan. 1–15	Jan. 1–15	Jan. 1–Feb. 15	Feb. 1–Mar. 1	Feb. 10–Mar. 10	Feb. 20–Mar. 15.
Onion, sets	Jan. 1–15	Jan. 1–15	Jan. 1–15	Jan. 1–Mar. 1	Jan. 15–Mar. 10	Feb. 1–Mar. 20	Feb. 15–Mar. 20.
Parsley	Jan. 1–30	Jan. 1–30	Jan. 1–30	Jan. 15–Mar. 1	Feb. 1–Mar. 10	Feb. 15–Mar. 15	Mar. 1–Apr. 1.
Parsnip			Jan. 1–Feb. 1	Jan. 15–Feb. 15	Jan. 15–Mar. 1	Feb. 15–Mar. 15	Mar. 1–Apr. 1.
Peas, garden	Jan. 1–Feb. 15	Jan. 1–Feb. 15	Jan. 1–Mar. 1	Jan. 15–Mar. 1	Jan. 15–Mar. 15	Feb. 1–Mar. 15	Feb. 10–Mar. 20.
Peas, black-eye	Feb. 15–May 1	Feb. 15–May 15	Mar. 1–June 15	Mar. 10–June 20	Mar. 15–July 1	Apr. 1–July 1	Apr. 15–July 1.
Pepper [1]	Feb. 1–Apr. 1	Feb. 15–Apr. 15	Mar. 1–May 1	Mar. 15–May 1	Apr. 1–June 1	Apr. 10–June 1	Apr. 15–June 1.
Potato	Jan. 1–Feb. 15	Jan. 1–Feb. 15	Jan. 15–Mar. 1	Jan. 15–Mar. 1	Feb. 1–Mar. 1	Feb. 10–Mar. 15	Feb. 20–Mar. 20.
Radish	Jan. 1–Apr. 1	Jan. 1–Apr. 1	Jan. 1–Apr. 1	Jan. 1–Apr. 1	Jan. 1–Apr. 15	Jan. 20–May 1	Feb. 15–May 1.
Rhubarb [1]							
Rutabaga				Jan. 1–Feb. 1	Jan. 15–Feb. 15	Jan. 15–Mar. 1	Feb. 1–Mar. 1.
Salsify	Jan. 1–Feb. 1	Jan. 10–Feb. 10	Jan. 15–Feb. 20	Jan. 15–Mar. 1	Feb. 1–Mar. 1	Feb. 15–Mar. 1	Mar. 1–15.
Shallot	Jan. 1–Feb. 1	Jan. 1–Feb. 10	Jan. 1–Feb. 20	Jan. 1–Mar. 1	Jan. 15–Mar. 1	Feb. 1–Mar. 10	Feb. 15–Mar. 15.
Sorrel	Jan. 1–Mar. 1	Jan. 1–Mar. 1	Jan. 15–Mar. 1	Feb. 1–Mar. 10	Feb. 10–Mar. 15	Feb. 10–Mar. 20	Feb. 20–Apr. 1.
Soybean	Mar. 1–June 30	Mar. 1–June 30	Mar. 10–June 30	Mar. 20–June 30	Apr. 10–June 30	Apr. 10–June 30	Apr. 20–June 30.
Spinach	Jan. 1–Feb. 15	Jan. 1–Feb. 15	Jan. 1–Mar. 1	Jan. 1–Mar. 1	Jan. 15–Mar. 10	Jan. 15–Mar. 15	Feb. 1–Mar. 20.
Spinach, New Zealand	Feb. 1–Apr. 15	Feb. 15–Apr. 15	Mar. 1–Apr. 15	Mar. 15–May 15	Mar. 20–May 15	Apr. 1–May 15	Apr. 10–June 1.
Squash, summer	Feb. 1–Apr. 15	Feb. 15–Apr. 15	Mar. 1–Apr. 15	Mar. 15–May 15	Mar. 15–May 1	Apr. 1–May 15	Apr. 10–June 1.
Sweetpotato	Feb. 15–May 15	Mar. 1–May 15	Mar. 20–June 1	Mar. 20–June 1	Apr. 1–June 1	Apr. 10–June 1	Apr. 20–June 1.
Tomato	Feb. 1–Apr. 1	Feb. 20–Apr. 10	Mar. 1–Apr. 20	Mar. 10–May 1	Mar. 20–May 10	Apr. 1–May 20	Apr. 10–June 1.
Turnip	Jan. 1–Mar. 1	Jan. 1–Mar. 1	Jan. 10–Mar. 1	Jan. 20–Mar. 1	Feb. 1–Mar. 1	Feb. 10–Mar. 10	Feb. 20–Mar. 20.
Watermelon	Feb. 15–Mar. 15	Feb. 15–Apr. 1	Feb. 15–Apr. 15	Mar. 1–Apr. 15	Mar. 15–Apr. 15	Apr. 1–May 1	Apr. 10–May 15.

(continued)

TABLE 1.8. *(Continued)*.

Crop	Planting dates for localities in which average date of last freeze is—						
	Apr. 10	Apr. 20	Apr. 30	May 10	May 20	May 30	June 10
Asparagus [1]	Mar. 10–Apr. 10	Mar. 15–Apr. 15	Mar. 20–Apr. 15	Mar. 10–Apr. 30	Apr. 20–May 15	May 1–June 1	May 15–June 1.
Beans, lima	Apr. 1–June 30	May 1–June 20	May 15–June 15	May 25–June 15			
Beans, snap	Apr. 10–June 30	Apr. 25–June 30	May 10–June 30	May 10–June 30	May–15–June 30	May 25–June 15	
Beet	Mar. 10–June 1	Mar. 20–June 1	Apr. 1–June 15	Apr. 15–June 15	Apr. 25–June 15	May 1–June 15	May 15–June 15.
Broccoli, sprouting [1]	Mar. 15–Apr. 15	Mar. 25–Apr. 20	Apr. 1–May 1	Apr. 15–June 1	May 1–June 15	May 10–June 10	May 20–June 10.
Brussels sprouts [1]	Mar. 15–Apr. 15	Mar. 25–Apr. 20	Apr. 1–May 1	Apr. 15–June 1	May 1–June 15	May 10–June 10	May 20–June 10.
Cabbage [1]	Mar. 1–Apr. 1	Mar. 10–Apr. 1	Mar. 15–Apr. 10	Apr. 1–May 15	May 1–June 15	May 10–June 15	May 20–June 1.
Cabbage, Chinese	(2)	(2)	(2)	Apr. 1–May 15	May 1–June 15	May 10–June 15	May 20–June 1.
Carrot	Mar. 10–Apr. 20	Apr. 1–May 15	Apr. 10–June 1	Apr. 20–June 15	May 1–June 1	May 10–June 1	May 20–June 1.
Cauliflower [1]	Mar. 1–Mar. 20	Mar. 15–Apr. 20	Apr. 10–May 10	Apr. 15–May 15	May 10–June 15	May 20–June 1	June 1–June 15.
Celery and celeriac	Apr. 1–Apr. 20	Apr. 10–May 1	Apr. 15–May 1	Apr. 20–June 15	May 10–June 15	May 20–June 1	June 1–June 15.
Chard	Mar. 15–June 15	Apr. 1–June 15	Apr. 15–June 15	Apr. 20–June 15	May 10–June 15	May 20–June 1	June 1–June 15.
Chervil and chives	Mar. 1–Apr. 1	Mar. 10–Apr. 10	Mar. 20–Apr. 20	Apr. 1–May 1	Apr. 15–May 15	May 1–June 1	May 15–June 1.
Chicory, witloof	June 10–July 1	June 15–July 1	June 15–July 1	June 1–20	June 1–15	June 1–15	June 1–15.
Collards [1]	Mar. 1–June 1	Mar. 10–June 1	Apr. 1–June 1	Apr. 15–June 1	May 1–June 1	May 10–June 1	May 20–June 1.
Cornsalad	Feb. 1–Apr. 1	Feb. 15–Apr. 15	Mar. 1–May 1	Apr. 1–June 1	Apr. 15–June 1	May 1–June 15	May 15–June 15.
Corn, sweet	Apr. 10–June 1	Apr. 25–June 15	May 10–June 15	May 10–June 1	May 15–June 1	May 20–June 1	
Cress, upland	Mar. 10–Apr. 15	Mar. 20–May 1	Apr. 10–May 10	Apr. 20–May 20	May 1–June 1	May 15–June 1	May 15–June 15.
Cucumber	Apr. 20–June 1	May 1–June 15	May 15–June 15	May 20–June 15	June 1–15		
Eggplant [1]	May 1–June 1	May 10–June 1	May 15–June 10	May 20–June 15	June 1–15		
Endive	Mar. 15–Apr. 15	Mar. 25–Apr. 15	Apr. 1–May 1	Apr. 15–May 15	May 1–30	May 1–30	May 15–June 1.
Fennel, Florence	Mar. 15–Apr. 15	Mar. 25–Apr. 15	Apr. 1–May 1	Apr. 15–May 15	May 1–30	May 1–30	May 15–June 1.
Garlic	Feb. 20–Mar. 20	Mar. 10–Apr. 1	Mar. 15–Apr. 15	Apr. 1–May 1	Apr. 15–May 15	May 1–30	May 15–June 1.
Horseradish [1]	Mar. 10–Apr. 10	Mar. 20–Apr. 20	Apr. 1–30	Apr. 15–May 15	Apr. 20–May 20	May 1–30	May 15–June 1.
Kale	Mar. 10–Apr. 1	Mar. 20–Apr. 10	Apr. 1–20	Apr. 10–May 1	Apr. 20–May 10	May 1–30	May 15–June 1.
Kohlrabi	Mar. 10–Apr. 10	Mar. 20–May 1	Apr. 1–May 10	Apr. 10–May 15	Apr. 20–May 20	May 1–30	May 15–June 1.
Leek	Mar. 1–Apr. 1	Mar. 15–Apr. 15	Apr. 1–May 1	Apr. 15–May 15	May 1–May 20	May 1–15	May 1–15.
Lettuce, head [1]	Mar. 10–Apr. 1	Mar. 20–Apr. 15	Apr. 1–May 1	Apr. 15–May 15	May 1–June 30	May 10–June 30	May 20–June 30.
Lettuce, leaf	Mar. 15–May 15	Mar. 20–May 15	Apr. 1–June 1	Apr. 15–June 15	May 1–June 30	May 10–June 30	May 20–June 30.
Muskmelon	Apr. 20–June 1	May 1–June 15	May 15–June 15	June 1–June 15			
Mustard	Mar. 10–Apr. 20	Mar. 20–May 1	Apr. 1–May 10	Apr. 15–June 1	May 1–June 30	May 10–June 30	May 20–June 30.
Okra	Apr. 20–June 15	May 1–June 1	May 10–June 1	May 20–June 10	June 1–20		
Onion [1]	Mar. 1–Apr. 1	Mar. 15–Apr. 10	Apr. 1–May 1	Apr. 10–May 1	Apr. 20–May 15	May 1–30	May 10–June 10.
Onion, seed	Mar. 1–Apr. 1	Mar. 15–Apr. 1	Mar. 15–Apr. 15	Apr. 1–May 1	Apr. 20–May 15	May 1–30	May 10–June 10.
Onion, sets	Mar. 1–Apr. 1	Mar. 10–Apr. 1	Mar. 10–Apr. 10	Apr. 10–May 1	Apr. 20–May 15	May 1–30	May 10–June 10.
Parsley	Mar. 10–Apr. 10	Mar. 20–Apr. 20	Apr. 1–May 1	Apr. 15–May 15	May 1–20	May 10–June 1	May 20–June 10.
Parsnip	Mar. 10–Apr. 10	Mar. 20–Apr. 20	Apr. 1–May 1	Apr. 15 May 15	May 1–20	May 10–June 1	May 20–June 10.
Peas, garden	Feb. 20–Mar. 20	Mar. 10–Apr. 10	Mar. 20–May 1	Apr. 1–May 15	Apr. 15–June 1	May 1–June 15	May 10–June 15.
Peas, black-eye	May 1–July 1	May 10–June 15	May 15–June 1				
Pepper [1]	May 1–June 1	May 10–June 1	May 15–June 10	May 20–June 10	May 25–June 15	June 1–15	
Potato	Mar. 10–Apr. 1	Mar. 15–Apr. 10	Mar. 20–May 10	Apr. 1–June 1	Apr. 15–June 15	May 1–June 15	May 15–June 1.
Radish	Mar. 1–May 1	Mar. 10–May 10	Mar. 20–May 10	Apr. 1–June 1	Apr. 15–June 15	May 1–June 15	May 15–June 1.
Rhubarb [1]	Mar. 1–Apr. 1	Mar. 10–Apr. 10	Mar. 20–Apr. 15	Apr. 1–May 1	Apr. 15–May 10	May 1–20	May 15–June 1.
Rutabaga			May 1–June 1	May 1–June 1	May 1–20	May 10–20	May 20–June 1.
Salsify	Mar. 10–Apr. 15	Mar. 20–May 1	Apr. 1–May 15	Apr. 15–June 1	May 1–June 1	May 10–June 1	May 20–June 1.
Shallot	Mar. 1–Apr. 1	Mar. 15–Apr. 15	Apr. 1–May 1	Apr. 10–May 1	Apr. 20–May 10	May 1–June 1	May 10–June 1.
Sorrel	Mar. 1–Apr. 15	Mar. 15–May 1	Apr. 1–May 15	Apr. 15–June 1	May 1–June 1	May 10–June 10	May 20–June 10.
Soybean	May 1–June 30	May 10–June 20	May 15–June 15	May 25–June 10			
Spinach	Feb. 15–Apr. 1	Mar. 1–Apr. 15	Mar. 20–Apr. 20	Apr. 1–June 15	Apr. 10–June 15	Apr. 20–June 15	May 1–June 15.
Spinach, New Zealand	Apr. 20–June 1	May 1–June 15	May 1–June 15	May 10–June 15	May 20–June 15	June 1–15	
Squash, summer	Apr. 20–June 1	May 1–June 15	May 1–30	May 10–June 10	May 20–June 15	June 1–20	June 10–20.
Sweetpotato	May 1–June 1	May 10–June 10	May 20–June 10				
Tomato	Apr. 20–June 1	May 5–June 10	May 10–June 15	May 15–June 10	May 25–June 15	June 5–20	June 15–30.
Turnip	Mar. 1–Apr. 1	Mar. 10–Apr. 1	Mar. 20–May 1	Apr. 1–June 1	Apr. 15–June 1	May 1–June 15	May 15–June 15.
Watermelon	Apr. 20–June 1	May 1–June 15	May 15–June 15	June 1–June 15	June 15–July 1		

Source: Wester (1972).
[1] Plants.
[2] Generally fall-planted (Table 1.7).

plant; it also gives spring and early summer dates *beyond which planting usually gives poor results* (Wester 1972).

Opposite each vegetable given in Table 1.8 are a series of columns. No gardener needs to use more than one of these columns. Each column contains two planting dates. The first date is the *earliest safe date* that a specific vegetable can be planted or transplanted by a person using that particular column. The second date is the last satisfactory date for planting that vegetable. All the times between these two are not equally satisfactory. Most of the vegetables grow better and yield more when planted near the earlier date given.

The proper time to plant any vegetable in nonsheltered locations in the spring can be determined as follows: (1) find the location of the garden in Fig. 1.6. Then find the solid line on the map (Fig. 1.6) that is closest to it. (2) Find the date shown on the solid line. This date is the average date of the last killing frost in the spring. The first number represents the month; the second number represents the day. Thus 4-30 is April 30. Once the gardener has found the date, the map is no longer needed. (3) Now turn to Table 1.8 and find the specific column that has your date over it. This is the only date column that the gardener will need. It can be outlined in ink for ready reference.(4) Find the dates in your column that are on the same line with the vegetable you wish to plant. The dates show the period during which the crop can be planted safely. The best time for planting is on the first date or shortly thereafter.

The Great Plains region warms up rapidly in the spring and is subject to dry weather. In these areas, very early planting is essential for the vegetables to escape the heat and drought. Most cool season crops grow poorly when planted in the spring in the southern part of the Great Plains and southern Texas (Wester 1972). These plants are generally fall-grown.

Fall Gardens

After the summer vegetables are harvested, a fall garden should be planted (Table 1.7). The fall garden is often more productive and fall-grown vegetables are usually of higher canning and freezing quality than those which mature during the hot dry periods of midsummer. Fall gardens usually contain hardy plants that can be planted in late summer or fall about 6–8 weeks before the average date of the first freeze in the fall. Rainfall, insects, and diseases are often more important than temperature for the successful growing of late summer and fall crops, particularly in the South and Southeast.

The season of the year in which a vegetable may be grown depends upon the region of production. In southern Florida and Texas, warm season crops can be grown during the winter. During the summer, cool season vegetables can be grown at high elevations. A good rule of thumb on planting time is that for every 100 miles distance from south to north, a week's delay in planting time is required. This generally does not hold if the climate is moderated by large bodies of water or adversely influenced by altitude.

In the extreme north, there is little opportunity for fall gardens. In the Deep South, spring is a short season. Cool-season crops such as collards, cauliflower, spinach, late cabbage, broccoli, turnips, and Chinese cabbage are usually planted in late summer for fall and winter harvest.

The area for the fall garden should be spaded and fertilized before planting, as for the spring garden. The area can then be seeded. However, the cool-season crops will need to be irrigated to establish the plants; and some vegetables such as lettuce seed will not germinate if soil temperatures are greater than 75°F. Some crops, such as Brussels sprouts, Chinese cabbage, cabbage, and cauliflower, can be planted as transplants for the fall garden. As these transplants can seldom be purchased, the gardener needs to seed these in a cooler area about 6 weeks before they are planted in the garden. A small portion of the garden can be used. The seed is planted and irrigated, and a board is placed over the row for several days until the seed has germinated. The board will help keep the ground cooler. Once the seeds have germinated, the board is removed and the plants are allowed to grow until transplanted (thin them if needed). Weeds grow quickly in a fall garden, but organic mulches may be applied soon after planting to control weeds. As the soil is warm, these mulches will also reduce the soil temperature, benefiting the cool season crops. Organic mulching materials require nitrogen for decomposition and will compete with the crop for the available soil nitrogen. About 2 lb of a complete fertilizer, such as 10-10-10, per 100 ft of row will usually be sufficient for fall-planted crops growing in an organic mulch.

In most areas of mid-America, a fall crop of green or wax beans can be grown. These plants will not withstand frost, however.

Frost-tolerant vegetables such as peas and radishes can be grown in a fall garden. Beets and carrots will withstand a frost (32°F) and can be harvested throughout the fall. Fall-grown beets and carrots are better for winter storage than summer-grown crops.

Many cool-season crops are very hardy and withstand temperatures of 25°F. These crops include Brussels sprouts, collards, kale, lettuce, leeks, onions, turnips, parsnips, salsify, cabbage, cauliflower, Chinese cabbage, and broccoli. The flavor of turnips, parsnips, collards, salsify, and Chinese cabbage improves after a frost.

Leeks have a long growing season and are usually planted as a spring or summer vegetable. They are very hardy, however, and can be left in the garden and harvested throughout the winter. Salsify and parsnips may be mulched and also harvested during winter and early spring. Onion top sets from winter onions can be planted for fall use. Those not used can be left in the garden and used as green onions in very early spring. Once winter onions have begun spring growth, the plants become very pungent.

Fall Planting Dates. To determine the dates for late summer and fall plantings, Fig. 1.7 and Table 1.9 are used. They are used in a similar manner to the method used to determine the spring and summer planting dates given previously. A planting date about in the middle of those given in Table 1.9 is usually best, but in most areas of the United States fair success can be achieved over the entire range of dates given. The late planting dates given in Table 1.9 are less exact and less dependable than the early date, particularly for gardens in the South where overwintered crops are grown.

In the Pacific Northwest, warm weather vegetables should not be planted quite so late as frost data (Table 1.9 and Fig. 1.7) indicate. Frost occurs late, but cool temperatures occur for some time before the frost. This reduces the growth rate of warm weather vegetables such as lima beans, sweet corn, and tomatoes.

CROPPING SYSTEMS

No one garden plan or arrangement of the plants within the garden will be suitable for all conditions. If space is at a premium and hand cultivation or mulching is practiced, the distance between rows can be kept to a minimum. The suggestions presented here can be adapted to the specific needs of individual gardeners.

Intercropping. Many gardeners prefer to mix several types of vegetables rather than plant them in blocks. Sweet corn must be planted in a block for proper pollination, however. Intercropping uses all the garden space efficiently. This technique also reduces insect activity and helps reduce problems with insect pests. Intercropping is planting fast-growing and slow-growing vegetables in alternate rows, alternating them within the same row, or planting early-maturing crops between rows of late-maturing ones.

An example of intercropping in alternate rows is to plant a row of radishes, then tomatoes, then onions, and then cauliflower. The radishes and onions can be harvested in time to make room for the later maturing crops. Pumpkins, winter squash, and gourds are often planted beside an early block of sweet corn. The vines can be trained to grow into the area where the sweet corn is growing. After the early crop of sweet corn is harvested, the stems are cut off about a foot from the soil surface and the vine crops left to utilize this space.

An example of intercropping by alternating different plants within the same row is to plant green onions, radishes, or lettuce between caged tomato plants or vine crops (cucumbers, pumpkins, winter squash). The early-maturing crops can be harvested before the tomatoes or vine crops require the space.

Several types of vegetables can be intercropped by planting them between rows. Green onions, lettuce, radishes, or spinach may be planted between rows of cabbage, peppers, sweet corn, and tomatoes. An early-maturing sweet corn can be planted in the middle of the growing season between rows of early planted potatoes. The early-maturing crops will not interfere with the growth of the late-maturing ones. (See Fig. 1.8.)

In dry-land areas where moisture is limited, intensive intercropping is generally not feasible. In these areas more space or irrigation water is required.

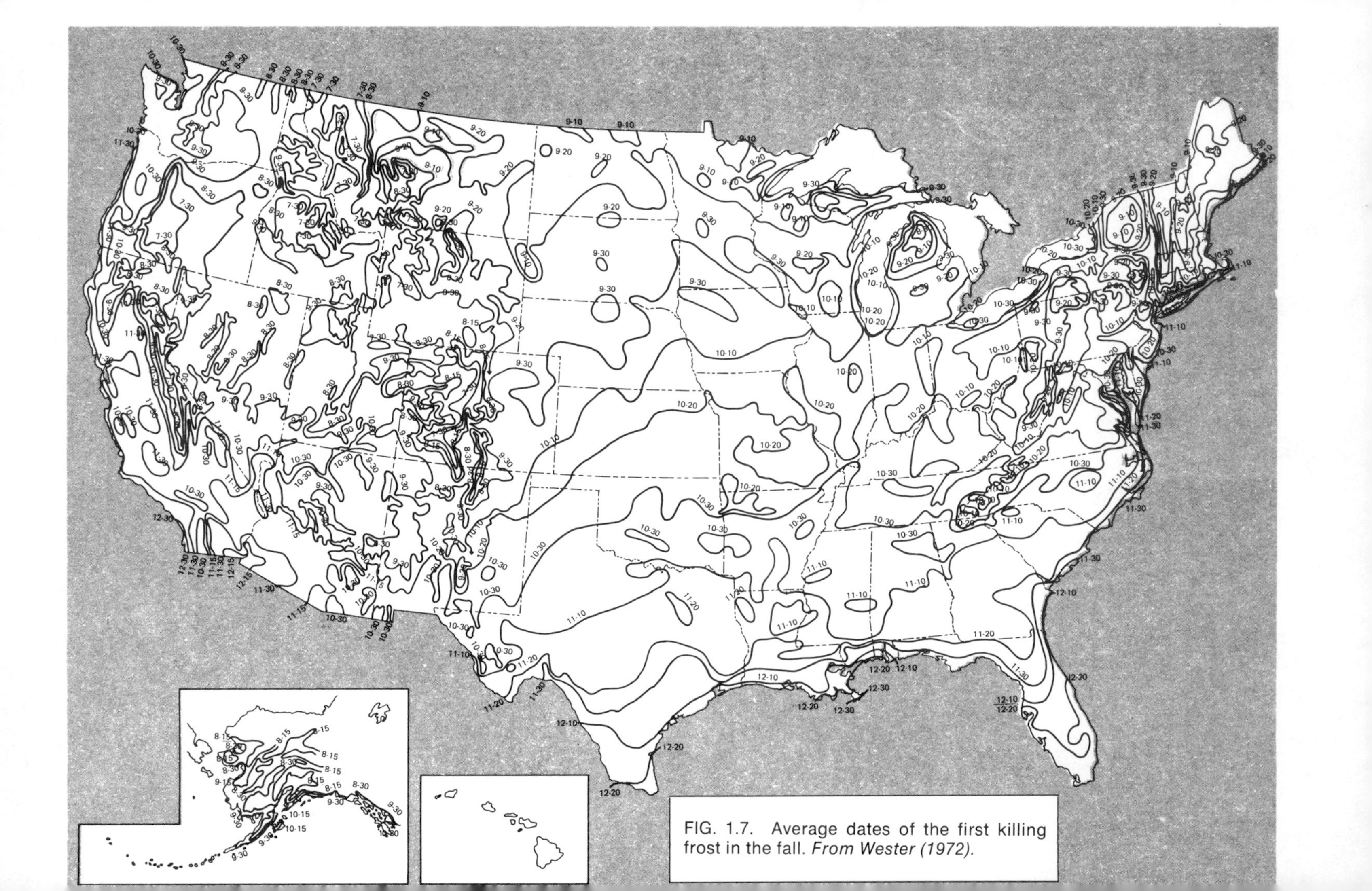

FIG. 1.7. Average dates of the first killing frost in the fall. *From Wester (1972).*

TABLE 1.9. Latest Dates, and Range of Dates, for Safe Outdoor Planting of Vegetables in the Fall.

Crop	Planting dates for localities in which average date of first freeze is—										
	Aug. 30	Sept. 10	Sept. 20	Sept. 30	Oct. 10	Oct. 20	Oct. 30	Nov. 10	Nov. 20	Nov. 30	Dec. 10	Dec. 20
Asparagus [1]					Oct. 20–Nov. 15	Nov. 1–Dec. 15.	Nov. 15–Jan. 1	Dec. 1–Jan. 1				
Beans, lima				June 1–15	June 1–15	June 15–30.	July 1–Aug. 1	July 1–Aug. 15	July 15–Sept. 1	Aug. 1–Sept. 15	Sept. 1–30	Sept. 1–Oct. 1.
Beans, snap		May 15–June 15	June 1–July 1	June 1–July 10	June 15–July 20	July 1–Aug. 1.	July 1–Aug. 15	July 1–Sept. 1	July 1–Sept. 10	Aug. 15–Sept. 20	Sept. 1–30	Sept. 1–Nov. 1.
Beet	May 15–June 15	May 15–June 15	June 1–July 1	June 1–July 10	June 15–July 25	July 1–Aug. 5.	Aug. 1–Sept. 1	Aug. 1–Oct. 1	Sept. 1–Dec. 1	Sept. 1–Dec. 15	Sept. 1–Dec. 31	Sept. 1–Dec. 31.
Broccoli, sprouting	May 1–June 1	May 1–June 1	May 1–June 15	June 1–30	June 15–July 15	July 1–Aug. 1.	July 1–Aug. 15	Aug. 1–Sept. 1	Aug. 1–Sept. 15	Aug. 1–Oct. 1	Aug. 1–Nov. 1	Sept. 1–Dec. 31.
Brussels sprouts	May 1–June 1	May 1–June 1	May 1–June 15	June 1–30	June 15–July 15	July 1–Aug. 1.	July 1–Aug. 15	Aug. 1–Sept. 1	Aug. 1–Sept. 15	Aug. 1–Oct. 1	Aug. 1–Nov. 1	Sept. 1–Dec. 31.
Cabbage [1]	May 1–June 1	May 1–June 1	May 1–June 15	June 1–July 10	June 1–July 15	July 1–20.	Aug. 1–Sept. 1	Sept. 1–15	Sept. 1–Dec. 1	Sept. 1–Dec. 31	Sept. 1–Dec. 31	Sept. 1–Dec. 31.
Cabbage, Chinese	May 15–June 15	May 15–June 15	June 1–July 1	June 1–July 15	June 15–Aug. 1	July 15–Aug. 15.	Aug. 1–Sept. 15	Aug. 15–Oct. 1	Sept. 1–Oct. 15	Sept. 1–Nov. 1	Sept. 1–Nov. 15	Sept. 1–Dec. 1.
Carrot	May 15–June 15	May 15–June 15	June 1–July 1	June 1–July 10	June 1–July 20	June 15–Aug. 1.	July 1–Aug. 15	Aug. 1–Sept. 1	Sept. 1–Nov. 1	Sept. 15–Dec. 1	Sept. 15–Dec. 1	Sept. 15–Dec. 1.
Cauliflower [1]	May 1–June 1	May 1–July 1	May 1–July 1	May 10–July 15	June 1–July 25	July 1–Aug. 5.	July 15–Aug. 15	Aug. 1–Sept. 1	Aug. 1–Sept. 15	Aug. 15–Oct. 10	Sept. 1–Oct. 20	Sept. 15–Nov. 1.
Celery [1] and celeriac	May 1–June 1	May 15–June 15	May 15–July 1	June 1–July 5	June 1–July 15	June 1–Aug. 1.	June 15–Aug. 15	July 1–Aug. 15	July 15–Sept. 1	Aug. 1–Dec. 1	Sept. 1–Dec. 31	Oct. 1–Dec. 31.
Chard	May 15–June 15	May 15–July 1	June 1–July 1	June 1–July 5	June 1–July 20	June 1–Aug. 1.	June 1–Sept. 10	June 1–Sept. 15	June 1–Oct. 1	June 1–Nov. 1	June 1–Dec. 1	June 1–Dec. 31.
Chervil and chives	May 10–June 10	May 1–June 15	May 15–June 15	(2)	(2)	(2)	(2)	(2)	Nov. 1–Dec. 31	Nov. 1–Dec. 31	Nov. 1–Dec. 31	Nov. 1–Dec. 31.
Chicory, witloof	May 15–June 15	May 15–June 15	May 15–June 15	June 1–July 1	June 1–July 1	June 15–July 15.	July 1–Aug. 10	July 10–Aug. 20	July 20–Sept. 1	Aug. 15–Sept. 30	Aug. 15–Oct. 15	Aug. 15–Oct. 15.
Collards [1]	May 15–June 15	May 15–June 15	May 15–June 15	June 15–July 15	July 1–Aug. 1	July 15–Aug. 15.	Aug. 1–Sept. 15	Aug. 15–Oct. 1	Aug. 25–Nov. 1	Sept. 1–Dec. 1	Sept. 1–Dec. 31	Sept. 1–Dec. 31.
Cornsalad	May 15–June 15	May 15–July 1	June 15–Aug. 1	July 15–Sept. 1	Aug. 15–Sept. 15	Sept. 1–Oct. 15.	Sept. 15–Nov. 1	Oct. 1–Dec. 1	Oct. 1–Dec. 1	Oct. 1–Dec. 31	Oct. 1–Dec. 31	Oct. 1–Dec. 31.
Corn, sweet			June 1–July 1	June 1–July 1	June 1–July 10	June 1–July 20.	June 1–Aug. 1	June 1–Aug. 15	June 1–Sept. 1			
Cress, upland	May 15–June 15	May 15–July 1	June 15–Aug. 1	July 15–Sept. 1	Aug. 15–Sept. 15	Sept. 1–Oct. 15.	Sept. 15–Nov. 1	Oct. 1–Dec. 1	Oct. 1–Dec. 1	Oct. 1–Dec. 31	Oct. 1–Dec. 31	Oct. 1–Dec. 31.
Cucumber			June 1–15	June 1–July 1	June 1–July 1	June 1–July 15.	June 1–Aug. 1	June 1–Aug. 15	June 1–Aug. 15	July 15–Sept. 15	Aug. 15–Oct. 1	Aug. 15–Oct. 1.
Eggplant [1]				May 20–June 10	May 15–June 15	June 1–July 1.	June 1–July 1	June 1–July 15	June 1–Aug. 1	July 1–Sept. 1	Aug. 1–Sept. 30	Aug. 1–Sept. 30.
Endive	June 1–July 1	June 1–July 1	June 15–July 15	June 15–Aug. 1	July 1–Aug. 15	July 15–Sept. 1.	July 15–Aug. 15	Aug. 1–Sept. 1	Sept. 1–Oct. 1	Sept. 1–Nov. 15	Sept. 1–Dec. 31	Sept. 1–Dec. 31.
Fennel, Florence	May 15–June 15	May 15–July 15	June 1–July 1	June 1–July 1	June 15–July 15	June 15–Aug. 1.	July 1–Aug. 1	July 15–Aug. 15	Aug. 15–Sept. 15	Sept. 1–Nov. 15	Sept. 1–Dec. 1	Sept. 1–Dec. 1.
Garlic	(2)	(2)	(2)	(2)	(2)	(2)	(2)	Aug. 1–Oct. 1	Aug. 15–Oct. 1	Sept. 1–Nov. 15	Sept. 15–Nov. 15	Sept. 15–Nov. 15.
Horseradish [1]	(2)	(2)	(2)	(2)	(2)	(2)	(2)	(2)	(2)	(2)	(2)	(2)
Kale	May 15–June 15	May 15–June 15	June 1–July 1	June 15–July 15	July 1–Aug. 1	July 15–Aug. 15.	July 15–Sept. 1	Aug. 1–Sept. 15	Aug. 15–Oct. 15	Sept. 1–Dec. 1	Sept. 1–Dec. 31	Sept. 1–Dec. 31.
Kohlrabi	May 15–June 15	June 1–July 1	June 1–July 15	June 15–July 15	July 1–Aug. 1	July 15–Aug. 15.	Aug. 1–Sept. 1	Aug. 15–Sept. 15	Sept. 1–Oct. 15	Sept. 1–Dec. 1	Sept. 15–Dec. 31	Sept. 1–Dec. 31.
Leek	May 1–June 1	May 1–June 1	(2)	(2)	(2)	(2)	(2)	(2)	Sept. 1–Nov. 1	Sept. 1–Nov. 1	Sept. 1–Nov. 1	Sept. 15–Nov. 1
Lettuce, head [1]	May 15–July 1	May 15–July 1	June 1–July 15	June 15–Aug. 1	July 15–Aug. 15	Aug. 1–30.	Aug. 1–Sept. 15	Aug. 15–Oct. 15	Sept. 1–Nov. 1	Sept. 1–Dec. 1	Sept. 15–Dec. 31	Sept. 15–Dec. 31.
Lettuce, leaf	May 15–July 15	May 15–July 15	June 1–Aug. 1	June 1–Aug. 1	July 15–Sept. 1	July 15–Sept. 1.	Aug. 15–Oct. 1	Aug. 25–Oct. 1	Sept. 1–Nov. 1	Sept. 1–Dec. 1	Sept. 15–Dec. 31	Sept. 15–Dec. 31.
Muskmelon			May 1–June 15	May 15–June 1	June 1–June 15	June 15–July 20.	July 1–July 15	July 15–July 30				
Mustard	May 15–July 15	May 15–July 15	June 1–Aug. 1	June 15–Aug. 1	July 15–Aug. 15	Aug. 1–Sept. 1.	Aug. 15–Oct. 15	Aug. 15–Nov. 1	Sept. 1–Dec. 1	Sept. 1–Dec. 1	Sept. 1–Dec. 1	Sept. 15–Dec. 1.
Okra			June 1–20	June 1–July 1	June 1–July 15	June 1–Aug. 1.	June 1–Aug. 10	June 1–Aug. 20	June 1–Sept. 10	June 1–Sept. 20	Aug. 1–Oct. 1	Aug. 1–Oct. 1.
Onion [1]	May 1–June 10	May 1–June 10	(2)	(2)	(2)	(2)		Sept. 1–Oct. 15	Oct. 1–Dec. 31	Oct. 1–Dec. 31	Oct. 1–Dec. 31	Oct. 1–Dec. 31.
Onion, seed	May 1–June 1	May 1–June 10	(2)	(2)	(2)	(2)			Sept. 1–Nov. 1	Sept. 1–Nov. 1	Sept. 1–Nov. 1	Sept. 15–Nov. 1.
Onion, sets	May 1–June 1	May 1–June 10	(2)	(2)	(2)	(2)		Oct. 1–Dec. 1	Nov. 1–Dec. 31	Nov. 1–Dec. 31	Nov. 1–Dec. 31	Nov. 1–Dec. 31.
Parsley	May 15–June 15	May 1–June 15	June 1–July 1	June 1–July 15	June 15–Aug. 1	July 15–Aug. 15.	Aug. 1–Sept. 15	Sept. 1–Nov. 15	Sept. 1–Dec. 31	Sept. 1–Dec. 31	Sept. 1–Dec. 31	Sept. 1–Dec. 31.
Parsnip	May 15–June 1	May 1–June 15	May 15–June 15	June 1–July 1	June 1–July 10	(2)	(2)	(2)	Aug. 1–Sept. 1	Sept. 1–Nov. 15	Sept. 1–Dec. 1	Sept. 1–Dec. 1.
Peas, garden	May 10–June 15	May 1–July 1	June 1–July 15	June 1–Aug. 1	(2)	(2)	Aug. 1–Sept. 15	Sept. 1–Nov. 1	Oct. 1–Dec. 1	Oct. 1–Dec. 31	Oct. 1–Dec. 31	Oct. 1–Dec. 31.
Peas, black-eye					June 1–July 1	June 1–July 1.	June 1–Aug. 1	June 15–Aug. 15	July 1–Sept. 1	July 1–Sept. 10	July 1–Sept. 20	July 1–Sept. 20.
Pepper [1]			June 1–June 20	June 1–July 1	June 1–July 1	June 1–July 10.	June 1–July 20	June 1–Aug. 1	June 1–Aug. 15	June 15–Sept. 1	Aug. 15–Oct. 1	Aug. 15–Oct. 1.
Potato	May 15–June 1	May 1–June 15	May 1–June 15	May 1–June 15	May 15–June 15	June 15–July 15.	July 20–Aug. 10	July 25–Aug. 20	Aug. 10–Sept. 15	Aug. 1–Sept. 15	Aug. 1–Sept. 15	Aug. 1–Sept. 15.
Radish	May 1–July 15	May 1–Aug. 1	June 1–Aug. 15	July 1–Sept. 1	July 15–Sept. 15	Aug. 1–Oct. 1.	Aug. 15–Oct. 15	Sept. 1–Nov. 15	Sept. 1–Dec. 1	Sept. 1–Dec. 31	Aug. 1–Sept. 15	Oct. 1–Dec. 31.
Rhubarb [1]	Sept. 1–Oct. 1	Sept. 15–Oct. 15	Sept. 15–Nov. 1	Oct. 1–Nov. 1	Oct. 15–Nov. 15	Oct. 15–Dec. 1.	Nov. 1–Dec. 1					
Rutabaga	May 15–June 15	May 1–June 15	June 1–July 1	June 1–July 1	June 15–July 15	July 10–20.	July 15–Aug. 1	July 15–Aug. 15	Aug. 1–Sept. 1	Sept. 1–Nov. 15	Oct. 1–Nov. 15	Oct. 15–Nov. 15.
Salsify	May 15–June 1	May 10–June 10	May 20–June 20	June 1–20	June 1–July 1	June 1–July 1.	June 1–July 10	June 15–July 20	July 15–Aug. 15	Aug. 15–Sept. 30	Aug. 15–Oct. 15	Sept. 1–Oct. 31.
Shallot	(2)	(2)	(2)	(2)	(2)	(2)	(2)	Aug. 1–Oct. 1	Aug. 15–Oct. 1	Aug. 15–Oct. 15	Sept. 15–Nov. 1	Sept. 15–Nov. 1.
Sorrel	May 15–June 15	May 1–June 15	June 1–July 1	June 1–July 15	July 1–Aug. 1	July 15–Aug. 15.	Aug. 1–Sept. 15	Aug. 15–Oct. 1	Aug. 15–Oct. 15	Sept. 1–Nov. 15	Sept. 1–Dec. 15	Sept. 1–Dec. 31.
Soybean				May 25–June 10	June 1–25	June 1–July 5.	June 1–July 15	June 1–July 25	June 1–July 30	June 1–July 30	June 1–July 30	June 1–July 30.
Spinach	May 15–July 1	June 1–July 15	June 1–Aug. 1	July 1–Aug. 15	Aug. 1–Sept. 1	Aug. 20–Sept. 10.	Sept. 1–Oct. 1	Sept. 15–Nov. 1	Oct. 1–Dec. 1	Oct. 1–Dec. 31	Oct. 1–Dec. 31	Oct. 1–Dec. 31.
Spinach, New Zealand				May 15–July 1	June 1–July 15	June 1–Aug. 1.	June 1–Aug. 1	June 1–Aug. 15	June 1–Aug. 15			
Squash, summer	June 10–20	June 1–20	May 15–July 1	June 1–July 1	June 1–July 15	June 1–July 20.	June 1–Aug. 1	June 1–Aug. 10	June 1–Aug. 20	June 1–Sept. 1	June 1–Sept. 15	June 1–Oct. 1.
Squash, winter			May 20–June 10	June 1–15	June 1–July 1	June 1–July 1.	June 10–July 10	June 20–July 20	July 1–Aug. 1	July 15–Aug. 15	Aug. 1–Sept. 1	Aug. 1–Sept. 1.
Sweet potato					May 20–June 10	June 1–15.	June 1–15	June 1–July 1	June 1–July 1	June 1–July 1	June 1–July 1	June 1–July 1.
Tomato	June 20–30	June 10–20	June 1–20	June 1–20	June 1–20	June 1–July 1.	June 1–July 1	June 1–July 15	June 1–Aug. 1	Aug. 1–Sept. 1	Aug. 15–Oct. 1	Sept. 1–Nov. 1.
Turnip	May 15–June 15	June 1–July 1	June 1–July 15	June 1–Aug. 1	July 1–Aug. 1	July 15–Aug. 15.	Aug. 1–Sept. 15	Sept. 1–Oct. 15	Sept. 1–Nov. 15	Sept. 1–Nov. 15	Oct. 1–Dec. 1	Oct. 1–Dec. 31.
Watermelon			May 1–June 15	May 15–June 1	June 1–June 15	June 15–July 20.	July 1–July 15	July 15–July 30				

Source: Wester (1972).

[1] Plants.

[2] Generally spring-planted (Table 1.6).

FIG. 1.8. Chinese cabbage growing abo crop of early potatoes. *From Anon. (1978*

Succession Planting. Succession planting is the growing of one crop, removing it, and then planting another crop in the same space. Except in dry-land areas, the garden space should be fully utilized throughout the growing season. In order to have a continuous supply of fresh vegetables, successive plantings are required. It is better to plant a small area that will yield a small amount that can be used than to have a large amount produced all at once. Two or three small plantings of leaf lettuce and radishes should be made a week apart during the cool season. Onions can be planted every 2 weeks to provide for green onions. An early planting of carrots, cabbage, and beets can be used fresh and a late planting can be used for freezing or storage.

It is best to follow one crop not with a similar crop, but with a different one to reduce insect and disease problems. Tomatoes can be planted after harvesting radishes, and cucumbers after spinach. Early peas or beans can be followed by beets, late cabbage, carrots, or celery. Early sweet corn or potatoes can be followed by spinach or turnips. Fast-maturing vegetables that can be harvested early and the space used by another crop include early beets, early carrots, early Chinese cabbage, kohlrabi, leaf lettuce, green onions, peas, radishes, and spinach.

In the South, south Atlantic, Gulf Coast, and some western regions, some vegetables can be grown in the garden every month of the year. In the north, late summer and fall plantings of vegetables should be made. In the extreme north, where the growing season is relatively short, succession plantings of early maturing vegetables only are possible. In the South, the late fall garden is as important as the late summer one. Late plantings of snap beans, lima beans, spinach, radishes, and beets can be grown.

Vertical Cropping. Many garden plants can be trained to use some type of support to use the space above the garden. This area is not otherwise used, and garden space can be increased without increasing the soil area. This is particularly important for growing plants in minigardens and container gardens. Peas and pole beans will climb naturally, but tomatoes, cucumbers, squash, melons, and gourds can be trained to climb upward (Fig. 1.9).

Only the gardener's imagination limits the use of various structures to support plants. Wire, fences, poles, trellises, and other structures can be used to support plants. Vines can be partially supported by loosely tying them to wooden fences. Tomatoes can be staked or grown in a wire cage. Peas and various pole beans can be grown on a fence made of string and closely spaced stakes. Cucumbers and other vines can be trained to grow into trees and shrubs at the edge of the garden.

It is important that the structures be sturdy enough so they will not tip over in the wind or from the weight of the plants growing on them. The fruit of pumpkins and some cultivars of winter squash and melons become very large. These fruits need to be supported to prevent them from breaking off

FIG. 1.9. Pole beans growing up strings supported by a center pole in the pot to use vertical space. *Courtesy of National Garden Bureau.*

from the vine and falling to the ground. These fruits can be supported by wrapping them with several layers of cheesecloth or other material and tying this to the support.

Cover Crops. A cover crop is grown, not for its edible plant parts, but to eventually incorporate it into the soil. When a vegetable has been harvested and it is too late in the season to plant another, a cover crop can be planted. Seed can be planted between the rows of all vegetables or broadcast over the entire area where no vegetables are growing. The most popular cover crop is an even mixture (3 lb) of winter rye and vetch. The vetch is a legume, and its bacteria fix nitrogen from the air. About 3 lb of annual rye or 1 lb of ryegrass or oat seeds can also be seeded per 1000 sq ft of garden. If the gardener ordered an excess of bush bean seeds in the spring, these can also be used as a cover crop. If there is a late frost, an additional picking of snap beans can be obtained as a bonus. If not, the beans will provide additional nitrogen as a legume and organic matter as a cover crop. The seed should be worked into the soil with a rake and irrigated to establish the plants, if it does not rain. Seeds should be planted about a month before the first frost.

The cover crops can be spaded into the soil as soon as the frost is out of the ground in the spring. Cover crops provide additional organic matter to the soil and return nutrients from the lower soil depths. Some of these nutrients would otherwise be leached from the soil by fall rains. Additionally, cover crops prevent erosion on sloping sites and in windy, sandy areas.

Permanent Mulch Systems. Some gardeners prefer to leave a mulch on the garden all year round and not plow or spade the soil. The decision to use a permanent mulch system depends upon the climate, the type of mulch, and the vegetables the gardener wishes to grow. Perennial vegetables such as asparagus and rhubarb can be organically mulched permanently and new mulch added each year. Organic mulches lower the soil temperature and conserve moisture. In northern climates or in rainy areas, a permanent mulch will prevent adequate drying and warming of the soil in the spring. The soil will remain cool and wet and planting dates for the vegetables will be delayed. The mulch will have to be moved aside for planting to occur; and small-seeded crops such as lettuce, carrots, beets, and radishes will have difficulty emerging through the mulch, particularly sawdust and wood chip mulches. Frequently, no small-seeded plants emerge through a 3 in. layer of mulch, and seeds of vine crops such as pumpkins and squash rot due to the cold, wet germination conditions.

A permanent mulch system can be used to an advantage in warm climates with low rainfall and sandy soils. A loose, porous organic mulch such as straw or hay should be used and a 3 to 6 in. layer of mulch spread over the area. The hay or straw is easily moved to one side in the spring to allow the planting of small-seeded crops; once they are established, the straw can be moved around the plants. The mulch can be placed around transplanted plants immediately. Sandy soils are well-drained and soil temperatures increase rapidly in southern areas, eliminating the cool, wet soil conditions.

As the soil temperatures increase rapidly, a permanent mulch will decrease the soil temperature in the spring increasing the time various cool season crops can be grown. The mulch will partially decompose during the year and less material will have to be moved aside to plant small-seeded crops. If legume (alfalfa, clover) hay is used, no additional nitrogen is needed, but straw will require about 2 lb of fertilizer (5-10-5; 6-10-4 or similar) per 100 sq ft to prevent any nitrogen deficiency occurring in the vegetable crops growing through the straw.

A variation of the permanent mulch system has been used successfully in cooler, wetter climates. The hay or straw is moved to one side of the garden in early spring, allowing the soil to dry out and warm up. The vegetables are seeded and once established, the mulch is replaced over the area.

Double Digging. This system is best used with deeply rooted crops that will be grown in one place for several years, such as asparagus and rhubarb; or with root crops.

The traditional double digging method entails first digging a trench (Area 1) one spade deep, and laying the soil aside. The soil in the bottom of the trench (Area 2) is dug one spade deep, mixed with organic matter, and returned to the trench. Compost or composted manure are best, but peat may be purchased and used. The final mixture contains 25–50% organic matter. Next, the soil behind Area 1 (Area 3) is dug one spade deep, mixed with organic matter, and placed in Area 1. The soil in the bottom of Area 3 is then dug one spade deep, and treated and returned as the soil in Area 2 was. The soil behind Area 3 is now dug, treated, and returned to Area 3, and so on. When the end of the garden is reached a trench will remain, and the the soil from Area 1 (mixed with organic matter) is placed in this trench. Typically, the garden soil has been dug to a 18-inch depth, and the organic matter has increased the soil volume 6–9 in. Most gardeners now outline the garden with wood ties or landscape timbers, making the garden a raised bed.

This method requires a lot of labor, but should be considered where sandy or poor soils exist. Root penetration is easier, there is more aeration, the nutrient holding capacity of the soil is increased, and the harvesting of root crops, such as potatoes, is made easier.

Hydroponics. The term hydroponics is generally used to describe any of several methods of growing plants without soil. A number of hydroponic systems are available for purchase, but all are expensive, complicated, and require considerable labor and maintenance.

An inexpensive hydroponic system (Imai 1987) can be made with a polystyrene container (pop cooler) (Fig. 1.10). Nutrient solution is added to a depth of 4–8 in. below the top. A layer of window screen is placed about one-half inch above the surface of the nutrient solution. This screen induces rapid lateral root growth and branching in this area, and thus promotes plant growth. It also anchors the plants. The polystyrene cover should then be added to exclude light which prevents algae from growing in the nutrient solution. Fifteen 3-inch round holes are made in the cover; and suitable plastic pots, with the bottoms removed, are placed through these holes and sit on the plastic screen. Transplants are placed in these pots.

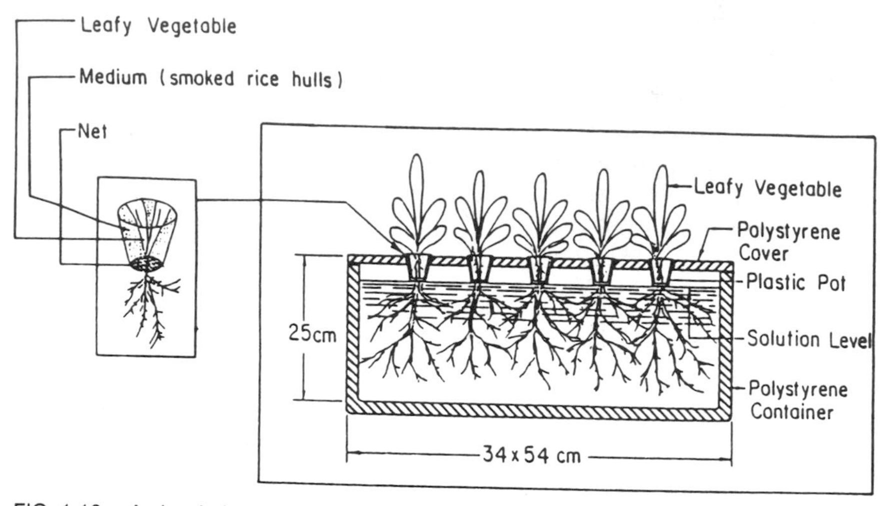

FIG. 1.10. A simple hydroponic system made from a pop cooler. *Courtesy of Asian Vegetable Research and Development Center.*

Seedlings are grown as transplants in peat (Jiffy-7), plastic pots with a layer of nylon net on the bottom, or a bag made of nylon netting and filled with peat and vermiculite. Rock wool cubes do not work well. After transplanting, additional peat may be added to fill the pots in the polystyrene container. Immediately after transplanting, care must be taken to insure that the seedlings have adequate water. Hand watering with nutrient solution is required or the nutrient level may be raised initially to help accommodate the water requirement of the seedlings.

For lettuce, which is grown seven weeks or less, it is usually not necessary to add water or nutrient solution for the duration of the crop. For crops of longer duration, such as tomatoes and cucumbers, additional nutrient solution may be added. However, the sudden addition of large volumes of water (from rain) or nutrient solution causes the plants to wilt and adversely affects the plant roots. The solution level should not be increased by more than 4 in. at any one time.

The level of nutrient solution will decrease due to evaporation, transpiration, and absorption by the root system. Plants have grown well with a 6-inch air space below the screen. In this system, the upper part of the root system absorbs oxygen while the lower portion in the nutrient solution absorbs water and nutrients. The plants have ample aeration.

The polystyrene container insulates the nutrient solution and maintains relatively stable solution temperatures. It holds a large water volume, which helps maintain pH, nutrient levels, and electrical conductivity. Complete nutrient solutions can be purchased from Hygroponics Inc., 3935 N. Palo Alto Avenue, Panama City, FL 32405; Plant Food Co., Wisner, LA 71378; or elsewhere.

If placed outside, the entire unit should be covered with a screened cage to prevent insect damage. A plastic cover should be placed over the cage during rains to protect the plants and to prevent rainwater from entering the polystyrene container.

AMOUNT TO PLANT

The amount of a specific vegetable that a gardener should plant depends upon the preference of the gardener (Table 1.10), the ultimate use of the vegetable, and the number of people consuming the vegetable. Table 1.11 gives the yield of some vegetables produced in a home garden. The amount individual gardens will produce depends upon the location of the garden within the United States and how much attention the vegetables receive. Good gardens may yield more; poor gardens will produce less.

GARDEN TOOLS AND SUPPLIES

Gardening does not require a lot of tools or supplies. The amount purchased depends upon the size of the garden, the types of jobs to be done, and the amount of money the gardener wishes to spend.

Tools

It is best to select good quality tools that can be used for many years rather than poorly designed or cheaply made tools. A small garden only needs a hoe, rake, shovel or spade, stakes and string, and a means to water

TABLE 1.10. Vegetable Consumption and Vegetables Grown in the Home Garden in the United States.

Crop	Pounds consumed per person[1]	Percent of gardeners growing that crop[2]
Potatoes	118.7	40
Tomatoes	81.4	96
Sweet corn	25.7	48
Lettuce	22.7	60
Onions	16.3	75
Carrots	11.3	59
Celery	7.1	<1
Broccoli	5.8	43
Green beans	5.7	71
Sweet potato	5.3	15
Peas	3.7	70
Cauliflower	3.6	30
Spinach	2.6	46
Asparagus	1.0	45

[1] From Tooley et al. 1989.
[2] From Neier and Marr 1988.

TABLE 1.11. Yield of Vegetables Produced in a Home Garden.

Vegetables	Average crop expected per 100 ft	Approximate planting per person: Fresh	Approximate planting per person: Storage, canning, or freezing
Asparagus	30 lb	10–15 plants	10–15 plants
Beans, snap bush	120 lb	15–16 ft	15–20 ft
Beans, snap pole	150 lb	5–6 ft	8–10 ft
Beans, lima bush	25 lb shelled	10–15 ft	15–20 ft
Beans, lima pole	50 lb shelled	5–6 ft	8–10 ft
Beets	150 lb	5–10 ft	10–20 ft
Broccoli	100 lb	3–5 plants	5–6 plants
Brussels sprouts	75 lb	2–5 plants	5–8 plants
Cabbage	150 lb	3–4 plants	5–10 plants
Cabbage, Chinese	80 heads	3–10 ft	—
Carrots	100 lb	5–10 ft	10–15 ft
Cauliflower	100 lb	3–5 plants	8–12 plants
Celeriac	60 lb	5 ft	5 ft
Celery	180 stalks	10 stalks	—
Chard, Swiss	75 lb	3–5 plants	8–12 plants
Collards and kale	100 lb	5–10 ft	5–10 ft
Corn, sweet	10 dozen	10–15 ft	30–50 ft
Cucumbers	120 lb	1–2 hills	3–5 hills
Eggplant	100 lb	2–3 plants	2–3 plants
Garlic	40 lb	—	1–5 ft
Kohlrabi	75 lb	3–5 ft	5–10 ft
Lettuce, head	100 heads	10 ft	—
Lettuce, leaf	50 lb	10 ft	—
Muskmelon (cantaloupe)	100 fruits	3–5 hills	—
Mustard	100 lb	5–10 ft	10–15 ft
Okra	100 lb	4–6 ft	6–10 ft
Onions (plants or sets)	100 lb	3–5 ft	30–50 ft
Onions (seed)	100 lb	3–5 ft	30–50 ft
Parsley	30 lb	1–3 ft	1–3 ft
Parsnips	100 lb	10 ft	10 ft
Peas, English	20 lb	15–20 ft	40–60 ft
Peas, Southern	40 lb	10–15 ft	20–50 ft
Peppers	60 lb	3–5 plants	3–5 plants
Potatoes, Irish	100 lb	50–100 ft	—
Potatoes, sweet	100 lb	5–10 plants	10–20 plants
Pumpkins	100 lb	1–2 hills	1–2 hills
Radishes	100 bunches	3–5 ft	—
Salsify	100 lb	5 ft	5 ft
Soybeans	20 lb	50 ft	50 ft
Spinach	40–50 lb	5–10 ft	10–15 ft
Squash, summer	150 lb	2–3 hills	2–3 hills
Squash, winter	100 lb	1–3 hills	1–3 hills
Tomatoes	100 lb	3–5 plants	5–10 plants
Turnip greens	50–100 lb	5–10 ft	—
Turnip, roots	50–100 lb	5–10 ft	5–10 ft
Watermelon	40 fruits	2–4 hills	—

Source: Abraham and Abraham (1977).

plants during dry weather. Larger gardens can effectively use some types of power equipment.

After each use, the tools should be cleaned and then stored in a clean, dry place. A light film of oil on shovels, hoes, and trowels will provide rust protection. At the end of the season, the tools should be sharpened and any repairs made. Power tools with engines should be winterized according to the manufacturer's recommendations.

Hand tools can be stored in the garage, basement, barn, or storage shed. If there is not storage available for power tools, they should be covered during winter to protect them.

Hoe. A hoe is needed for weed control and to cover seeds after planting. It can also be used to make a small trench into which seeds can be planted.

Irrigation Equipment. If water is available, irrigation equipment is desirable in nearly all gardens. Furrow irrigation requires careful planning of the garden and precise handling of the soil to ensure even water distribution. Trickle irrigation is used in many gardens, and this equipment can be purchased during the winter. Most gardeners will use a garden hose with or without a sprinkler to provide water to their garden. (For a discussion of plant water and the systems used to supply it, see Chapter 3 on Soils and Plant Nutrition).

Rake. A rake is used to smooth out the soil and to prepare the seedbed. It can also be used to loosen the crust that forms on some soils after a heavy rain.

Shovel or Spade. A spade or shovel is needed to turn the soil over and incorporate organic material into the soil. A shovel can also be used to harvest large root crops such as potatoes and sweet potatoes.

Trowel. An inexpensive trowel is very handy in the garden. It can be used to dig holes for transplants, to clean equipment free of soil, and to loosen soil around plants, such as onions, for easier harvest.

String, Stakes, and Measuring Stick. These can be used to lay out evenly spaced rows in straight lines. String is also needed to help train plants such as cucumbers onto a trellis or as a guide for pole beans. When the garden is planted, each row should have a stake labeled with the vegetable that was planted.

Sprayer or Duster. Some method of insect and disease control is needed. For most small gardens, the use of the spray or dust in an aerosol can or a self-contained duster is the most economical and convenient method of applying pesticides. In large gardens, an inexpensive sprayer or duster can be used. Both are effective. When finished with the sprayer, it should be rinsed in the garden area, placed upside down, and allowed to dry completely.

Hand Cultivator. A hand-held or wheel-supported cultivator saves time and effort in weed control in large gardens. It can also be used to loosen the soil crust so water can penetrate some soils more easily. A hand cultivator is useful to dig a deep trench for planting of some crops such as potatoes.

Seeder. For large gardens, some type of seeder will take the backache out of planting. Some seeders open the furrow, plant, cover the seed, and then pack the soil in one operation. Most seeders are adaptable to various sized seed.

Fertilizer Spreader. Fertilizer is easily applied by hand in a small garden. For a large garden a spreader will make the job quicker and easier and provide a more uniform application. It is particularly useful for organic gardeners who apply large amounts of various natural deposits.

Rotary Tiller. There are a number of rotary tillers adapted for large gardens. They can plow the garden and prepare it for planting or be adjusted to shallow cultivation for weed control. If a tiller is to be used, the garden rows need to be wide enough to easily operate the tiller between rows. Common tillers need 16–30 in. between rows of large gardens.

There are two general types available. The common and least expensive type uses the rotating tiller shaft to provide forward motion. The engine weight helps provide penetration into the soil.

The second type of tiller is more expensive but easier to use. The engine provides the power to the rotating tiller shaft and the wheels separately. The tiller speed may be regulated and is not greatly affected if stones are encountered. These tillers frequently have a reverse gear that allows the gardener to work in cramped areas.

Chipper-Shredder. A chipper-shredder is extremely valuable for all gardens. It is particularly useful for homeowners and organic gardeners. Although expensive, a machine that will grind 2 or 3 in. limbs and used Christmas trees is best. Most rotary mowers can be used to chop leaves and succulent branches. A chipper-shredder will grind garden debris for compost, prepare and mix material for soil mixes, and grind leaves, newspapers, and material pruned from trees and shrubs for use as a mulch. These mulches can be used not only in the garden, but around trees and shrubs as well (Fig. 1.11).

Chemicals and Fertilizer

Supplies. Various gardening supplies should be purchased during the winter and early spring. The materials will then be available when they are needed.

Chemicals. Various insect and disease control chemicals should be ordered in advance. This is particularly important if botanical insecticides are

FIG. 1.11. A chipper-shredder (background) can be used to produce mulch (foreground) from tree limbs.

used. Some botanical and biological control agents are readily available but must be ordered from seed catalogues. All pesticides should be stored in a locked box or cabinet to keep them away from pets and children.

Fertilizer. Fertilizer can be purchased in the spring and easily stored in a dry place. Many fertilizers become solid when exposed to wet conditions. Various organic fertilizers and natural deposits should be ordered early. The natural deposits should be added at the end of the growing season rather than in the spring. Many garden centers only stock these items in the spring, however.

Mulch. Many gardeners use a mulch to control weeds, conserve moisture, and keep the fruits clean. Organic materials such as peat moss, leaves, wood chips, or straw should be on hand. Leaves can be collected in the fall, and straw is also available. The grass in a lawn can be allowed to grow quite tall and then mowed but not collected. The lawn clippings are allowed to dry, raked up, and used as a mulch. Organic mulches are normally added after the plants are established.

Various nonorganic types of mulches, such as black plastic or aluminum foil, need to be on hand before planting. These materials are normally applied to the soil before planting.

Plant Growing Aids. Many types of devices are available which allow plants to utilize the vertical space in a garden. Various stakes, bean poles, fencing, tomato cage wire, and trellises are often used. Hot caps may be needed in early spring. Various fences to keep out unwanted animals may be required; and nylon or other types of netting may be needed to prevent bird damage. Plant blankets, row covers, and plant temperature protectors can be used to lengthen the growing season by preventing frost damage.

Seeds and Plants

Good quality seeds and plants are required to produce quality vegetable plants with high yields. The seeds are often ordered from seed catalogues in the winter. Some cultivars are frequently sold out by spring. Not all cultivars recommended for your area will be available from one seed source. Several seed catalogues should be obtained. (See Table 1.12)

TABLE 1.12. Seed Companies.

Retail company	Comments on catalogue
Alberta Nurseries & Seeds, Ltd. P.O. Box 29 Bowden, Alberta, Canada TOM OKO	14 of 48 pages on vegetables. Listings of short season, hardy types.
Arcnias Seed Store Corp. 106-108 E. Main Street Sedalia, MO 65301	13 of 42 pages on vegetables. Includes flowers, fruits, and garden aids.
Burgess Seed & Plant Co. 905 Four Seasons Road Bloomington, IL 61701	26 of 44 pages on vegetables. Many unusual items, herbs.
W. Atlee Burpee & Co. 300 Park Avenue Warminster, PA 18974	54 of 182 pages on vegetables. Includes flowers, herbs, general garden supplies.
D. V. Burrell Seed Growers Co. Box 150 Rocky Ford, CO 81067	97 pages with special emphasis on melons, peppers, tomatoes, and cultivars for California and the Southwest.
DeGiorgi Co., Inc. 1529 No. Saddle Creek Rd. Omaha, NE 68104	112 pages including unusual cultivars. Catalogue for sale.
Farmer Seed & Nursery Co. 1706 Morrissey Drive Bloomington, IL 61704	30 of 82 pages on vegetables. Includes midget, unusual, and early maturing cultivars; herbs; flowers; fruits.
Henry Field Seed & Nursery Co. 407 Sycamore Street Shenandoah, IA 51602	116 pages for general gardening. Many cultivars and suggestions for growing vegetables.
G. S. Grimes 201 W. Main Street Box 398 Smethport, PA 16749	6 of 46 pages on vegetables. Listing of cultivars and price. No descriptions. Includes flowers and vegetable transplants.

TABLE 1.12 (*Continued*)

Retail company	Comments on catalogue
Glecklers Seedmen Metamora, OH 43450	4 pages of unusual vegetables listed with brief descriptions.
Gurney Seed & Nursery Co. Yankton, SD 57079	22 of 76 pages on vegetables. Many unusual items, hardy plants, and general garden aids.
Harris Seeds 961 Lyell Avenue Rochester, NY 14606	45 of 70 pages on vegetables. Includes descriptions. Considered authority in many part of the country.
J. L. Hudson P.O. Box 1058 Redwood City, CA 94604	16 of 112 pages on vegetables. Many unusual vegetables and herbs.
Le Jardin du Gourmet Box 75 St. Johnsbury, CT 05863	16 pages of hard-to-find vegetables and herbs.
Johnny's Selected Seeds Foss Hill Road Albion, ME 04910	28 pages including hard-to-find Oriental vegetables.
Earl May Seed & Nursery Co. Shenandoah, IA 51603	82 pages of vegetables, flowers, fruits, general garden aids. Many cultivars, All-American selections.
Meyer Seed Co. 600 S. Carolina Street Baltimore, MD 21231	23 pages of vegetables. Includes All-American selections.
Nichols Garden Nursery 1190 N. Pacific Highway Albany, OR 97321	64 pages of rare and unusual vegetables, herbs, and recipes.
Park Seed Co., Inc. Cokesbury Road Greenwood, SC 29647-0001	22 of 122 pages on vegetables. Includes midget and unusual vegetables, herbs, flowers, general garden aids.
Porter & Son, Seedsmen 1510 E. Washington St. Box 104 Stephenville, TX 76401-0104	14 of 32 pages on vegetables. Includes trickle irrigation systems.
Spring Hill Nurseries 110 West Elm Street Tipp City, OH 45371	8 of 67 pages on vegetables. Includes flower and fruit trees.
Stokes Seeds Box 548 Buffalo, NY 14240	62 of 158 pages on vegetables. Includes many cultivars, unusual vegetables, and herbs.
Otis W. Twilley Seed Co. P.O. Box 65 Trevose, PA 19047	44 of 64 pages on vegetables. Many cultivars for various climates, disease-resistant cultivars.

Seeds can also be purchased at garden centers and department stores, but these sources seldom have unusual cultivars or many types that are suited for container gardens. Seeds for spring, summer, and fall gardens should be purchased, as well as early and late cultivars. The length of time required for a vegetable to produce the edible plant parts, from planting to maturity, is usually given in catalogues and on the seed package. The maturity time is usually given as a range of dates, such as "Best Bet" tomato: 75–80 days. In a warm climate with ample sunlight, the tomato fruit may be mature in 70 days. With cool, cloudy spring conditions, the tomato fruit may require 90 days to reach the edible stage. In the case of vegetables that are transplanted, such as tomatoes, peppers, eggplant, and melons, the maturity time usually refers to the number of days required after the plant is placed in the garden as a 6-week-old transplant. With vegetables that are seeded directly into the garden, the maturity time usually refers to the number of days from planting to harvest. Early-maturing cultivars should be considered for fall gardens.

Small seeds, such as lettuce and carrots, are easily planted too deep and too close together. Seed tapes are designed for precise spacing and easy handling (Fig. 1.12). The tapes may be placed in a straight line or any other formation desired. The tapes may be cut at any point for multiple row planting or part of the tape saved for a later planting. The tapes dissolve after planting.

Seeds should be stored in a cool, dry location such as in a glass jar or tin container. Frequently, not all the seeds purchased are used and some seeds may

FIG. 1.12. Seed tapes dissolve after planting and result in rows of evenly spaced vegetables. *Courtesy of W. Atlee Burpee Co.*

be stored for several years (see Table 2.1). It is sometimes more economical to purchase a large amount of seeds that are viable for several years than to purchase small amounts yearly.

SELECTED REFERENCES

Abraham, G. and Abraham, K. 1977. Planning your vegetable garden—Pots, pyramids and planters. *In* Growing Your Own Vegetables. U.S. Dep. Agric. Bull. *409*.

Anon. 1978. Asian Veg. Res. Dev. Cent. Progr. Rep. 1977. Asian Vegetable Research and Development Center, Taiwan.

Anon. 1979. California Assembly Bill No. 443, Chapter 914, Section 4, September.

Anon. 1980. USDA Report and recommendations on organic farming.

Arthurs, K.L. 1977. Vegetables in containers require enough sun, space, drainage. *In* Growing Your Own Vegetables. U.S. Dep. Agric. Bull. *409*.

Blackwell, C. 1977. Why folks garden and what they face. *In* Gardening for Food and Fun. J. Jayes (Editor). U.S. Dep. Agric. Yearbook of Agriculture, Washington, DC.

Carbonneau, M.C. 1969. Gardening in containers. Ill. Agric. Exp. Stn. Circ. *997*.

Gomez, R. 1974. Home vegetable gardening. N.M. Agric. Exp. Stn. Circ. *457*.

Imai, H. 1987. AVRDC non-circulating hydroponics system. Proceedings of a symposium on horticultural production under structures. Taiwan Ag. Res. Institute, Taichung, Taiwan.

Mansour, N.S. and Bagget, J.R. 1977. Root crops more or less trouble-free, produce lots of food in a small space. *In* Growing Your Own Vegetables. U.S. Dep. Agric. Bull. *409*.

Marr, C.W. 1977. End of one season is start of another. *In* Gardening for Food and Fun. J. Hayes (Editor). U.S. Dep. Agric. Yearbook of Agriculture, Washington, D.C.

Neier, R. and Marr, C. 1988. Vegetable preferences of home gardeners. *HortScience* **23**:916.

Sims, W.L. *et al.* 1977. Home vegetable gardening. Calif. Agric. Exp. Stn. Leafl. *2989*.

Staff Liberty Hyde Bailey Hortorium. 1976. Hortus Third. Macmillan Co., New York.

Tooley, J., Carle, L.A. and Knight, M.I. 1989. Vegetable Scoop. *U.S. News and World Reports*. Feb. 13, p. 84.

Utzinger, J.D. and Connolly, H.E. 1978. Economic value of a home vegetable garden. HortScience *13*, 148–149.

Vandemark, J.S., Hopen, H.J. and Courter, J.W. 1975. Eat well and save money by growing your own vegetables. Ill. Res. *17* (2) 8–9.

Vandemark, J.S. and Splittstoesser, W.E. 1978. Size and composition of pots affect vegetable transplants' growth. Ill. Res. *20* (1) 5.

Wester, R.E. 1972. Growing vegetables in the home garden. U.S. Dep. Agric. Home Gard. Bull. *202*.

Wilson, J.W. 1977. Where to garden—setting your sites. *In* Gardening for Food and Fun. J. Hayes (Editor). U.S. Dep. Agric. Yearbook of Agriculture, Washington, D.C.

2

Plant Growth

SEEDS

A seed is a dormant undeveloped plant. It usually contains its own food supply and is protected by a seed coat. The seed carries the genetic material from its two parents and controls the maximum quality and performance of the developed plant. The cost of most vegetable seed is a small part of the cost of producing the edible vegetable product and considerable care should be taken in purchasing the seed. Good vegetable seeds are clean, disease-free, and viable and produce plants typical of the cultivar listed on the seed packet.

Diseases. Vegetable seed should be disease-free. Some diseases are carried on the seed coat, such as black rot of cabbage, and can be controlled by seed treatment. Other diseases, such as blackleg, are carried within the seed of cabbage and cauliflower and can be fairly well controlled with a hot-water treatment. Such treatment must be precise in temperature and time or the seed will be injured; the gardener should not attempt this procedure. Instead the gardener should purchase seed that has been commercially treated. The seed packet will state "Hot Water Treated." Some diseases, such as anthracnose of beans, are not controlled successfully by seed treatment. These types of diseases are particularly a problem if the gardeners save their own seed and live in a humid climate. Commercial seed is produced under hot, dry conditions where anthracnose is not a problem.

Some seed will also be treated with a fungicide to control fungi that attacks the seedling during germination. This is particularly true with corn, cucumbers, and melons. These seed packages will carry the warning "Do Not Use for Food, Feed, or Oil." Unwanted seed that has been treated with a fungicide should not be used as birdseed.

Vitality. Vegetable seed should have enough vitality to germinate, emerge from the soil, and produce the plant. A good stand is important, as a partial stand results in wasted space. Overseeding results in a costly thinning operation or, frequently, reduced yields due to overcrowding. Much vegetable seed a year or two old is still usable, but one should know its vitality to ensure a good stand. National seed companies test their seeds and sell only seed with a high germination percentage. Minimum germination requirements for most vegetables are controlled by the federal government and some seed packets carry the germination percentage of the seed and the date the seed was tested. If the seed is below the minimum germination, the package will be labeled "Below Federal Standard." The percentage of germination must be stated. Some seeds such as carrot are usually low in germination percentage, while others such as cabbage frequently give 99 to 100% germination. Table 2.1 gives the federal standard for germination of the principal kinds of seed and the expected longevity of the seed. The longevity figures are for seeds stored under favorable conditions from the

TABLE 2.1. Vegetable Seed: Federal Minimum Germination; Average Number of Seeds per Ounce and Relative Longevity.

Vegetable	Minimum germination (%)	Average number of seeds per oz	Relative longevity (years)
Asparagus	60	1,400	3
Bean, lima	70	20–70	3
Bean, snap	75	100	3
Beet	65	2,000	4
Broccoli	75	8,100	3
Brussels sprouts	70	8,500	4
Cabbage	75	7,700	4
Carrot	55	22,000	3
Cauliflower	75	8,600	4
Celeriac	55	50,000	3
Celery	55	76,000	3
Chard, Swiss	65	1,500	4
Chicory	65	20,000	4
Chinese cabbage	75	7,000	3
Cucumber	80	1,100	5
Eggplant	60	7,200	4
Endive	70	17,000	5
Kale	75	10,000	4
Kohlrabi	75	9,200	3
Leek	60	9,900	2
Lettuce	80	26,000	5
Muskmelon	75	1,100	5
New Zealand spinach	40	430	3
Okra	50	500	2
Onion	70	8,500	1
Parsley	60	18,000	1
Parsnip	60	6,800	1
Pea	80	50–230	3
Pepper	55	4,500	2
Pumpkin	75	200	4
Radish	75	3,100	4
Rutabaga	75	11,000	4
Salsify	75	2,000	1
Spinach	60	2,900	3
Squash	75	180–380	4
Sweet corn	75	140	2
Tomato	75	10,000	3
Turnip	80	14,000	4
Watermelon	70	320	4

Source: Harrington and Minges (1954).

time of harvest (not time of purchase). Under proper conditions many seeds are viable longer than indicated, but the author found that some kinds of lettuce, when stored under hot, humid conditions, were viable for only 18 months.

Cultivar. Good vegetable seeds produce plants typical of the cultivar listed on the seed packet. A horticultural classification recognizes the gradations into kind, cultivar, and strain. This information will be listed on the seed package.

A "kind" of vegetable seed includes all plants recognized as a single vegetable such as cucumber, tomato, or pumpkin. This is not the same as the genus or species of botanical classification. The genus and species *Brassica oleracea* includes many kinds of vegetables (cabbage, collards, kale, cauliflower, and others).

A vegetables cultivar includes all those plants of a given kind that are practically alike in their important characteristics. Each cultivar should be different from all others in one or more prominent and significant features. On the seed package the kind is listed, such as tomato, and then the cultivar, such as Roma or Jet Star.

A strain includes those plants of a given cultivar that have the general characteristics of the cultivar but differ from it in one important aspect or two or three minor respects. These differences are not great enough to justify a new cultivar name, and vegetables strains are not commonly grown by gardeners. For example, Dickinson Field Pumpkin is the pumpkin used in over half of the commercial pumpkin pie mixes; the strain "Libby Select" is one strain selected by a commercial company for proper color and thickness of the edible flesh.

Gardeners who save their own seed may not have a pure cultivar. Some vegetables are self-pollinated and do not require the plants to be isolated from other cultivars. These include beans, peas, lettuce, and nonhybrid tomatoes. Beets, sweet corn, spinach, and Swiss chard are wind-pollinated, and their cultivars must be isolated from each other by 1 mile. Most other vegetables are pollinated by insects and are entirely or readily cross-pollinated. These vegetables must be separated by at least one-quarter mile. Gardeners should not save seed from wind- or insect-pollinated vegetables. Sweet corn will cross with a nearby field of field corn, and it is unwise to save this seed. Broccoli, cabbage, cauliflower, collards, kale, and kohlrabi all intercross readily, and their seed should not be saved. Seed may be saved from watermelon if isolated from other watermelons and citrons. Muskmelon seed may be saved if isolated from other melons, even if it was grown near cucumbers, as these two plants do not cross.

GERMINATION REQUIREMENTS

When a seed receives the proper amount of air, water, and heat, it will begin to grow. With a few vegetable seeds, notably celery, light may also be required. Some germination will usually occur over a wide range of each of these factors. However, as the extremes are approached, above or below the optimum range for each factor, the total germination is reduced, the rate is slower, and abnormal seedlings are more frequent. Beyond the minimum and maximum, no germination occurs.

Temperature. The effect of temperature on seed germination has received much attention. Some vegetable seeds, such as onion, will germinate at freezing temperatures, but the rate of growth is extremely slow. At 32°F, it takes 135 days for the appearance of an onion seedling when planted one-half inch deep. This emphasizes that some seeds can withstand cold temperatures and the seed will be alive when the soil temperature increases. Some seeds, such as sweet corn and beans will rot if planted at low temperatures and left for long periods of time. Table 2.2 gives the minimum temperature for germination of some vegetable seeds. This minimum temperature can be used as an indication of when to plant the crops in the spring. When the temperatures have reached the minimum, the vegetable may be planted. It is assumed that the soil temperatures will continue to increase, however. Planting vegetables before the minimum temperature required for germination is reached does not produce an earlier crop. Instead, the seed will not germinate. Early planting gives more time for soil organisms to destroy the seed, resulting in reduced stands. Standard sweet corn may be planted when temperatures have reached 50°F (Table 2.2). However, this is too cold for the supersweet types, and seeds often rot, resulting in very reduced stands. Satisfactory results were obtained when the supersweet corn was planted at temperatures of 60°F or above.

Some seed companies sell seeds by the number of degree days required to reach maturity instead of the number of days from planting to harvest. If we know the number of degree days for a cultivar to reach maturity, we can determine when the crop should be harvested. There is a minimum temperature below which the vegetables will not grow (see Table 2.2). This minimum is 50°F for sweet corn (Hortik and Arnold 1965). Above 50°F, the higher the temperature, the more rapidly the corn plant grows and approaches maturity. If the temperature becomes too high, the corn growth will slow down, but within reasonable limits, this error is not too important. The number of degree days required for Golden Cross Bantam sweet corn to reach maturity is 1875. This number signifies the number of degrees by which the mean temperature was above 50°F (the minimum temperature for corn germination). Thus, if the mean (average of the highest and lowest temperature for that day) temperature was 74°F, the corn crop was exposed to 24 degree days (74 minus 50). Any temperature below 50°F represents 0° of effective temperature. These degree days are added together beginning with the date of germination. This number of degree days does not indicate how many calendar days are involved. If your sweet corn crop was exposed to a 56°F mean temperature for 4 days, this would result in 24 degree days (56 minus 50 equals 6 degree days for each of 4 days). This is the same number of degree days calculated when the mean temperature was 74°F for 1 day. It is important to recognize that the sweet corn plant will be just as far

TABLE 2.2. Soil Temperature for Vegetable Seed Germination.

Minimum				
32°F (0°C)	40°F (4.4°C)	50°F (10°C)	60°F (15.6°C)	65°F (18.3°C)
Endive	Beet	Asparagus	Bean, lima	Eggplant
Lettuce	Broccoli	Sweet corn	Bean, snap	Muskmelon
Onion	Cabbage	Tomato	Cucumber	Pumpkin
Parsnip	Carrot	Turnip	Okra	Squash
Spinach	Cauliflower		Pepper	Watermelon
	Parsley			
	Pea			
	Radish			
	Swiss chard			
	Celery			

Optimum				
70°F (21°C)	75°F (24°C)	80°F (26.7°C)	85°F (29.4°C)	95°F (35°C)
Celery	Asparagus	Bean, lima	Bean, snap	Cucumber
Parsnip	Endive	Carrot	Beet	Muskmelon
Spinach	Lettuce	Cauliflower	Broccoli	Okra
	Pea	Onion	Cabbage	Pumpkin
		Radish	Eggplant	Squash
		Tomato	Parsley	Watermelon
		Turnip	Pepper	
			Sweet corn	
			Swiss chard	

Maximum			
75°F (24°C)	85°F (29.4°C)	95°F (35°C)	105°F (40.6°C)
Celery	Bean, lima	Asparagus	Cucumber
Endive	Parsnip	Bean, snap	Muskmelon
Lettuce	Pea	Beet	Okra
Spinach		Broccoli	Pumpkin
		Cabbage	Squash
		Carrot	Sweet corn
		Cauliflower	Turnip
		Eggplant	Watermelon
		Onion	
		Parsley	
		Pepper	
		Radish	
		Swiss chard	
		Tomato	

along toward maturity in 1 day at 74°F as it will be in 4 days at 56°F. Thus, if a gardener wants successive crops of sweet corn during the summer, the time between planting the first and second crop will be far apart in the spring when temperatures are cool and there are few degree days accumulating. As the temperature becomes hotter, the harvest dates become closer and closer together.

There is a maximum temperature at which vegetable seeds will germinate (Table 2.2), and high temperatures may kill many vegetable seeds outright. High soil temperatures are also damaging at the time of seedling emergence, when the seedling may die due to heat injury at the soil surface. Lettuce and endive, however, will not germinate at high temperatures. If the soil temperature has not reached the maximum and killed the seed, the seeds will germinate when the soil temperature has cooled. This is particularly important in summer plantings for a fall crop. Lettuce and endive may be germinated between wet paper towels in the house, or if still too warm, in the refrigerator; and then planted into the fall garden.

Water. Vegetable seeds require water to germinate. They can be divided into five groups, depending on how much water their seeds need in order to germinate (Table 2.3). It is important to recognize that this is the amount of water required for the seeds to germinate, not the amount of water required to produce the crop. The rate of germination is faster at higher moisture levels than at the minimum. The amount of water in soils ranges from field capacity to the permanent wilting percentage. Field capacity is the maximum amount of water your particular soil will hold. Any additional water will drain out of the soil. Permanent wilting percentage is that small amount of water remaining in the soil when the plant can no longer remove any more water and the plant wilts. This permanently wilted plant will not recover unless water is supplied to the soil.

Some bean varieties are susceptible to having the cotyledons crack when the seeds are very dry and the seeds absorb water rapidly (Jones 1971). The

TABLE 2.3. Soil Moisture Required for Vegetable Seed Germination.

Group		
Group 1.	Seeds that give nearly as good germination at the permanent wilting percentage as at higher soil moisture contents	
	Cabbage	Pumpkins
	Broccoli	Radish
	Brussels sprouts	Sweet corn
	Cauliflower	Squash
	Kohlrabi	Turnip
	Muskmelon	Watermelon
	Mustard	
Group 2.	Vegetable seeds that require a soil moisture content at least 25% above the permanent wilting percentage	
	Bean, snap	Peanut
	Carrot	Pepper
	Cucumber	Spinach
	Leeks	Tomato
	Onion	
Group 3.	Vegetable seeds needing a soil moisture content at least 35% above the permanent wilting percentage	
	Lima bean	Pea
Group 4.	Vegetable seeds needing a soil moisture content above 50% of the permanent wilting percentage	
	Beet	Endive
	Chinese cabbage	Lettuce
Group 5.	Vegetable seeds needing a soil moisture content close to field capacity	
	Celery	

bean cotyledon is the food storage organ for the new bean plant and comprises over 95% of the dry seed. If the seed is very dry, planted, and immediately irrigated, the outer layers of the bean cotyledon absorb water rapidly and begin to expand. However, the inner layers of the cotyledon are still dry, and when the outer layer expands, the dry inner layers crack. This disrupts the food transport system; the newly developing plant cannot receive any of the food beyond the crack, and the plant frequently does not emerge from the soil. The gardener can overcome this problem by storing the beans for several days at 50–60% humidity to raise the bean seed moisture content to above 14% or by planting the seed at a low soil moisture (see Table 2.3) and not irrigating the garden for several days.

Planting Depth and Rate. Vegetable seeds should be planted at a depth equal to about four times the diameter of the seed. This rule or the specific depth chart (Table 2.4) must be used with care. In wet weather or heavy soils, seeds should generally be planted shallower. In dry weather or light and sandy soils, seeds should be planted deeper. Some small seeds like tomato, pepper, and eggplant germinate slowly and are often planted in the garden as transplants. If they are planted directly in the garden, care must be taken to prevent the soil from drying out and forming a crust on the soil surface. This crust is often so hard that the germinating seed does not have the strength to break through the crust, resulting in poor stands. The gardener can overcome this problem by making a small trench, about 1 inch deep, planting the seed at the bottom of this trench, and covering the seed with vermiculite or sawdust. This will prevent crusting, and the gardener can reduce the frequency of irrigation.

Oxygen. Oxygen is required for germination of a vegetable seed but is usually limiting only when the soil around the seed is saturated with water. This is particularly a problem where the garden is located at the low end of the homeowner's lot and rainwater from a neighbor's house floods the garden. This condition removes all the oxygen from the soil, and if the area remains flooded for any period of time the seed will be killed. It is not too much water that has killed the seed but a lack of oxygen. Some seeds such as the cucurbits (muskmelon, cucumber, watermelon, pumpkin) are particularly sensitive to low levels of oxygen. These plants are frequently grown on sandy soils, not because they require less water, but because they require more oxygen. Sandy soils contain more air spaces than heavy soils.

TRANSPLANTS

The term "transplanting" means shifting of a plant from one soil or culture medium to another. Some vegetables grow best when they are planted indoors and then transplanted into a garden. There are a number of advantages to producing your own transplants including (1) permitting the

TABLE 2.4. Vegetable Planting Depth and Rate.

Vegetables	Depth to plant seeds (in.)	Plants or seed per 100 ft	Spacing (in.)		Number of days until ready to use
			Rows	Plants	
Asparagus	½	60 plants or 1 oz	36–48	18	(2 years)
Beans, snap bush	1–1½	½ lb	24–36	3–4	45–60
Beans, snap pole	1–1½	½ lb	36–48	4–6	60–70
Beans, lima bush	1–1½	½ lb	30–36	3–4	65–80
Beans, lima pole	1–1½	¼ lb	36–48	12–18	75–85
Beans, fava, broad bean	2½	½ lb	18–24	3–4	80–90
Beans, garbanzo-chickpea	1½–2	½ lb	24–30	3–4	105
Beans, yardlong or asparagus	½–1	½ lb	24–36	12–24	65–80
Beets	½	1 oz	15–24	2	50–60
Broccoli	¼	[1]40–50 pl or ¼ oz	24–36	14–24	60–80
Brussels sprouts	¼	[1]50–60 pl or ¼ oz	24–36	14–24	90–100
Cabbage	¼	[1]50–60 pl or ¼ oz	24–36	14–24	60–90
Cabbage, Chinese	¼	[1]60–70 pl or ¼ oz	18–30	8–12	65–70
Carrots	¼	½ oz	15–24	2	70–80
Cauliflower	¼	[1]50–60 pl or ¼ oz	24–36	14–24	70–90
Celeriac	⅛	200 pl	18–24	4–8	120
Celery	⅛	200 pl	30–36	6	125
Chard, Swiss	1	2 oz	18–30	6	45–55
Chicory, witloof, or French endive	¼	½ oz	18–24	8–12	90–120
Collards	¼	¼ oz	18–36	8–16	50–80
Corn, sweet	2	3–4 oz	24–36	12–18	70–90
Cress	¼	¼ oz	12–16	2–3	25–45
Cucumber	1	½ oz	48–72	24–48	50–70
Dandelion	½	½ oz	12–16	8–10	70–90
Endive	½–1	¼ oz	12–24	9–12	60–90
Eggplant	¼–½	⅛ oz	24–36	18–24	80–90
Fennel	½	¼ oz	18–24	6	120
Garlic (cloves)	1	1 lb	15–24	2–4	140–150
Ground cherry or *Physalis*	½	¼ oz	36	24	90–100
Horseradish	4–6	100 roots	24	10–18	(6–8 months)
Jerusalem artichoke	4	50–70 tubers	30–60	15–24	100–105
Kale	¼	¼ oz	18–36	8–16	50–80
Kohlrabi	½	½ oz	15–24	4–6	55–75
Leeks	½–1	½ oz	12–18	2–4	130–150
Lettuce, head	¼–½	¼ oz	18–24	6–10	70–75
Lettuce, leaf	¼–½	¼ oz	15–18	2–3	40–50
Muskmelon (cantaloupe)	1	[1]50 pl or ½ oz	60–96	24–36	85–100
Mustard	½	¼ oz	15–24	6–12	30–40
Okra	1	2 oz	36–42	12–24	55–65

TABLE 2.4. *(Continued).*

Vegetables	Depth to plant seeds (in.)	Plants or seed per 100 ft	Spacing (in.)		Number of days until ready to use
			Rows	Plants	
Onions	1–3	400–600 plants or sets	15–24	3–4	80–120
Onions (seed)	½	1 oz	15–24	3–4	90–120
Parsley	¼–½	¼ oz	15–24	6–8	70–90
Parsnips	½	½ oz	18–30	3–4	120–170
Peas, English	2	1 lb	18–36	1	55–90
Peas, Southern	½–1	½ lb	24–36	4–6	60–70
Peppers	¼	⅛ oz	24–36	18–24	60–90
Potatoes, Irish	4	6–10 lb of seed tubers	30–36	10–15	75–100
Pumpkins	1–2	½ oz	60–96	36–48	75–100
Radishes	½	1 oz	14–24	1	25–40
Rhubarb	4	20 pl	36–48	48	(2 years)
Salsify	½	½ oz	15–18	3–4	150
Soybeans	1–2	1 lb	24–30	2	120
Shallot	1–2	700 bulbs	12–18	2–4	60–75
Spinach	¾	1 oz	14–24	3–4	40–60
Spinach, New Zealand	1½	½ oz	24	18	70–80
Squash, summer	1	1 oz	36–60	18–36	50–60
Squash, winter	1	½ oz	60–96	24–48	85–100
Sunflower	1	4 oz	36–48	16–24	80–90
Sweet potato	4	75–100 pl	36–48	12–16	100–130
Tomatoes	½	[1]50 pl or ⅛ oz	24–48	18–36	70–90
Turnip greens	½	½ oz	14–24	2–3	30
Turnip, roots or rutabaga	½	½ oz	14–24	2–3	30–60
Watermelon	1	1 oz	72–96	36–72	80–100

[1] As transplants.

plant to grow before the danger of frost is over and the soil is dry enough to prepare; (2) maximum numbers of plants can be obtained from costly seed; (3) the problem of soil crusting can be avoided; (4) depth for planting is more easily controlled; (5) growing conditions can be controlled to produce suitable plants when needed, such as peppers, which germinate slowly, or pumpkins, which require high germination temperatures; (6) the gardener can grow special cultivars of vegetable plants that may not be available locally, such as cherry or paste-type tomatoes, and (7) the hazard of importing diseases with purchased transplants is eliminated.

Success in growing good transplants depends on four growing requirements: (1) an insect, weed, and disease-free growing medium; (2) adequate heat and moisture for growing the plants; (3) enough light to ensure a stocky growth of the plant; and (4) an adjustment or "hardening" period to prepare the plant grown indoors to grow successfully in the outdoor environment.

Containers. Many types of containers can be used to grow vegetable plants for transplanting. Almost anything is satisfactory as a container as long as there is a hole in the bottom to allow adequate drainage. Some gardeners use large containers made from wood called flats. They may be made of metal or other material but are usually constructed from cedar, cypress, or redwood. If the flats are made from other woods, they should be treated with a wood preservative such as copper naphthenate. Pentachlorophenol or creosote should not be used, as these materials will retard plant growth.

Clay or plastic pots can be used but are expensive and must be sterilized before re-use. However, cut-off milk cartons, plastic jugs, tin cans, freezer boxes, cottage cheese, or margarine tubs can also be used. There are a number of commercially prepared organic-composition pots, such as peat-pots, that can also be used. These pots minimize root disturbance when the plants are placed in the field, as the entire pot is planted. However, the pots must be completely covered with soil and watered thoroughly so that the plant roots can grow through them. If the pots are not covered completely, they act as a wick that draws the water from around the plants. All these containers require some type of growing medium.

Seeds for transplanting may be sown into small blocks or pressed peat blocks. Kys-Kubes and BR8 Blocks are small cubes or blocks containing fertilizer. These are thoroughly watered, and the seeds are then placed in them. Jiffy-7 pellets are compressed peat enclosed in a plastic container. These pellets are soaked in water and expand 3–4 times in size. The seed is then planted into the expanded pellet. These cubes, blocks, and pellets, containing the vegetable transplant, are placed directly into the soil and little root disturbance occurs (Fig. 2.1).

The size of the container is more important than the kind of container. The size should be equivalent to the minimum size suggested for trans-

FIG. 2.1. Jiffy-7 used for transplanting tomatoes as an individual container. Left to right: Dry; soaked in water and planted with a seed; a tomato plant grown for transplanting. *Courtesy of J.S. Vandemark II.*

plants in Table 2.5. Avoid close spacings, which cause plants to compete with each other for light, nutrients, and moisture. Crowded seedlings tend to become weak and spindly and are more susceptible to disease. Wider spacings or larger containers generally give superior results. If you want to produce the highest quality plants, space them so the leaves of one plant do not touch those of another plant. However, it may not be economical to provide the space required for growing a volume of plants in this manner.

Plants are frequently overcrowded when seeded directly into hotbeds where they remain undisturbed until ready for transplanting into the field or garden. Such vegetable plants (cabbage, onion, pepper, and tomato) seeded at the recommended rates should be thinned to stand one-half inch or more apart in the row.

Growing Media. The soil medium for growing transplants should not be soil from the garden or field. This type of soil usually lacks proper structure (too much clay), contains insect pests, disease organisms, and weed seeds, and may contain some residual herbicides. A good growing medium should be sterile, uniformly fine, well aerated, and well drained.

Composted soil and soil mixes are commonly used as media for growing transplants. Frequently, artificial media are also added to the soil mix and include vermiculite, perlite, peat, and sand. Vermiculite is a lightweight expanded mica that absorbs large amounts of water (3–4 gal./cu ft) and

TABLE 2.5. Seed and Plant Spacing Chart for Transplants.

Crop	Planting depth (in.)	Seeds per in. of row (number)	Row width (in.)	Minimum space for transplants (in.)
Broccoli	¼–½	10	2–3	3 × 3
Brussels sprouts	¼–½	10	2–3	3 × 3
Cabbage				
for re-transplanting	½	10	2	3 × 3
for planting	½	3–6	4–6	—
Cauliflower	¼–½	10	2–3	3 × 3
Cucumber[1]	¾–1	—	—	3 × 3
Eggplant	¼–½	10	2–3	4 × 4
Lettuce (leaf, bibb, head)	¼–½	10–15	2–3	2 × 2
Muskmelon[1]	¾–1	—	—	3 × 3
Onion (for planting)	½	10	3–4	—
Pepper				
for re-transplanting	¼–½	10	2–3	3 × 3[2]
for planting	½	3–6	4–6	—
Tomato				
for re-transplanting	¼–½	10	2–3	3 × 3[2]
for planting	¼–½	3–6	4–8	—
Watermelon[1]				
regular	¾–1	—	—	3 × 3
seedless	½–1	—	—	3 × 3

Source: Courter *et al.* (1972).
[1] These crops should be seeded directly in individual containers. Refer to following section on Transplanting Techniques for Specific Crops for seeding rates.
[2] For growing in flats, plants may be spaced 2 × 2 in.

contains a considerable amount of available magnesium and potassium. Perlite is a lightweight expanded volcanic material mined from lava flows that has no fertility value but has an excellent water-holding capacity.

Soil mixes for growing transplants should contain adequate nutrients and be well drained and well aerated. Mixtures of (1) equal parts of soil, sand, and peat; (2) equal parts of soil, peat, and perlite; or (3) two parts soil, one part sand, and one part peat are good media. The soil provides the nutrients; the sand, perlite, or vermiculite provides the aeration; and peat, vermiculite, or perlite provides the water-holding capacity. Sand alone is not recommended because it has low water-holding capacity.

The growing media should be sterile. Small amounts of compost or garden soil may be sterilized by baking the soil in the oven. All of the soil should be heated to 180°F for 30 min. In a shallow pan, soil may reach this temperature in 45 min with the oven at 350°F. Overcooking releases toxic materials and kills various helpful microorganisms in the soil. These helpful microorganisms will degrade some of the organic matter in the soil and release nutrients, particularly nitrogen, for the plants' growth. Without these microorganisms, the soil will usually be deficient in nitrogen.

There are a number of commercially prepared media suitable for starting and growing vegetable plants, such as Pro-mix, Jiffy Mix, Redi-Earth, and others.

Sowing. The date to sow vegetables seeds for transplanting depends upon the date the transplants are to be planted into the field and the desired age of the transplant. How early you may plant the transplant depends upon the hardiness of the vegetable and the climate in your area. Some plants can withstand frost while others cannot. Table 2.6 gives the recommended growing periods and frost tolerance of the plants. Plants grown under less than optimum conditions will require longer growing times than those listed.

A common practice is to sow the seed rather thickly in flats or large containers and then, when the first true leaves are fairly well developed, to transplant the seedlings spaced further apart into new flats. Suggested seed and plant spacings are given in Table 2.5. After seeds are sown in flats, they are generally covered with a fine layer of sand, soil, or vermiculite. Sprinkle the flat with water and then cover with a clear plastic film. This plastic film seals in moisture and air around the seeds and increases the temperature. The flat will require no further attention until the seedlings have begun to emerge.

When the seedlings have developed their first true leaves, they are ready to transplant into other containers. When transplanting, the seedling should be held by a leaf, not the stem. The slighest injury to the stem may cause permanent damage. Do not transplant weak, damaged, or malformed seedlings. After transplanting, the seedlings will become established faster if they are kept shaded and moist for one or two days. Transplanting is not beneficial, and may be harmful, to the plants. The degree to which trans-

TABLE 2.6. Transplanting Tolerance and Time Required from Seeding to Transplanting.

Vegetable	Transplanting tolerance	Time to grow (weeks)	Frost susceptibility
Broccoli	Survive well	5–7	Tolerant
Brussels sprouts	Survive well	5–7	Tolerant
Cabbage	Survive well	5–7	Tolerant
Cauliflower	Survive well	5–7	Tolerant
Celeriac	Survive well	8–10	Tolerant
Celery	Require care	8–10	Very susceptible
Cucumber[1]	Seed in container	3–4	Very susceptible
Eggplant	Require care	6–8	Very susceptible
Endive	Survive well	3–4	Tolerant
Kale	Survive well	5–7	Tolerant
Lettuce	Survive well	5–7	Moderately tolerant
Muskmelon[1]	Seed in container	3–4	Very susceptible
Onions, dry	Survive well	8–10	Very tolerant
Parsley	Survive well	5–7	Tolerant
Pepper	Require care	6–8	Susceptible
Pumpkin[1]	Seed in container	4–6	Very susceptible
Squash[1]	Seed in container	4–6	Susceptible
Tomato	Survive well	4–7	Susceptible
Watermelon[1]			
regular	Seed in container	4–6	Susceptible
seedless	Seed in container	6–8	Susceptible

Source: Courter *et al.* (1972).
[1] Crops not generally successfully transplanted unless planted in individual containers, cubes, blocks, or pellets. Any root disturbance hinders growth.

planting checks growth depends on the amount of injury to the root system, the age and size of the seedling, environmental conditions following transplanting, and the kind of vegetable seedling involved.

The seedlings of some vegetables, such as cucumbers, melons, squash, and pumpkin (Table 2.6), are difficult to transplant successfully and should be seeded directly in individual containers. Thus, many gardeners initially sow seeds directly into individual containers (peat pots, Jiffy-7, and the like) so that additional transplanting is not required.

Light, Temperature, and Water for Transplants. It is seldom possible to keep transplanted plants in house windows without the plants becoming spindly and weak. They should be grown in a hotbed, cold frame or other place where they will receive enough sunlight, ventilation, and suitable temperature.

There is no economical method of providing adequate artificial lighting to grow a volume of good plants. However, strong, vigorous seedlings can be started under 40 watt, 48 in. long fluorescent tubes, 6–8 in. above the seedlings. Best results are obtained if the fluorescent fixture is next to a window to increase the amount of light reaching the young plants and if the plant receives 12 hr of total light (natural sunlight and artificial light). It is best to add one or two incandescent light bulbs with each four-tube fluorescent bank of lights to provide light that is more nearly like natural sunlight. "Plant-growth" type of lamps designed for growing plants indoors may also be used. These types of lamps emit a light that simulates sunlight, but the color they emit may not be pleasing in a household environment. A timer to turn the lights on and off is very helpful.

Optimum day and night temperature ranges for growing plants are given in Table 2.7. During times of low light intensity, the lower temperatures should be used.

High temperatures (above the recommended range) at any time, especially during conditions of low light, will cause plants to become spindly and weak. Temperatures lower than recommended will reduce growth and delay plant development, and they may also cause rough fruit in tomatoes and premature seeding of cabbage and cauliflower.

Use accurate thermometers to frequently check temperatures if the plants are being grown in a hotbed or cold frame. The thermometers should be located at the level of the plants being grown. Uniform temperatures are essential for adequate control of plant development and production of uniformly sized transplants.

Daytime temperatures and humidity are primarily controlled by ventilation. This air exchange supplies carbon dioxide, which is used in photosynthesis, and helps to minimize disease problems. Ventilation at night, except during excessively windy or cold periods, ensures adequate air circulation and reduced humidity. Avoid direct cold drafts, which may be harmful to the plants.

Plant growth can be regulated by careful watering. Water the plants only when moisture is needed and then wet the soil thoroughly. Too frequent watering, a common error in plant growing, results in soft, succulent plants with restricted root growth. It may also promote damping-off diseases. The

TABLE 2.7. Recommended Seed-Germination and Plant-Growing Temperatures.

Crop	Optimum soil temperature range for germination (°F)	Days to emerge[1]	Plant-growing temperature[2] (°F)	
			Day	Night
Broccoli	70–80	5	60–70	50–60
Brussels sprouts	70–80	5	60–70	50–60
Cabbage	70–80	4–5	60–70	50–60
Cauliflower	70–80	5–6	60–70	55–60
Cucumber	70–95	2–5	70–80	60–70
Eggplant	75–85	6–8	70–80	65–70
Lettuce	60–75	2–3	55–75	45–55
Muskmelon	75–95	3–4	70–80	60–70
Onion	65–80	4–5	60–70	45–55
Pepper	75–85	8–10	65–80	60–70
Tomato	75–80	6	60–75	60–65
Watermelon				
regular	70–95	4–5	70–80	60–70
seedless	85–95	5–6	70–80	60–70

Source: Courter *et al.* (1972).
[1] At optimum soil temperature range.
[2] The lower temperatures are recommended during cloudy weather.

less the light and the cooler the temperature, the less water is required by the plant. Plants should be watered in the morning to allow the foliage and soil surface to dry before night when the temperatures near the window or in the cold frame are cooler. Wet, cool conditions encourage damping-off diseases.

Fertilizer. Fertilizers can be easily supplied as the plants are watered. This allows a practical means of adjusting nutrient levels according to the stages of plant development and existing environmental conditions. You can control plant growth by the amount and concentration of the fertilizer solution used and the frequency of application.

Many soluble fertilizers are available for supplemental feeding. Starter fertilizers of various analyses, such as 10-52-17, 10-50-10, 20-20-20, 5-25-15, or 16-32-16, have been used with good results. These are high analysis, water-soluble fertilizers that are primarily mixtures of diammonium phosphate and monopotassium phosphate. Potassium nitrate (14-0-46 analysis) and ammonium nitrate (33-0-0 analysis) have also been used successfully.

Many gardeners prefer to fertilize once each week. A rate of 1 tbsp of soluble fertilizer per gallon of water is suggested. After the plants are three weeks of age, the strength can be increased to 2 tbsp of the fertilizer per gallon of water. To remove any fertilizer that might burn the foliage, give the plants a light watering with clear water.

Soluble salts can be a problem in plant-growing beds as a result of using too much fertilizer or improper fertilizer. Because it is easy to overfertilize a small area, be careful not to use rates higher than those suggested. Also, avoid the use of muriate fertilizers, which contain large amounts of chlorides. Symptoms of soluble salt injuries are poor seed germination, stunted plant growth, small dark leaves, and wilting. Wilting may occur even when the soil is sufficiently moist. Thus, it is a good practice to use plain water between fertilizer applications to prevent the accumulation of salts that could injure the transplants.

Cold Frames and Hotbeds. In determining the type of equipment for growing plants for later transplanting, the gardener must consider the climate of the area and the type of plants being grown. Frost-tolerant plants (see Table 2.6) such as cabbage need only inexpensive, simple facilities while tender seedlings such as tomatoes or peppers require more elaborate facilities.

Cold frames are structures which depend on the sun for heat. In warmer parts of the United States and in protected locations, a coldframe or plastic covered pit on the sunny side of a building usually is sufficient. In cooler sections or in exposed areas, some additional heat is needed as a protection against cold damage and these heated cold frames are called hotbeds. The heat is provided by hot air, steam or hot water pipes, electric soil cables, infrared lights, or fermenting manure.

The growing structure should be on well-drained land, free from danger of flooding in the spring. Covers for hotbeds and cold frames may be glass fiberglass, plastic, canvas, or muslin. The amount of covering is determined by the outside weather, the heat generated inside the structure, and the type of plants being grown. In cold climates a tight, well-glazed structure is necessary.

Large plant hoods that resemble a small greenhouse may be made from tubular aluminum or steel pipe and plastic. Upright semicircles of pipe are placed in the ground and a double layer of plastic film is placed over the pipes. This insulates against 5°–10°F frost temperatures. Clear plastic film transmits about as much visible light as glass and more ultraviolet and infrared (heating rays) than glass.

Electrically heated hotbeds are convenient for gardeners. A complete unit may be purchased, complete with frame, heating cables, switches, and thermostats (Fig. 2.2). Their cost is relatively low, depending on electric rates, and the soil temperatures are easily controlled.

Gardeners may construct a hotbed by building a cold frame on the south side of their house in front of a heated basement window. The heated basement will heat the hotbed, and the temperature of the hotbed can then be controlled by opening and closing the basement window.

Hardening. Hardening is a physiological process whereby plants accumulate carbohydrate reserves and produce additional cuticle on the leaves. This allows the plants to withstand such conditions as chilling, drying winds, shortage of water, or high temperatures. Plants can be hardened by any method that stops growth, such as lowered temperature, withholding water, or limiting fertility. During the marketing of vegetable transplants,

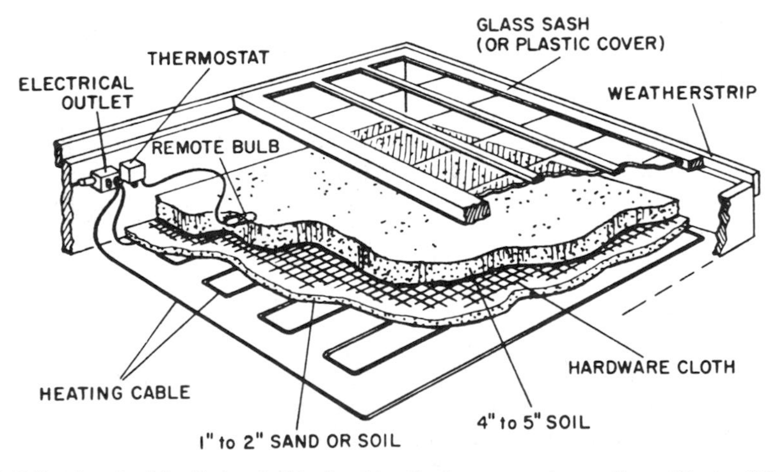

FIG. 2.2. An electrically heated hotbed for the home gardener. *From Wester (1972).*

a combination of these factors occurs, hardening the plants. Gardeners often place the plants outside in a protected area and expose them to lower temperatures for two weeks before planting. The plants should be brought inside or covered in a cold frame if frost is likely. The plants should be exposed to more sunlight, which causes the plant to produce more cuticle, thereby reducing water loss. Cabbage, lettuce, onion, and some other plants can be hardened to withstand frost; tomatoes, peppers, and eggplant cannot.

Hardening is not necessarily helpful to the plant, and it may be detrimental. It is not recommended for most kinds of vegetable transplants. With the exception of tomatoes, plants that are susceptible to frost should not be hardened (see Table 2.6). Overly hardened plants may survive rigorous garden conditions, but they begin growth slowly and they may never fully recover, reducing yields and delaying maturity. Some biennial crops such as cauliflower and cabbage can receive enough cold temperature below 50°F that, if they are large enough plants, they will be induced to produce a seed stalk instead of the edible part.

When hardening plants, it is better to reduce the growth of the plants than to stop it completely. Hardening reduces the growth rate, thickens the cuticle, increases the wax on leaves of some plants, increases the carbohydrate reserve, and induces a pink color, especially in the stems, petioles, and veins of the leaves. Usually hardened plants contain leaves that are smaller and darker green than nonhardened plants. Hardened plants produce new roots faster than nonhardened plants, which is important for plants not grown in individual transplanting containers. The hardened plants have accumulated a food reserve that is used in the formation of new roots.

TRANSPLANTING TECHNIQUES FOR SPECIFIC CROPS

Although all vegetables can be grown as transplants, they are not all grown in an identical manner. Some can be transplanted only if they are planted in an individual container so their roots are not disturbed. The best plants to transplant are those which do not contain fruit and still have the cotyledons (initial leaves) remaining on the plant.

Cabbage and Broccoli. These plants germinate and grow best at soil temperatures of 70°–80°F. If the seeds are sown in a hotbed, it may be necessary to transplant the seedlings to a cold frame, because it is sometimes impractical to lower the hotbed temperature enough to grow good, stocky cabbage or broccoli plants. Although cabbage and broccoli thrive on relatively low temperatures, repeated chilling or exposure of the plants to low temperature (50°F) for a period of 2 weeks or more may cause them to prematurely form seedstalks after they are set in the field. In warm climates, cold frames can be used for growing cabbage and broccoli plants.

Cauliflower and Brussels Sprouts. These vegetables are similar to cabbage, except that cauliflower requires slightly higher night temperatures. Cauliflower plants are delicate and require careful handling. The seedlings should not be crowded, exposed to low temperatures, or inadequately watered, as this will slow or stunt growth of the plants. If the stems become woody or tough due to poor growing conditions, they will form small inferior cauliflower heads in the garden. Cauliflower plants should be grown in individual containers or, when removed from the flat for transplanting, the plants should be lifted out with as much soil on the roots as possible.

Eggplant. Eggplant seeds require about a week to germinate in a flat and should be ready to transplant to other containers in an additional 10–15 days. Seedlings should be grown uniformly from germination onwards. If growth stops, the stems become hard and woody and the plants do not produce satisfactory fruits. Eggplant is susceptible to low temperatures and should not be hardened. Plants should be planted in the garden when they have 4–5 true leaves, the soil is warm, and there is no danger of frost.

Lettuce. Lettuce seeds germinate in two or three days and can be transplanted into other containers after an additional week. Bibb lettuce has a higher optimum growing temperature than head or leaf lettuce and should be grown at temperatures higher than those listed in Table 2.3. Lettuce seedlings are spaced 2 in. apart and should be grown in individual containers. If they are grown in flats, as much soil as possible should be left on the roots when planted in the garden.

Onions. Onions are grown from sets or seeds. The Spanish type is usually grown as a transplant. Seeds should be planted 90 days before they are to be planted in the garden. Onions do better if moderately hardened by watering less frequently and exposing the plants to night temperatures of 40°–50°F for 7–10 days before transplanting. The tops of the plants are often clipped to a 6 in. height to reduce water loss and for ease of transplanting. Excessive clipping will delay onion growth and decrease yields.

Pepper. The seed germinates slowly and requires about 8 days. Seedlings should be ready for planting into other containers in an additional 10–15 days when the first true leaf has appeared. Peppers are grown similar to tomatoes.

Tomato. Tomato seeds germinate in about 6 days, but the time is influenced by the viability of the seed and the germination temperature. Lowered growing temperatures (see Table 2.7) and adequate light and water will produce short stocky plants. The seedlings should be transplanted into individual containers or flats 10–15 days after sowing. In general, container-grown seedlings produce earlier yields than plants grown in flats and pulled out bare-root prior to transplanting. When grown under good conditions, tomatoes should have stocky, thick stems, well-developed root systems, and visible flower buds on the first cluster. Flowers should not be in bloom, and fruit should not be set. If fruit is set before the plant becomes established in the field, the plant will not develop properly and the fruit should be removed. Plants should not be pruned or trimmed to slow down growth. By removing the leaves, less food will be available for the plants' growth. Pruning will increase branching of the plants but this has not been found to be beneficial. Hardening of the plants often results in delayed establishment of the plants in the field, delaying harvest. Hardening frequently results in rough tomato fruit on the first two clusters. The plants can be held a week in good conditions if weather prevents transplanting into the garden.

Vine Crops. Watermelons, muskmelons, cucumbers, summer squash, pumpkins, and gourds are generally planted directly into the garden. These crops do not transplant easily if grown in flats, and they must be grown in individual containers about 3 in. or larger in size. This allows transplanting of the plant with minimum disturbance to the root system. Usually 3 to 5 seeds are sown and then thinned to the 2 strongest seedlings after emergence. Plants should be thinned by pinching off or cutting the stems with a knife. Pulling the excess plants out can result in injury to the roots of the remaining plants. The cost of some seedless and hybrid seed is such that many gardeners plant only 1 or 2 seeds per container.

Vegetative Propagation

Vegetative propagation of plants is asexual propagation and does not involve seeds. Asexual propagation involves the regeneration of the missing plant parts. A root cutting develops a new shoot, and a stem develops new roots. There are several advantages to vegetative propagation. Seeds do not always produce identical plants, but vegatatively propagated plants are always identical with the mother plant. Some vegetables produce few viable seeds or are not commercially available; vegetative propagation allows the production of these crops, such as garlic, horseradish, potato, and sweet potato. The following list summarizes various vegetables and the plant part that is vegetatively propagated.

Asparagus (crowns)
Rhubarb (crowns)
Potato (tuber)
Jerusalem artichoke (tuber)
Sweet potato (slips or shoots)
Horseradish (root)
Garlic (bulbs)
Shallots (bulbs)

The crown of a plant is that part just above or below ground level and may be thought of as a very short stem. Crowns produce buds that produce new plants, as occurs with rhubarb and asparagus. In these vegetables the crown also serves as a food storage organ.

Vegetables such as garlic and shallots are propagated by bulbs, which consist of many fleshy leaves and a very short stem. Bulbs produce new bulbs at the base of these leaves that may be used to produce new plants.

Potatoes and Jerusalem artichokes are propagated from tubers, which are enlarged stems grown underground. This tuber contains a number of "eyes" or buds, each of which is capable of producing a new plant. Commonly, 3 oz of tuber containing at least 2 eyes are used as the seed piece. If the entire tuber is planted, the terminal "eye" or bud inhibits the other buds from growing. This is an advantage, as all of the tuber's energy is channeled into one vigorous shoot. A whole potato about the size of a chicken's egg is ideal for planting, as the tuber has a large food reserve and only one plant is produced.

Sweet potatoes are propagated from the edible part, the fleshy, swollen root (fibrous root). These roots are grown usually in sand and produce a number of new rooted plants. These plants (slips) are removed from the fibrous root and planted in the garden. Horseradish is commonly propagated vegetatively from a small root about 3 to 5 in. long. This root will produce a new plant.

GERMINATION

Germination of a seed involves four phases: (1) the absorption of water, (2) the formation of enzyme systems and the breakdown of the food reserves, (3) the growth of the new root and shoot, and (4) the growth of the seedling up to the time it has emerged from the soil.

When seeds absorb water, active metabolism begins and respiration (the breakdown of sugars into carbon dioxide and water with the release of energy) and protein synthesis begin. The growth of the small root and shoot (embryo) require large amounts of energy for new cell material. At the beginning of germination, the embryo has enough carbohydrate, fat, and protein food reserve, but this is soon depleted. The bulk of the seed is composed of food reserves, which must be broken down by various enzymes made of protein. The food reserves start to dissolve shortly after the embryo begins to grow. The cell walls are degraded, reserve protein is hydrolyzed to soluble amino acids, starch is hydrolyzed to sugar, and fats are converted to sugar or energy. These events are controlled by various enzymes, which have been synthesized, and in about 4 days the food reserves have been liquified. Much of the food reserve has been transported to and used by the growing embryo.

The radicle or root emerges first from the seed. The young root can now absorb water and nutrients from the soil. Shoot development begins after the root has emerged. The seedling now grows in the absence of light in the soil and is dependent upon the stored food reserves until the shoot emerges.

If small seeds are planted deep in the soil, the seedling will run out of food reserve before it emerges from the soil. When plants are grown in the dark or in the soil, the shoot continues to elongate until it has emerged into the light. Plants have various ways in which they protect the shoot from damage as it moves through the soil toward the surface. In grasses such as sweet corn, the tip of the shoot is enclosed in a leaf cylinder called the coleoptile. The area between the seed and the coleoptile grows until light hits the coleoptile, and then secondary roots are formed. As the secondary roots form only when the light hits the coleoptile, these roots are always at the same depth in the soil, regardless of how deep the seed was planted.

In dicotyledonous plants, the stem forms a hook; as the shoot moves through the soil, it pulls the top of the shoot behind it. This prevents damage to the cotyledons and/or leaves, which are pulled rather than pushed upward. Once the seedling is exposed to light, the shoot becomes unhooked and the leaves begin to expand. The cotyledons of some vegetables such as tomatoes, peppers, beans, and squash emerge from the soil and function as leaves. The cotyledons of peas remain underground, however. The cucurbits (cucumber, squash, pumpkins) develop a "foot" on the shoot as it develops, and this foot pries off the large seed coat from the cotyledons as they emerge or shortly thereafter.

VEGETATIVE GROWTH

Once the seedling has emerged from the soil, it is capable of continuous and uninterrupted growth until flowering. During this period of growth when the plant cannot readily be induced to flower, it is considered to be juvenile. Specific environmental conditions are also required (such as day length or low temperature) for some plants to flower.

There are a number of advantages of the plant for having a period of rapid vegetative growth. The plant can be better compete with weeds and other plants if it is large. It maintains its competitive position and is able to receive more sunlight by growing rapidly. As a result, the plant will have more food produced from photosynthesis, and a large plant with more food reserve will produce a higher yield of the edible part. By maintaining a vegetative condition, the food reserve of beets, carrots, onions, and radish is used to produce the edible part. When these plants flower, however, the food reserve is used to produce flowers and seeds. These plants are not capable of developing the edible part and also producing flowers and seeds. In those vegetables in which fruit is desired, a larger plant can withstand the stresses of flowering much easier than a small plant. A small plant often produces seed and then dies. A buildup of food materials occurs in the leaves during vegetative growth, and these materials are used in the flowering process. In many plants, once flowering begins, vegetative growth stops and no more food accumulates. After flowering and fruit set, the plant often stops growing; but if the fruit is removed, the plant continues to grow and produce new flowers and fruit, as occurs in pepper and beans. In some plants, such as sweet corn, the vegetative part of the plant becomes the flowering part (the tassel on sweet corn) and the plant cannot continue to grow or produce new flower parts even if the flower or fruit (sweet corn ear) is removed.

Often vegetative and reproductive growth proceed together. In peanuts, flower parts often develop in the seed before it is harvested. In sweet corn, the plant has formed the ear and tassel before the plant has reached the sixth leaf stage.

Vegetative growth is influenced by the environment. Pungency of onion and radish develop rapidly at high temperatures, and most gardeners do not want "hot" radishes. If radishes are grown in the greenhouse, high temperatures frequently occur, resulting in radishes with a high pungency content. The amount of solids and starch of potato tubers is also affected by temperature.

FLOWERING

Once the plant has reached a certain developmental stage and the plant has overcome juvenility and when certain environmental conditions have been met, the plant will flower. Many environmental factors affect flowering. Water, nutrition, light, and temperature play a vital role. The actual environmental conditions that promote flowering do not have to be present during the flowering process. The gardener is interested in producing two types of vegetables; those in which the edible part is the flower bud, fruit, or seed; and those in which the edible part is the leaf, root, stem, or petiole. The plants in the latter group may, however, produce flowers instead of a satisfactory edible part.

Temperature. Temperature is one critical environmental factor that affects flowering. Many vegetables are induced to flower by low temperatures, particularly biennial and perennial plants. Some, such as spinach, are induced to flower at high temperatures. Tomatoes will produce more flowers if grown at 80°F during the day and 65°F at night than at higher or lower night temperatures.

The induction of flowering by low temperature is called vernalization. Biennial plants normally grow one year, are induced to flower during the winter, and flower the next year. They may, however, be induced to flower the first year; if this happens, the plant does not produce a satisfactory edible plant part. As the plants become older, they respond more easily to low temperature flower induction. However, the gardener can manipulate the plants to prevent flowering. The vernalization process requires water, oxygen, growing point or bud, cold temperature, and a food reserve. If any of these requirements are lacking, the plant will not flower.

Horseradish is commonly grown from a small root cutting obtained one year and planted the next year. This root cutting is a section of a side root. When this root is planted early in the spring, it receives a cold treatment but does not flower that year. The root appears to have enough food reserve but lacks an active growing point, a requirement for vernalization. Thus, the root cutting produces a new fleshy root instead of flowers. If the plant is left in the ground over the winter, it will flower the next year as the plant now has an active growing point and is vernalized.

Broccoli and cauliflower require very little chilling for vernalization. The gardener can transplant these plants when they are small and do not have enough food reserve for vernalization to occur or can transplant them as larger plants after the soil temperatures have warmed up. If the plants receive a cold treatment and have enough food reserve, they will flower and produce a head so early that it will be very small.

Carrot cultivars are highly variable in their vernalization response. The large root of carrot is easily vernalized at 60°F to induce it to flower.

Onions are vernalized if they undergo 2 weeks at 40°–50°F. Onion sets are normally stored during the winter at temperatures just above freezing to prevent spoilage. During this storage the sets are vernalized, and if they were planted immediately they would flower and would not form a bulb. However, the sets can be converted back to their original nonflowering condition by high temperatures. Exposing the sets to 80°F for 2 or 3 weeks will devernalize the sets, and they will form bulbs instead of flowers. Most gardeners do not need to be concerned with converting onion sets back to a nonflowering condition. Onion sets should be planted very early in the spring, and these sets will usually be exposed naturally to 40°–50°F temperatures. To ensure that onion bulbs rather than flowers are produced, medium-sized sets about one-half inch in diameter should be selected for use as dry onions or bulbs. These sets are small enough that, even if they are exposed to low temperatures when planted in the spring, the sets do not have enough food reserve to allow them to become vernalized. The larger sets will be vernalized and should be used to produce green onions.

Beets, cabbage, and celery may also be vernalized by cold temperature. To prevent this, it is best to plant beets when the temperature is warming up rapidly in the spring and the seedling does not have enough food reserve to be vernalized and the cold temperatures do not remain long enough. Small cabbage transplants with low food reserves or larger plants planted late will produce satisfactory heads instead of flowers.

Nutrition. When some vegetables are given a large amount of nitrogen fertilizer, the plants produce large amounts of vegetative growth but few flowers and fruit. This is particularly a problem when large amounts of manure are applied to a garden and tomatoes are produced.

Light. A large number of vegetables respond to the variations in the length of the light and dark periods, or day lengths. Some vegetables initiate flowers during long days (12 hr or more of light) such as radish, lettuce, spinach, potato, and beets. Other vegetables initiate flowers during short days, such as some gourds and beans. Still other vegetables, such as tomatoes and peppers, produce flowers under either long or short days. As a result, different cultivars of vegetables are grown in the south from those grown in the north. Some types of sweet corn are adapted to short days and, if grown under long day conditions, the plants will grow vegetatively but will not flower and produce an ear. Some gourds and beans are short day plants and will not flower under long days. In northern climates, they grow vegetatively under the long days of summer and flower in the fall. Frequently, a frost occurs before the fruit has reached an acceptable size.

Flower Types. Most vegetables, of which the edible part is the fruit or seed, produce complete flowers. Complete flowers are those containing both male and female parts. The cucurbits (cucumber, squash, pumpkins), however, have separate male and female flowers (Fig. 2.3). These plants produce a number of male flowers and, as the vine grows, they begin to produce both male and female flowers. The fruit is produced only by the female flowers, and the gardener should not be concerned with a lack of fruit on the first part of the vine. The production of male flowers ensures a larger amount of vegetative growth before fruit set, thereby providing more food material to produce more and larger fruit.

FRUIT SET

Pollination. The changes that mark the transition of the flower into a young fruit are called fruit set. The first event a gardener notices is pollination. Peas, beans, and some other legumes are already pollinated when the flowers emerge. In others like tomato, the flowers are self-pollinated but must be shaken by the wind to ensure pollination. Beets, sweet corn, spinach, and Swiss chard are pollinated by airborne pollen, and crossing between cultivars readily occurs. Most vegetables are pollinated by insect-borne pollen, including those given in the following list.

asparagus
broccoli
Brussels sprouts
cabbage
carrot
cauliflower
celeriac
celery
Chinese cabbage
collard
cucumber
eggplant
gherkin
gourds
kale
kohlrabi
muskmelon
mustard
onion
parsley
peppers
pumpkin
radish
rutabaga
squash
turnip
watermelon

FIG. 2.3. Pumpkin vine showing male flower (left) and female flower (right) with its ovary (immature pumpkin fruit).

Gardeners cannot control the flight path of insects. Although your garden may not contain any plants that will cross with each other, bees and other insects from other gardens may pollinate your plants. If seed is to be saved from your plants, the various cultivars must be separated from each other by 400 yards. This distance is impractical in the garden, and it is advisable to purchase new seeds yearly.

Gourds, pumpkin, and squash have separate male and female flowers (Fig. 2.3), and bees or some other insect must transfer the pollen from the male to the female. Commercial seeds of these plants are produced in isolated areas where no other cultivar is grown. The gourds, pumpkins, and squash are all members of the genus *Cucurbita*. Cultivars within a species will cross-pollinate with each other, even though they may be called a gourd or a pumpkin or a squash (Table 2.8). This cross-pollination does not affect the shape, color, or taste of the fruit presently produced. However if the seed from these fruits is saved and planted the next year, a vast array of colors, shapes, sizes, textures, and tastes of fruit will be produced.

Parthenocarpy. Some vegetables can have fruit set without the production of seeds, and this is called parthenocarpy. Fruit development by this means can occur (1) without any pollination, (2) with pollination but without fertilization, or (3) by pollination and fertilization but the embryo aborts.

Fruit development without pollination is often noted by gardeners who use auxin-like herbicides to kill weeds in their lawn. A small amount of this material will stimulate production of tomato fruit with no seeds inside the fruit. Frequently, the inside of the fruit will be hollow. Parthenocarpic fruit produced without pollination occasionally occurs naturally with tomato, pepper, pumpkins, and cucumbers (Fig. 2.4). There are several cultivars of cucumbers for greenhouse production that produce fruit without pollination. Most of these cultivars will produce seed if grown outdoors and pollinated.

Fruit development with pollination but without fertilization can occur in tomato. Below 59°F, pollination of the flower can occur, but the pollen tube does not grow so that fertilization does not occur. In this case, various hormones (such as "Blossomset") can be applied by the home gardener to induce fruit set.

Production of parthenocarpic fruit by embryo abortion occurs in seedless watermelons. Although no seeds are produced, the fruit still contains small seed coats, which are sometimes confused with true seeds.

FRUIT DEVELOPMENT AND GROWTH

The control of fruit development is extremely complex. Many growth hormones are involved, and there is considerable competition among various growth substances. It is obvious that fruits that contain a large amount

TABLE 2.8. Cultivars Within the *Cucurbita* Genus That Are Frequently Cross-Pollinated Within a Given Species.

		Cultivar		
		Squash		
Genus and species	Gourds and ornamentals[1]	Summer	Winter	Pumpkin
Cucurbita pepo	Apple Bicolor Bird nest Crown of Thorns Miniature Miniature bottle Orange Pear Spoon Warted Other small hard-shelled types	Yellow elongated types Butterbar Crook Neck Eldorado Goldbar Golden Girl Golden Zucchini Straight Neck Flat-shaped types Green Tint Pattie Pan Scallopini White Scallop Green elongated types Caserta Cocozelle Zucchini	Acorn types Acorn Ebony Table Ace Table King Table Queen Novelty types Vegetable spaghetti Vegetable gourd Edible gourd	Naked seed types Eat-All Lady Godiva Triple Treat Standard types Big Tom Cinderella Connecticut Field Early Sweet Sugar Funny Face Halloween Howden's Field Jack-o'-Lantern Jackpot Luxury Small Sugar Spirit Sugar Pie Tricky Jack Young's Beauty
Cucurbita moschata			Butternut types Butternut Hercules Hybrid butternut Patriot Ponca Waltham	Standard types Cheese Dickinson Field Golden Cushaw Kentucky Field
Cucurbita maxima	Aladdin Turk's Turbin		Baby Blue Hubbard Banana Boston marrow Buttercup Delicious Emerald Gold Nugget Golden Turban Hubbard Hybrid R Kindred Marblehead Sweet Meat	Standard types Big Max King of the Mammoths Mammoth Chili Mammoth Prize
Cucurbita mixta				Green-Striped Cushaw Japanese Pie Tennessee Sweet Potato White Cushaw

Source: Adapted from Doty (1973) and Vandemark and Courter (1978).
[1] Although listed as gourds and ornamentals, some of these are eaten, particularly Turk's Turbin.

FIG. 2.4. Parthenocarpic cucumbers produced in a garden where no other cucumbers were grown nearby. *Courtesy of J.S. Vandemark.*

of food material must obtain this from the rest of the plant. The seeds in the developing fruit produce new growth substances that induce the production of a conducting system between the fruit and the plant. Usually this conducting system is among certain leaves nearest the developing fruit. In peas and cucumbers, about 80% of the sugar is provided by the leaves nearest the fruit. When the fruit is removed, very little sugar moves out to the rest of the plant. Thus, old leaves may fall off or be shaded by new growth without any yield reduction. These old leaves are not providing much food material to the new, developing fruit anyway. When the food material is being rapidly transported to the fruit, a nutrient deficiency sometimes occurs in the leaves because of this rapid transport.

Without the production of growth hormones by the developing seed, the fruit usually falls off. In cucumber, however, if two flowers at the same node are pollinated at the same time, both fruits develop simultaneously. If there is one day difference in pollination, the one pollinated first develops first. The other fruit remain small and does not develop until the first fruit is developed or picked. Sometimes the food material may move from the second fruit into the first fruit, causing the later pollinated fruit to fall off the plant.

Fruit may stop growing without falling off the plant. If the young developing fruit is removed, other fruits continue to grow. In okra, if the old fruit is allowed to mature and remain on the plant, yield is drastically reduced. In tomato, it normally takes about 45 days to go from pollinaton to incipient color (just turning red). If fruit growth is stopped, it may take as long as 130 days to get color. The seeds in these fruit are still viable, and these fruits begin to grow when other fruits are removed from the plant.

In sweet corn and peas, the young seeds are watery and high in sugar content. As they continue to develop, they make starch, which is not sweet tasting, and form woody-like material in the seed coat, making the seeds tough. In peas, development and enlargement of the pod begins first. Development of the embryo is second, and seed enlargement is the last development stage. We can make use of this knowledge and harvest edible-podded peas when pod enlargement is complete and before embryo and seed development begins.

Fruit size depends upon (1) the number of leaves, (2) the amount of sunlight received by these leaves, (3) an ample supply of water, (4) the right temperature, and (5) the competition between fruit. The plant has only so much food material available for the developing fruit. If the plant is large and has ample food, it is capable of producing more and larger fruit than a small plant. If a tomato transplant is planted with a small fruit on the plant, the fruit will be small. This transplant cannot grow vegetatively and develop an adequate-sized fruit at the same time. The plant will use the available energy to produce this fruit and will have limited energy for new growth and additional fruit. Fruit on a tomato transplant should be removed before planting in the garden. Gardeners prefer smaller fruit on cucumbers grown for pickles. The cucumber plant is capable of producing many medium-sized fruit or a few large fruit. However, total weight of the fruit produced by the plant is about the same. Gardeners frequently prefer one large pumpkin instead of several small ones. To produce the largest pumpkin fruit, the first and second female flowers that the plant produces should be removed and the next female flowers pollinated. Total numbers of fruit will be reduced, but the weight of the fruit produced will be increased as all the plant's energy is being directed into this one fruit.

FRUIT RIPENING

When a fruit has reached its maximum size, it is mature. A tomato fruit beginning to turn red is mature and will not increase further in size and may be picked. Ripening of fruit refers to the processes that change in the mature fruit. The general changes during ripening are (1) softening of the fruit flesh, (2) hydrolysis of stored materials into soluble material, (3) changes in the pigment content (color) of the fruit, (4) changes in flavor, and (5) changes in respiration.

During ripening, cell walls are broken down, making the fruit soft. Various insoluble starches are converted to sugar. Ripe bell peppers are sweeter than green ones due to this increased sugar. However, the hydrolysis of starch in pumpkin or winter squash makes the product watery, which is not desirable.

Most gardeners consider a fruit ripe when it has changed to a characteristic color. Tomato and bell peppers become red. Winter squash turns light yellow or orange, starting at the area which was in contact with the ground. Usually this color is due to new pigments being synthesized. In summer beets, there is a high sugar content and a small amount of red color. In the fall, under cooler temperatures, the red color increases greatly; as less photosynthesis is occurring, the sugar content declines. A summer-grown red beet has about the same percentage of sugar as a typical sugar beet.

During ripening, various flavor components appear. In snap beans, over 40 volatile flavor components appear, although only 6 or 7 of these are important. Sweetness and acidity are often important in flavor, and sugar usually accumulates during ripening and acid levels decline. Flavor components in tomato are affected by temperature, and cold temperatures result in less fresh tomato flavor. Tomatoes should not be stored in the refrigerator for fresh use.

It is not desirable to have all the ripening changes occur at the same time. Color development and softening of tomato fruit, for example, are greatly influenced by temperature. At high temperatures (85°F), the fruit becomes soft but color development is slow. The fruit becomes partially colored or becomes orangish instead of red in color and is soft. Tomato fruit should be exposed to temperatures in the 70s (°F). At these temperatures, color development will be rapid, and softening will be slow, resulting in a red, firm fruit. This can be achieved by picking the fruit at the first appearance of red color and allowing the fruit to ripen in the light in an air-conditioned house (or another cooler place).

During fruit ripening, the formation and release of ethylene occur. Once ethylene is formed, it induces more of itself, and the ripening changes occur more rapidly. To hasten ripening, fruit such as tomatoes, pears, bananas, or avocado can be placed in a "fruit-ripening bowl" or another closed container with a few holes for air exchange. Little ethylene will be lost, and fruit will be induced to ripen faster. Some air exchange is needed because the ripening changes require a large amount of energy and oxygen is needed to allow the fruit to increase respiration.

BULB AND TUBER FORMATION

A tuber is an inflated stem produced by a swelling type of growth. This tuber has a number of buds, called "eyes" in potatoes. In a bulb, the stem does not increase in size but the base of the leaves swells. Bulb formation results from an increase in cell size, while tubers increase because of an increase in both cell size and cell number.

Both bulb and tuber formation are affected by day length. Tuber formation of potatoes is accelerated by short days, but some cultivars produce tubers over a wide range of day lengths. Potato tuber formation is greatest at 70°F. Where summer temperatures are greater than this, the crop must be planted as early as possible and mulched with about 8 in. of straw to keep temperatures as low as possible.

Onions require cool temperatures during early growth, and high temperatures and long day lengths for bulb formation. Therefore, if onion seedlings are grown in the house under artificial light for 18 hr per day, the seedlings will not grow faster; instead, these seedlings will form very small bulbs and the tops will die.

Tubers and bulbs do not ripen as fruits do. Tubers should be stored 2 weeks at warm temperatures after harvest to allow a thicker skin and suberin layers to form. This protects the tubers from a high rate of water loss and covers bruises that may have occurred.

SELECTED REFERENCES

Courter, J.W., Vandemark, J.S., and Shurtleff, M.C. 1972. Growing vegetable transplants. Ill. Coll. Agric. Cir. *884.*

Doty, W.L. 1973. All About Vegetables. Chevron Chemical Co., San Francisco.

Harrington, J.F. and Minges, P.A. 1954. Vegetable seed germination. Univ. Calif. Coll. Agric. Leafl.

Hortik, H.J. and Arnold, C.Y. 1965. Temperature and the rate of development of sweet corn. Proc. Am. Soc. Hortic. Sci. *87,* 303–312.

Jones, T.L. 1971. Injury to bean *(Phaseolus vulgaris L.)* in relation to imbibition. Ph.D. Thesis. University of Illinois, Urbana.

Vandemark, J.S. and Courter, J.W. 1978. Vegetable gardening for Illinois. Univ. Ill. Agric. Circ. *1150.*

Wester, R.E. 1972. Growing vegetables in the home garden. U.S. Dep. Agric. Home Gard. Bull. *202.*

3

Soils and Plant Nutrition

SOIL

An ideal garden soil is fertile, deep, friable, well drained, and high in organic matter. The exact type of soil is not as important if the soil is well drained, adequately supplied with organic matter, and retains moisture. The kind of subsoil is also important. Hard shale, rock ledges, gravel beds, deep sand, or a hardpan under the surface is particularly undesirable. They make the development of a good garden soil difficult or impossible. However, if the soil has good physical properties but is just unproductive, this soil can be made productive by adding organic matter and fertilizer and adjusting the pH.

Good water drainage is essential. The garden should not contain low places where water would stand after heavy rains. Good air drainage is also important. The garden may be on a slight slope that allows the air to move downward to lower levels on the hill. This allows the garden to escape early and late frosts.

Soil Composition. Soil is made up of inorganic substances, decaying organic material, air, water, and various amounts of insects, earthworms, bacteria, fungi, and microorganisms. The living organisms degrade the organic material into a residual material called humus. Humus is dark in color and can absorb large amounts of water and nutrients.

Soil Type. Soils that contain 20–65% organic matter are called muck soils. Soils containing over 65% organic matter are peat soils. These soils retain large amounts of moisture but are deficient in several plant nutrients.

Most soils contain less than 20% organic matter and are mineral soils. These soils contain various sized particles of decreasing size: sand, silt, and clay. These materials are bound together into soil particles and the coarseness or fineness of these particles is referred to as soil texture. Texture is important because the area between the particles is filled with organic matter, air, and water. The total area is not as important as the size of the individual spaces. Clay soils have more total space than sandy soils, and clays can hold more water than sands. However, the individual space between particles is so small in the clays that air and water movement is slow; and when it is filled with water, it prevents air movement essential for root growth. Sands allow this water to drain out and the spaces become filled with air. Although this allows for good aeration, without water, plant nutrients are not kept in solution to be absorbed by the plant. The best garden soils contain a mixture of sand (for aeration), silt, and clay (holds water and nutrients).

Mineral soils may be separated into sandy soils, loam soils, and clayey soils. Sandy soils contain less than 15% silt or clay. They are well aerated, dry out, and warm up rapidly. They are low in fertility and cannot hold much added nutrients or water.

Loam soils contain a relatively even mixture of sand, silt, and clay. They contain less than 20% clay, 30–50% silt, and 30–50% sand. These soils are ideal for most vegetables.

Clay soils contain 20–30% clay. Clay soils retain water and dry out slowly making them difficult to cultivate and work properly. Root growth is poor due to the small spaces between particles. These soil surfaces crust easily, reducing stands of some vegetables.

Sand and clay soils may be modified in the garden by adding organic matter and various soil amendments. The relative composition of desirable soils is 45% mineral matter, 5% organic matter, 25% air, and 25% water (Bartelli *et al.* 1977).

Cation Exchange. Cation exchange is the ability of clay and humus, which are negatively charged, to attract and exchange positively charged ions called cations. These cations, such as calcium, magnesium, potassium, and ammonia, are thus held in the soil and are not lost by leaching. These cations become dissolved in the soil solution and are absorbed by the plant. The cations are replenished by the application of fertilizer or released from the degradation of organic matter or decomposition of rocks. The amount of material able to exchange cations is important. The more cation exchange capacity available, the more nutrients the soil can hold and fewer that are lost. Cation exchange capacity is equivalent to the milligrams of hydrogen ions that will combine with 100 g of dry soil. The actual amount varies with the amount of humus and the amount and type of clay present. The cation exchange capacity range for typical garden soils is for sands, 2–10; loams, 2–40; and clays, 5–60.

Soil Preparation. Vegetables grow best on fairly deep soils. If the garden soil is shallow it should be deepened gradually by increasing the depth of spading an inch or two a year until the desired depth is reached. Organic matter should be incorporated into these shallow soils to make the subsoils more productive (see later section on Soil Amendments). The deeper the soil is prepared, the greater is its capacity for holding air and moisture. One of the purposes of turning over the soil is to separate soil particles and allow air to come into contact with as many particles as possible and thereby provide a good growing media for the growth of roots and soil microorganisms.

The garden can be plowed, tilled, or spaded in the spring or fall. Fall preparation has several advantages over spring preparation: (1) organic matter decomposes more rapidly, (2) insect and disease problems are reduced by burying them in the soil or exposing them to the weather, (3) more water is absorbed, (4) the physical condition of clay soils is improved by exposing them to frost action, (5) the soil can be worked and planted earlier in the spring, and (6) the trapped air acts as insulation for increasing the survival of earthworms. Fall preparation is particularly desirable when sod, manure, or a large amount of organic material is to be turned under. These materials will decompose during the fall and early spring and be of value to the crops planted the next season.

Spring preparation of soils is desirable where soil erosion occurs, where shallow tilling is practiced, or on sandy soils. In the south, where conditions are hot and dry, the gardener may wish to consider a type of mulch system (see earlier section on Cropping Systems—Chapter 1) instead of spring soil preparation.

Garden soils should not be worked when they are wet. When a handful of soil is squeezed in the hand, it should readily crumble and not feel sticky when the pressure is released. If the soil forms a compact, muddy ball, it is too wet to be worked. When examining the soil to determine if it is dry enough to work, samples should be taken both at and a few inches below the surface. The soil may be dry at the surface, but the lower layers may be wet. Soil that sticks to a shovel or other tools is usually too wet. Shiny unbroken surfaces of spaded soil is another indication of a wet soil condition. Clay soils low in organic matter lose their crumbly texture if they are worked when wet and become hard, compact, and unproductive.

Fertilizer can be added before soil preparation, and organic fertilizers and natural deposits particularly should be added at this time. If fertilizer is added after the soil is prepared, the fertilizer should be worked into the soil to a depth of two or three inches.

Before planting, the seedbed can be smoothed with a harrow or rake. A freshly prepared seedbed will prevent weeds from emerging before the vegetable plants. Each soil has its own characteristics that determine the best physical condition or tilth suitable for planting. Soils should not be worked to such a fine consistency that crusting occurs after rains. Clay soils particularly crust easily and prevent emergence of seedlings. These soils should be left comparatively rough and cloddy to reduce crusting of the soil surface.

WATER

Adequate soil moisture is essential to the production of vegetables. Water dissolves plant nutrients in the soil, plays an important role in biological activities, keeps the plant cool, and transports food and nutrients in the plant. Many vegetables are about 80% water. Plants require hydrogen (H) and obtain this from water (H_2O). Plants cannot use hydrogen from the air. Water for plant growth comes from rain or snowfall, surface drainage water and underground water. There are usually dry periods during the growing season when additional water will be required to begin germination, keep vegetables growing rapidly, and ensure continual fruit production. The critical periods to ensure that vegetables have adequate water are given in Table 3.1. Rain showers that provide less than one-fourth inch of water barely wet the soil surface and most of this water is lost by evaporation. Whenever the rainfall is less than 1 in. during the week, the garden should be irrigated. About 1 in. of water a week, including rainfall, is desirable. It is better to thoroughly soak the soil to a depth of at least a foot (2 ft in southern and western soils) than to lightly sprinkle the area frequently. Watering in the morning or before 2:00 P.M. will allow the plants to become completely dry and help reduce diseases from mildew, blight, damping off, and others. An easy way to determine how much water has been received by the garden is to place four or five cans at various spots in the garden and measure the amount of irrigation water received in these cans.

Water is often required to establish a garden, particularly summer or fall gardens. If the soil is dry when it is time to plant, about one inch of water should be applied to the area. The surface should be allowed to dry out and the surface then raked or very lightly cultivated to prepare the seedbed. This entire procedure often takes one day. The vegetables should then be planted but no additional water applied for two days. If no rainfall occurs, the area should receive one-half inch of water every other day until the seedlings have emerged. This procedure is particularly important for green, wax, and lima beans as heavy watering results in reduced stands and yields (see Germination Requirements, Water, in Chapter 2).

Sprinkle and Furrow Irrigation. Most gardeners will apply water with overhead sprinklers. Sprinkler irrigation can be applied at a slow rate to allow the water to be absorbed in compacted soils and to reduce run-off on sloping gardens. Winds may disturb the sprinkler pattern, and it is best to water when the air is still. Sprinklers should be placed so the spray pattern

TABLE 3.1. Critical Periods of Water Needs of Vegetables.

Critical period	Vegetables
Germination	All, particularly summer or fall crops
Pollination	Lima beans
Pod enlargement	Lima and snap beans Edible-podded peas
Head development	Broccoli, cabbage, cauliflower
Root, bulb, and tuber enlargement	Carrot, onion, parsnip, potato, radish, turnip
Flowering and ear, fruit, and seed development	Sweet corn, cucumber, pea, squash
Fruit set and early development	Melons
Uniform supply from flowering to harvest	Eggplant, pepper, tomato

overlaps with each nearby sprinkler. Less water is received at the edges of the sprinkler spray than at the center (Fig. 3.1).

In the West and Southwest, many gardeners use furrow irrigation. The beds growing the vegetables should be about 6 in. high and 3 ft apart. The irrigation water is applied in furrows between the beds (Fig. 3.1). Raised beds are also useful in rainy areas as they allow the excess rain water to drain off the beds.

Trickle Irrigation. Trickle irrigation or drip irrigation is a method of applying small amounts of water directly to the growing plant. (Fig. 3.2). The system consists of a network of water-conducting plastic tubes, which allow water to move through the walls of the tube at a slow rate. The tubes are placed at one side of the row and frequently buried 1 or 2 in. deep in the soil. Trickle irrigation is best used with no-till mulch systems (see the section on Cropping Systems—Chapter 1), as the plastic tubes can be left permanently on the soil. About 50% of the water used in the U.S.A. is for irrigation, and some areas experience water shortages. Trickle irrigation uses about 50% less water than overhead irrigation; applies water more uniformly to the plants; keeps the plant foliage dry, thereby reducing plant diseases; and leaches high salt concentrations from the plant roots. The soil between the rows is dry as only about

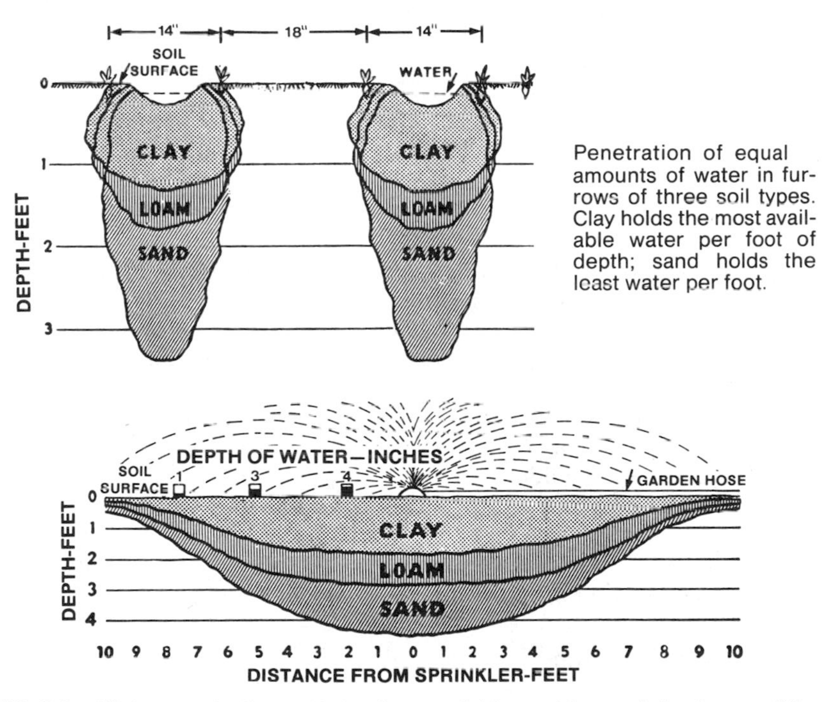

FIG. 3.1. Water penetration patterns from sprinkler and furrow irrigation on different soil types. *From MacGillivray et al. (1968).*

FIG. 3.2. A trickle irrigation system used to irrigate sweet potato plants. *Courtesy of J. Gerber.*

20–40% of the area occupied by roots is irrigated. The plant takes water from the wetted soil area and transfers it to all parts, preventing moisture stress.

If city water is used, a filter system will not be required. Trickle irrigation systems are frequently sold at garden centers. The systems are relatively expensive, but in some areas with water shortages, they are required. The system should be flushed before and after the growing season to prevent the emitter holes from plugging up.

AIR

Plant Growth

All plants require air to provide oxygen (O_2) and carbon dioxide (CO_2). Oxygen is required by the aboveground parts at night for respiration and is required all the time for the roots. When too much water is applied and the drainage is poor, the plant roots die, not from too much water, but from a lack of oxygen.

All plants require CO_2 to manufacture food. If the plants are being grown outdoors there is little way to increase the amount of CO_2 to the plant. In a greenhouse, CO_2 can be added from a CO_2 generator and a dramatic increase in plant growth can be noted. In the winter time, CO_2 is used in the greenhouse for plant growth and CO_2 levels inside are usually lower than outside. Frequently the vents are opened during the warm part of the day to increase air circulation and increase CO_2 levels.

Air Pollution Damage

Injury to vegetation induced by exposure to air pollutants has become more important to gardeners. When plants are injured by an air pollutant, symptoms characteristic of that specific pollutant usually develop. However once the pollutant contacts the plant tissue, the pollutant is itself changed, and the damage symptoms are frequently the only remaining evidence of air pollution. Air pollution injury is frequently confused with nutrient deficiency symptoms and pest damage.

Vegetables may be injured by one exposure at a high concentration or a prolonged exposure to a lower concentration of an air pollutant. The amount of damage depends upon (1) meterological factors that lead to air stagnation, (2) the air pollutant, and (3) the susceptibility of the vegetable plant. Table 3.2 gives the relative sensitivity of some vegetables to various air pollutants.

Ozone and PAN. Ozone probably causes more air pollution injury to vegetables than any other air pollutant. Hydrocarbons and nitrogen oxides are emitted from automobile engines and industrial plants, and a photochemical reaction occurs between these chemicals, producing ozone and PAN (peroxyacetyl nitrate). Concentrations of ozone high enough to cause damage to susceptible vegetables have been found more than 70 miles from large metropolitan areas (Jacobson and Hill 1970).

Ozone damage appears as very small, irregularly shaped spots on the upper leaf surface. These spots are dark brown to black and after 24 hr become light tan and then white several days later (Hindawi 1970). Recently matured leaves are most susceptible, with very young and old leaves being the most resistant. Injury is usually greatest at the leaf tip and margins. Onions and melons grown in a garden are the vegetables most likely to be damaged.

PAN usually affects the underside of newly matured leaves; the area becomes bronzed, glazed, or silvery in appearance. PAN injury develops on only three or four rapidly expanding leaves of sensitive plants. Very young and mature leaves are highly resistant (Jacobson and Hill 1970). Pale green to white areas may appear on the leaf surfaces.

Nitrogen Dioxide. Nitrogen dioxide is produced by the combustion of coal, fuel oil, natural gas, and gasoline used in power-generating operations. Vegetable cultivars vary widely in their susceptibility to nitrogen dioxide. Injury symptoms appear as irregular white or brown collapsed lesions on leaf tissue between the veins and near the leaf margin. The concentration of nitrogen dioxide influences plant injury more than the duration of exposure. Damage is seldom seen in a garden.

Sulfur Dioxide. Major sources of sulfur dioxide come from coal and fuel oil with a high sulfur content. When these are burned for heating purposes or to generate power, or used in petroleum refineries and some chemical plants, sulfur dioxide is produced. Both acute and chronic plant injury occurs. Acute injury is characterized by dead tissue between the veins or on the margins of leaves. The dead tissue may be white, silver, tan, brown, orange, or red in color, depending upon the plant species and time of year exposed. Chronic injury is marked by brownish-red, turgid, or bleached white areas on the blade of the leaf. Fully expanded leaves are most sensitive, while young leaves seldom show damage (Hindawi 1970).

TABLE 3.2. Relative Susceptibility of Selected Vegetables to Damage by Some Air Pollutants.

Air pollutants	Most susceptible	Intermediate	Least susceptible
Ozone	Bean Broccoli Muskmelon Onion Peanut Potato Radish Spinach Sweet corn Tomato	Carrot Endive Parsnip Turnip	Beet Cucumber Lettuce
PAN	Bean Celery Chard Endive Lettuce Mustard Pepper Spinach Tomato	Beet Carrot Sweet corn	Broccoli Cabbage Cauliflower Cucumber Onion Radish Squash
Nitrogen dioxide	Lettuce Mustard Sunflower	Dandelion	Asparagus Bean
Fluoride	Sweet corn	Sunflower	Asparagus Squash Tomato
2,4-D	Tomato	Sweet corn Potato	Bean Cabbage Eggplant Rhubarb
Chlorine	Mustard Onion Radish Sweet corn Sunflower	Bean Cucumber Southern pea Dandelion Squash Tomato	Eggplant Pepper
Sulfur dioxide	Bean Beet Broccoli Brussels sprouts Carrot Chard Endive Lettuce Okra Pepper Pumpkin Radish Rhubarb Spinach Squash Sweet potato Turnip	Cabbage Pea Tomato	Cucumber Onion Sweet corn

Source: Adapted from Jacobson and Hill (1970) and Hindawi (1970).

Fluoride. Fluoride is widespread in the earth's crust as a natural component of soil, rocks, and minerals. When these materials are heated to a high temperature, fluoride may be released. The production of aluminum, steel, ceramics, phosphorus chemicals, and fertilizers may emit significant quantities of fluoride. Most potential sources of large quantities of fluoride have emission control equipment, and vegetable injury is confined to a few sensitive cultivars in limited areas.

2,4-D. There are a number of "auxin" type herbicides on the market, including 2,4-D. These herbicides are frequently used by homeowners to kill various broadleaved weeds in their lawn. Most cultivars of sweet corn are not affected by 2,4-D. Tomatoes, however, are sensitive, causing the plant to produce distorted leaves (see Fig. 3.3), and heart-shaped fruit instead of round fruit. Sometimes the fruit may be hollow. The extent of the injury depends upon the dose received and the rate of plant growth at the time of exposure. Injury from auxin-type herbicides occurs frequently on garden vegetables. This problem is increasing, as most lawn-care companies use a weed killer. On windy days, the material can drift into the garden and result in injury.

Chlorine. Chlorine injury probably occurs frequently to vegetables, but the damage occurs close to the source. Chlorine as an air pollutant has been associated with accidental release during transportation and storage, water treatment plants, sewage disposal plants, swimming pools that are chlorinated, and burning of chlorine-containing plastics. Injury symptoms are

FIG. 3.3. Characteristic moderate to severe 2,4-D damage to a tomato leaf. The downward curvature of young leaflets, prominent light-colored veins, rolled edges, distorted leaf surfaces, and sharp points on the leaflets are all evident.

similar to those caused by sulfur dioxide. Middle-aged leaves are most susceptible to chlorine injury, followed by the oldest and then the youngest leaves. Chlorine-injured plants are most often observed in gardens near swimming pools and sewage disposal systems.

LIGHT

Sunlight is absolutely necessary to produce high-quality vegetables. A garden should not be planted near shrubs, trees, or the north side of a building where shading occurs. Leafy vegetables can be grown in partial shade but vegetables producing fruit need full sun. The rate of photosynthesis (food manufacture) of green plants is proportional to the intensity of the sunlight up to about 1200 foot-candles. Full sunlight is about 10,000 foot-candles, while overcast, cloudy conditions allow about 1000 foot-candles to reach the plant. Thus, on sunny days a little over 10% of the available sunlight is used in photosynthesis, while on cloudy days all the light is utilized.

ESSENTIAL PLANT ELEMENTS

Many different chemical elements are found in plants, but only 16 have been shown to be essential. Table 3.3 lists the 16 nutrients required by vegetables, in order of greatest accumulation. Carbon, oxygen, and hydrogen comprise 96% of the total dry weight of the plant, and these are obtained from air and water. The remaining 13 elements are nitrogen, potassium, calcium, magnesium, phosphorus, sulfur, iron, chlorine, manganese, boron, zinc, copper, and molybdenum. These are absorbed from the soil as inorganic salts. Any of these elements in a high enough concentration will be toxic to the plant and reduce yields.

TABLE 3.3. Relative Amounts of the Essential Nutrients Required by Most Vegetables.

Element	Percentage of total dry weight
Carbon	45.0
Oxygen	45.0
Hydrogen	6.0
Nitrogen	1.5
Potassium	1.0
Calcium	0.5
Magnesium	0.2
Phosphorus	0.2
Sulfur	0.1
Iron	0.01
Chlorine	0.01
Manganese	0.005
Boron	0.002
Zinc	0.002
Copper	0.0006
Molybdenum	0.00001

Source: Gerber and Swiader (1981).

TABLE 3.4. General Deficiency Symptoms of Nutrients That Usually Limit Vegetable Production in Home Gardens.

Nutrient	Symptoms
Nitrogen	General yellowing (chlorosis) of the older leaves; slow growth; small leaves; spindly plants; older leaves drop. Most common on highly leached soils or high organic matter soils at low temperatures.
Phosphorus	Dark blue-green or red-purple leaves, especially on the underside; slow growth. Most common on acidic or alkaline soils and organic soils under cold, dry conditions.
Potassium	Margins of lower leaves appear yellow and develop brown regions; slow growth. Most common late in growing season when nutrient moves to storage organs. Most common on highly leached acidic soils and organic soils.

Nutrient Deficiency Symptoms. Inexperienced gardeners often mistake insect, disease, or cold-weather damage for nutrient deficiency symptoms. The diagnosis of a nutrient deficiency, based on visual symptoms, is difficult (Gerber and Swiader 1981). Symptoms vary according to the plant species, soil type, climate, and age of the plant. This is further complicated by the fact that several of the nutrient deficiency symptoms are similar in appearance. The best practice with a home garden is to make an annual application of a balanced garden fertilizer. Table 3.4 describes the general deficiency symptoms for the three nutrients most often found to limit the productivity of vegetable gardens.

Carbon, Hydrogen, and Oxygen. These three elements make up the bulk of the plant tissue. Much of the live plant is made up of water (hydrogen and oxygen), which is used to keep the plant cool, keep materials in solution, transport minerals from the soil through the plant, and transport food from the leaves. About 90% of the dry weight of the plant is cellulose and various sugars, much of this being in the cell walls. These materials are composed of carbon, hydrogen, and oxygen. Plants receive carbon and oxygen from the air and hydrogen from water. All other nutrients are obtained from the soil solution.

Nitrogen. Nitrogen is usually more responsible for increasing plant growth than any other element. It is used in the formation of various proteins needed by the plant to do useful work. Although nitrogen is about 80% of the earth's atmosphere, it cannot be used in this form. It must be converted by various microorganisms or industrial processes into a form the plant can use. The two main forms are ammonium (NH_4^+) and nitrate (NO_3^-) ions. If various organic nitrogen forms are added as a fertilizer, this material must be broken down into these ions by soil microbes before the plant can use the nitrogen. Plants cannot use the nitrogen in the organic form. The organic material (whether added as a fertilizer or plowed under plant debris) is converted to the ammonium ion by various soil bacteria, and this ion moves little in the soil. However, the ammonium ion is coverted by other soil bacteria into the nitrate ion, which is soluble and easily leached from the soil. These conversion processes are dependent upon air, temperature, and misture. When air is lacking in the soil, various bacteria convert nitrates to atmospheric nitrogen and the nitrogen is lost. Under cool temperatures, little ammonium is converted to nitrate but warm temperatures cause some of the ammonia to be lost to the atmosphere by volatilization and also hasten its conversion to nitrate. Large amounts of moisture will result in leaching of these nitrates.

Four fates await any nitrogen applied to the soil (Luckhardt 1976), whether the nitrogen is found in a commercial fertilizer, an organic fertilizer, or crop residue: (1) it escapes into the air as a gas, (2) it is leached by water, (3) it is incorporated into new organic matter, and (4) the crop can use the remainder. Some 15–20% of the applied nitrogen is lost by volatility and about 50% of the nitrogen in manure is frequently lost by volatility. Leaching of the nitrates amounts to a lot on sandy soils and less on heavy clay soils. However, on clay soils saturated with water, the nitrates are converted to atmospheric nitrogen and are lost. Thus, heavy clay soils may lose as much nitrogen as sandy soils. The added nitrogen may be used by microorganisms involved in the breakdown of plant debris. These microbes use whatever amounts of nitrogen they require and the crop gets what is left. Thus, 30–60% of the applied nitrogen is tied up in organic matter. The nitrogen available to the vegetable crop is 15–75% of that applied and averages about 50%.

Snap beans and peas are legumes and these crops require a comparatively small amount of nitrogen compared with other vegetable crops (Janssen 1976). These legumes are associated with symbiotic bacteria (Fig. 3.4) that convert

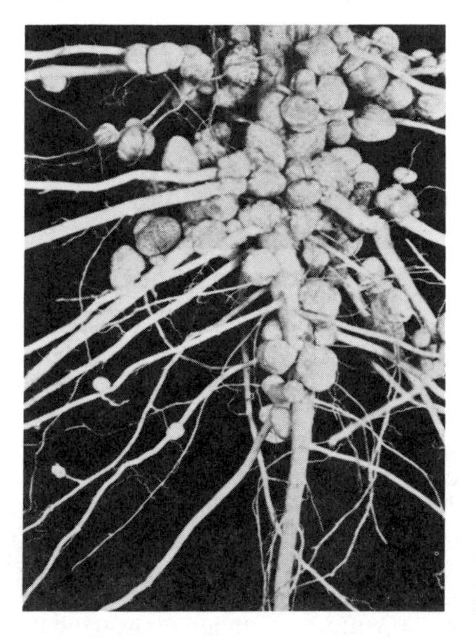

FIG. 3.4. Bean plant with root nodules containing bacteria that fix atmospheric nitrogen. *From Shanmugam and Valentine (1976) (Univ. Calif. Div. Agric. Sci.).*

atmospheric nitrogen into forms the plant can use. These bacteria depend directly upon the plant for their energy supply and cannot convert atmospheric nitrogen into a usable form unless they are associated with a specific plant. The amount of nitrogen converted into a usable form depends upon the vegetable involved, and the estimated pounds of nitrogen fixed per acre for some vegetables are as follows: peas 72 lb, peanuts 42 lb, beans 40 lb, and soybeans 58 lb (Delwiche 1970).

Usable nitrogen is also added to the soil by nonsymbiotic types of bacteria and by lightning and rainfall. The amount of nitrogen added by these methods is unreliable for the typical gardener and can be ignored in fertilizer recommendations. The amount of nitrogen released from decaying organic matter for plant use depends upon the amount of organic matter in the soil.

Large amounts of nitrogen are used by plants when they are growing vegetatively and developing their roots, stems and leaves. Nitrogen stimulates the production of these parts at the expense of the fruiting and food storage parts. Leafy vegetables such as cabbage, kale, lettuce, and spinach require more nitrogen than other garden crops. They may be stimulated to produce more leaves by side-dressings of nitrogen.

Phosphorus. Phosphorus promotes root growth, flower, fruit and seed development, and stimulates stiffer stems. Many soils contain large amounts of phosphorus in a form not available to the plant. The availability of phosphorus is related to soil pH and is most readily available at pH 5–7. Phosphorus is not very soluble and little is removed by leaching. However, movement of phosphorus in the soil is low and added phosphorus fertilizer remains where it is placed. Thus, phosphorus should be worked into the soil to make it available to be absorbed by plant roots. When transplants are planted in the spring, a starter fertilizer is frequently applied around the plants. This fertilizer contains a large amount of water-soluble phosphorus to stimulate root growth and provide available phosphorus directly to the roots that are there.

Potassium. Potassium is soluble in the soil and its loss by leaching is controlled by the amount of organic matter and the type of clay in the soil. Potassium is attracted to these types of materials and is held in reserve by them for later plant use. In soils high in organic matter such as peat and muck soils, potassium is usually limiting.

Potassium contributes heavily to the growth of root crops and has a stimulating effect on plant vigor and health. Root and tuber crops such as carrots, beets, parsnips, potatoes, sweet potatoes, and turnips require larger amounts of potassium than other vegetables.

Calcium. Calcium is seldom lacking in the soil as a plant nutrient. It is frequently added as limestone to adjust the pH of the soil. In many mixed fertilizers calcium is combined with phosphorus and added incidentally to the soil. Calcium is used by the plant for the formation of new cell walls.

Magnesium. Magnesium is frequently deficient in sandy, well drained soils. It is soluble in the soil solution but held to the clays and organic matter. Magnesium is important in chlorophyll formation and is involved in photosynthesis.

Sulfur. Sulfur is not present in large amounts in soil and is easily removed from the soil by leaching. Sulfur makes the soil acidic and has been used to control soil pH. It is added to the soil in various organic material and in association with superphosphate types of fertilizer. Sulfur is seldom limiting in a garden soil.

Other Essential Elements. Manganese, iron, zinc, boron, molybdenum, copper, and chlorine are usually not limiting in garden soils. Manganese and iron, however, may be unavailable if the soil pH is alkaline. These two elements must then be applied in a water soluble form or the soil pH made slightly acid.

Soil Test and pH

In order to determine what elements are deificient in your soil, a soil test should be taken. This test will indicate the amount and availability of nitrogen, potassium, and phosphorus and the soil pH. A small amount of soil from six to eight representative areas in the garden should be collected, dried, and taken to the nearest county soil-testing laboratory, or contact the local county agent or farm adviser.

The acidity or alkalinity of the soil is expressed as pH, with pH 7 being neutral. The pH of the soil is regulated by the amount of hydrogen ions and mineral cations found attached to the soil colloids. A large number of hydrogen ions makes the soil acid(below pH 7) while a large number of mineral cations makes the soil alkaline (above pH 7). A change of one pH unit is a 10-fold increase in acidity or alkalinity from pH 7. Orange juice is acid (about pH 4), pure water is neutral (pH 7), and most soap solutions are alkaline (about pH 9). A pH above 9 or below 4 is toxic to plant roots. Plants grow best at a slightly acid pH of 6.5. Soil pH determines the availability (not total amount) of many nutrients. Calcium, phosphorus, and magnesium may be unavailable at an acid pH below 6, and at this low pH, manganese, boron, zinc, iron, and aluminum may become so available to the plant that they are toxic. An alkaline pH frequently renders iron unavailable for plant growth. The optimum pH for vegetable crops is given in Table 3.5. When a number of crops are grown in the same area, a common pH must be chosen. If commercial fertilizers are used, a pH of 6.5–6.8 is best. If organic fertilizers (such as rock phosphate) are used, they will not dissolve and will not become available to the plant at this pH level. A pH of 6.0 is recommended for organic gardeners, but below a pH of 6.0 most plants will not grow as well (Table 3.5). Potatoes are often grown at a pH of 4.5

TABLE 3.5. Optimum pH for Vegetables.

pH	Crops
6–8	Asparagus
6–7.5	Beet, cabbage, muskmelon, peas, spinach, summer squash
6–7.0	Celery, chives, endive, horseradish, lettuce, onion, radish, cauliflower
5.5–7.5	Sweet corn, pumpkin, tomato
5.5–7	Snap beans, lima beans, carrots, cucumbers, parsnips, peppers, rutabaga, Hubbard squash
5.5–6.5	Eggplant, watermelon
4.5–6.5	Potato

Source: Anon. (1968).

or 5 to prevent damage from potato scab disease, as this disease increases with increasing pH. The club root disease thrives in acid soils, and cabbage is sometimes grown about pH 7 to reduce damage from this organism.

Soil pH is adjusted with limestone (calcium) to increase alkalinity or sulfur to increase acidity. The material can be applied the same time fertilizer is added in the fall and the soil then plowed or spaded. Spring applications are of little value that growing season. The amount of limestone or sulfur needed to adjust soil pH is given in Table 3.6.

Dolomitic limestone contains about 5% magnesium oxide and should be used to adjust soil pH if the soil is deficient in magnesium. Wood ashes contain about as much calcium as dolomitic limestone and may be used to adjust soil pH in place of limestone (Fletcher and Ferretti, undated).

TABLE 3.6. Amounts of Limestone or Sulfur Needed to Adjust Soil pH.

	Raising pH to pH 6.5 (lb of limestone per 1000 sq ft)[1]		
Change in pH desired	Sandy soil	Loamy soil	Clayey soil
From 6.2 to 6.4	40	80	80
From 5.8 to 6.4	120	200	200
From 5.4 to 6.4	320	360	400
From 4.8 to 5.3	160	240	280
From 4.2 to 4.7	200	320	—
	Lowering pH (lb of sulfur per 1000 sq ft)		
Change in pH desired	Sandy soil	Loamy soil	Clayey soil
From 8.5 to 6.5	40	50	60
From 7.5 to 6.5	10	15	20
From 6.5 to 5.5	10	15	20
From 7.5 to 5.5	20	30	40
From 8.5 to 5.5	50	65	80

Source: Adapted from Vandemark (1973).
[1] A similar amount of wood ashes or oyster shells may be used in place of limestone.

FERTILIZER

A fertilizer improves plant growth directly by providing one or more necessary plant nutrients. Organic fertilizers are directly derived from plant and animal sources. Inorganic fertilizers are not derived from plant and animal sources, although many such materials come from naturally occurring deposits. If these materials are not refined they are often referred to as "organic." Refined or inorganic fertilizers are referred to as chemical fertilizers. Both forms can be equally good. Organic materials are degraded by soil microorganisms into inorganic, water-soluble forms. Chemical fertilizers are supplied in this form. Plants can only use the inorganic, water-soluble form and cannot determine if the nutrient came from organic matter, a chemical fertilizer, a natural deposit or natural weathering of rocks (Gowans and Rauschkolb 1971; Utzinger *et al.* 1973).

Analysis. By law, a commercial fertilizer must show the analysis of the material. Usually this is shown as: first, the percent of nitrogen; second, the percent of phosphorus expressed as phosphorus pentoxide (P_2O_5); and third, the percent of potassium expressed as potassium oxide (K_2O). Only nitrogen is expressed in the elemental form. There is 43.7% phosphorus in phosphorus pentoxide and 83% potassium in potassium oxide. A fertilizer showing an analysis of 12-5-10 would contain 12% nitrogen, 5% phosphorus pentoxide and 10% potassium oxide. When the total amount of nutrients is below 30%, the fertilizer is considered a low analysis fertilizer and above 30%, a high analysis fertilizer. Transportation costs per amount of nutrient are higher with low analysis fertilizer. However, with low analysis fertilizers, gardeners have less problem with applying the fertilizer evenly, have less problem with the fertilizer burning the plant, and frequently the mineral impurities in low analysis fertilizers are of value.

Natural Deposits as Fertilizers

Natural deposits of material are usable as fertilizers. Limestone and sulfur are used to correct soil pH and often not thought of as fertilizers. However, dolomite limestone contains about 9% magnesium and about 25% calcium. Other natural deposits are used as fertilizer but the rock must be pulverized. Grinding of the material is the only change that is made and the smaller the particle size, the more readily available are the nutrients. The value of these natural materials depends upon the availability of the nutrients to the plant, not the total amount of nutrients present. This can be demonstrated with rock phosphate. Raw rock phosphate has been ineffective on soils with a pH higher than 6.0, even though it was finely ground (Fletcher and Ferretti, undated). Above pH 6.0 the phosphrous in rock phosphate is only slightly soluble and is released from the rock very slowly. Rock phosphate needs acid to release the phosphate for plant use, and is therefore more effective on acid soils. Rock phosphate contains from 20 to 32% phosphorous pentoxide but the amount which is available for plant use is less than 5%. However, in acid soils, the phosphorus may be slowly released for plant use in subsequent years after application.

Rock Phosphate. Rock phosphate is best used for reinforcing manure and compost piles. If used as a garden fertilizer, large amounts of rock phosphate must be applied to provide enough phosphorus for plant growth. Rock phosphate comes from marine deposits of apatite. Over 40% of the rock phosphate in North America comes from Florida and North Carolina (Anon. 1988).

Granite Meal and Greensand. Granite meal and greensand are used as a source of potassium. Granite meal is ground up granite rock, often a by-product of the granite building stone industry, and often contains some phosphorus. Greensand is obtained from sandy deposits. The potassium in greensand dissolves rapidly into the soil solution but the total amount of potassium in greensand is quite small. Because of this low concentration of nutrients, substantially more greensand is required than chemical fertilizer. The potassium in granite meal is not readily available, and granite meal is better used as a soil amendment than a fertilizer. Granite dissolves slowly, which is one of the reasons it is used as headstones in cemeteries.

Sodium Nitrate. Sodium nitrate is a source of nitrogen that is rapidly available for plant growth. The availability of this nitrogen is similar to that in a chemical fertilizer. Most sodium nitrate, as a natural deposit, comes from bat guano in caves. It may be found as "Chilean nitrate."

Natural deposits vary widely in their composition of plant nutrients. They also vary widely in the nutrient availability. Table 3.7 lists the average amount of nutrients in some natural deposits.

Organic Fertilizers

Organic fertilizers are usually low in nitrogen, the one nutrient that must be added in large amounts. When organic matter is added to the soil and is low in nitrogen, the microorganisms that degrade this organic matter use the available soil nitrogen. The plants receive what is left, and deficiencies often develop.

TABLE 3.7. Average Nutrient Value of Some Natural Deposits[1].

Material	Percentage			Relative availability
	N	P_2O_5	K_2O	
Granite meal	0	0	3	Very slow
Greensand	0	1	7	Medium
Rock phosphate	0	30	0	Very slow
Sodium nitrate	16	0	0	Rapid

[1] The amount of plant nutrients is highly variable, depending upon the material's origin. Their availability depends upon soil conditions and the fineness to which the material was ground.

The cost per unit of nutrient in organic fertilizers is almost always higher than chemical fertilizers. Organic fertilizers do have a number of advantages. Their nutrients are released slowly, making them available to the plant over the entire growing season; the nutrients are less likely to be lost from the soil; many organic fertilizers also act as soil amendments, improving the physical condition of the soil and indirectly improving plant growth. The nutrient content of some common organic fertilizers is given in Table 3.8. These values are averages, as the percentage of plant nutrients is highly variable. From an ecology and conservation viewpoint, bonemeal, dried blood, cottonseed, and soybean meal are more useful and valuable as a livestock feed than as a fertilizer. These products should first be used as an animal feed and the manure recycled and used as an organic fertilizer. Manure will be discussed in a separate section later.

Bonemeal and Blood. Steamed bonemeal is a by-product of the meat industry and is an excellent source of phosphorus, calcium, and trace elements. Dried blood is also a by-product of the meat industry and is sold for its nitrogen content. The supply is governed by the number of cattle being slaughtered. It is used mainly as a feed, making it expensive for gardeners to use. Fish meal is also sold for its nitrogen and phosphorus content, which becomes slowly available to the plant.

Gardeners can provide their own bonemeal by adding various cooked bones, without meat, to the compost pile. As only 5 lbs per 1,000 sq. ft. are needed (see Table 3.12), this amount can be accumulated in compost over a year's time. The bones will not decompose rapidly, so the compost material should be run through a chipper-shredder to grind them.

TABLE 3.8. Nutrient Value of Some Organic Fertilizers.

	Percentage		
Material	N	P_2O_5	K_2O
Animal tankage	9	10	1
Bonemeal, steamed	2	22	0
Blood, dried	13	1	0
Compost	2	1	1
Cottonseed meal	6	3	1
Fish meal	10	5	0
Guano	10	4	2
Mushroom compost	1	1	1
Wood ashes[1]	0	1	5
Sawdust	0.2	0.1	0.2
Seaweed	1	1	11
Sewage sludge	2	2	1
Soybean meal	6	1	2
Manure	See later section on Manures		

[1] Burning removes all nitrogen. The K_2O content depends upon the tree species burned. Wood ashes contain about 26% calcium and are alkaline.

Cottonseed and Soybean meal. Cottonseed and soybean meal are good sources of nitrogen that is moderately available to the plant. They are also used in cattle feed, making them expensive to use as a fertilizer.

Compost. Compost is composed of decayed organic material and the nutritive value depends upon the material added to the compost pile. The major value of compost is as a soil amendment as discussed later, and not as a fertilizer.

Mushroom compost is often available in areas where commercial mushrooms are produced. It is inexpensive and an excellent source of organic matter. For soils low in organic matter, about 4 in. may be spaded into the soil. Sawdust is a common material, but a nitrogen deficiency is almost inevitable unless other fertilizers are applied. Sawdust contains too much carbon for the soil microorganisms to decompose without additional nitrogen. For each bushel of sawdust added, about one-half pound of a 33-0-0 fertilizer is needed.

Wood Ashes. Wood ashes contain no nitrogen, as this is lost during burning. The amount of potassium in wood ashes depends upon the tree species burned, and varies from 5% in white oak to 13% in red oak ashes. As only 25 lbs per 1000 sq. ft. are needed (see Table 3.12), this amount can be accumulated from a fireplace. The wood ashes can be added to the compost pile. Wood ashes average about 26% calcium, are alkaline, and can be used to neutralize the natural acidic conditions in compost. The compost can be added to the garden in the fall (Lerner and Utzinger, 1986). Under these conditions, little change in soil pH should occur.

Seaweed. Seaweed (kelp) is a good source of trace elements. During World War I, kelp was harvested to produce explosives and for fertilizer. Because kelp grows in the oceans, it contains large amounts of soluble salts including common table salt. Heavy applications of seaweed should be avoided, as these salts may reduce plant growth.

Sewage Sludge. Sewage sludge varies greatly in composition. Raw sewage or untreated or improperly treated sewage sludge should not be applied to the garden soil, as these sludges often carry human diseases that are absorbed by the plant and found in the edible part. Heat-treated sludges are normally safe from a sanitary viewpoint. Another potential hazard is toxicity from nonessential elements such as cadmium and possibly selenium. Other toxic elements like nickel, lead, chromium, and arsenic are less hazardous than cadmium, selenium, and mercury, as they are less readily absorbed by plants and transferred to man. Most municipal sludges from industrial areas contain some or all of the toxic elements; it is not possible to use sludge as a fertilizer without adding toxic elements to the soil. Leafy vegetables like lettuce, cabbage, spinach, asparagus, and Swiss chard are effective in transporting a toxic element like cadmium between the soil and man. It has been recommended that vegetables not be consumed if soil cadmium levels exceed 5 ppm. But the recommendation implies that a relationship exists between soil tests levels of cadmium and plant uptake. Other factors such as pH, soil phosphorus and zinc, the chemical form of cadmium, plant type, and growth rates also affect the accumulation of cadmium in the plant. When the only variable is the level of soil cadmium, a relationship does exist between soil and plant cadmium. Given a single crop, grown at one location, the soil test for cadmium may have some value. But when many soils, crops, and garden practices were examined (Gerber *et al.* 1981), there was no meaningful relationship between soil and plant cadmium. In fact, 44% of the samples of leafy greens grown on soil with less than 5 ppm cadmium had more than 1 ppm in the edible parts. A steady diet of vegetables with more than 1 ppm cadmium is not recommended. Until a method of treatment is developed to remove potentially hazardous heavy metals from sewage waste, sludge should not be applied to vegetable gardens (Gerber *et al.* 1981). Most sludges may be used on lawns, flowers, shrubs, and trees if the area is not to be converted to a vegetable garden in the near future.

Chemical Fertilizers

Many chemical fertilizers are derived from natural deposits. The nutrients are concentrated and converted into a form that is readily available in the soil for the plant to absorb. This is particularly true of phosphorus and potassium. Much of the nitrogen fertilizer is synthesized from the nitrogen in the air. The three major forms are ammonia, nitrate, and urea. Urea is an organic compound, so technically it is an organic fertilizer. Because it is manufactured commercially it is referred to here as a chemical fertilizer.

Chemical fertilizers are usually concentrated and so only a small amount is added to the soil to provide the needed nutrients. Since chemical fertilizers do not contain a lot of nonnutrient material, they are usually more economical than organic fertilizers or natural deposits. Chemical fertilizers are immediately available for plant use. The use of a concentrated, readily available fertilizer can also become a disadvantage instead of an advantage. Many gardeners apply too much chemical fertilizer since only small amounts are needed. Because they are soluble and readily available, there may be large amounts of nutrients in the root zone, which results in salt injury to the plant. Some nutrients in chemical fertilizers are very water soluble and can be leached from the root zone and lost. Table 3.9 gives some common chemical fertilizers and their nutrient value.

Unmixed fertilizers contain only one element. The nitrogen carriers are the most important. Urea is a common chemical fertilizer that is rapidly broken down by soil microorganisms into ammonia. Ureaform types of chemical fertilizers are broken down slowly to release nitrogen for plant use. The nitrogen is made available to the plant for a longer period of time during the growing season. In addition, these ureaform fertilizers do not burn the plant if applied to the foliage and there is less problem of salt injury. These slow-release types of chemical fertilizers are similar in many

TABLE 3.9. Approximate Composition of Chemical Fertilizers.

Material	Percentage		
	N	P_2O_5	K_2O
Urea	42–46	0	0
Ureaform	30–40	0	0
Ammonium nitrate	33.5	0	0
Ammonium sulfate	20.5	0	0
Superphosphate	0	16–20	0
Triple superphosphate	0	46	0
Muriate of potash	0	0	48–62

respects to organic fertilizers but lack the soil amendment properties of the organic fertilizers.

Mixed fertilizers contain nitrogen, phosphorus and potassium in various amounts. Various analysis suitable for general garden use are 5-10-5; 3-12-12; 5-10-10; 8-16-8; 10-10-10; 12-12-12. Several highly water soluble, high phosphate fertilizers are available for use when transplanting to promote faster initial growth of the plants, such as 10-55-10 or 11-52-17.

Fertilizer Recommendations

The natural fertility of a soil is dependent upon the factors that built the initial soil: the initial rock formation, climate, time, topography, and the native vegetation. Soil that has not been disturbed reaches an equilibrium. The amount of nutrients released into the soil by the breakdown of organic matter, weathering of the rocks, and so on equals the amount of nutrients removed from the soil by leaching, formation of new plant material, and the like. However, when the soil is used to grow vegetables, some of the nutrients are tied up in the formation of new plants and some are removed in the edible parts. The amount removed varies with the vegetable and the yield. The approximate amount of the major nutrients removed by vegetables is shown in Table 3.10. The major nutrients that need replenishment are nitrogen, phosphorus, and potassium. These three nutrients are the major ingredients of most fertilizers.

The amount of fertilizer to add to the soil depends upon many factors. These include (1) the amount removed in the edible part (shown in Table 3.10); (2) the disposition of the plant debris. About one-third the nitrogen and half of the phosphorus and potassium absorbed by a sweet corn plant remains in the nonedible part of the plant. If this material is returned to the soil, these nutrients are not lost; (3) the type of plant. Bacteria on legumes (peas, beans) have the capacity to convert atmospheric nitrogen (see Fig. 3.4) into a form the plant can use, and these plants do not need as much nitrogen fertilizer; (4) the type of soil. Soils differ in the amount and availability of nutrients and some Midwest soils contain adequate amounts of phosphorus; and (5) the organic matter content of the soil. If corn stalks, sawdust, unfinished compost, manure, or straw is added to the soil, soil microorganisms will use the available nitrogen to break down these materials, and additional nitrogen is required. Dark-colored soils contain organic matter that is continuously being broken down,

TABLE 3.10. Nutrients Removed by the Edible Plant Part of Vegetables.

Crop	Yield (lb/1000 sq ft)	lb removed/1000 sq ft		
		N	P_2O_5	K_2O
Asparagus	90	0.5	0.1	0.2
Beans, snap	90	0.3	0.1	0.3
Beans, dry	54	1.8	0.1	0.7
Beets (without tops)	450	1.1	0.5	1.9
Cabbage	450	1.4	0.3	1.3
Carrots	580	1.4	0.6	3.4
Cauliflower	420	1.4	0.5	1.1
Corn, sweet	135	0.6	0.1	0.2
Cucumbers	430	0.5	0.2	1.1
Eggplant	260	0.5	0.2	1.1
Horseradish	90	0.9	0.4	0.7
Lettuce	270	0.9	0.1	0.5
Muskmelon	150	0.3	0.1	0.6
Onion, bulbs	450	1.0	0.05	0.5
Parsnips	570	1.4	1.0	5.0
Peppers	240	0.6	0.1	0.5
Potatoes	410	1.4	0.6	2.0
Pumpkin	580	1.4	0.6	3.9
Spinach	135	0.7	0.1	1.3
Squash, summer	455	0.5	0.1	0.9
Squash, winter	580	1.4	0.6	3.9
Sweet potato	320	0.9	0.3	2.0
Tomato	570	1.1	0.3	1.7
Turnip	510	1.1	0.5	1.4
Watermelon	225	0.2	0.07	0.3

and the nitrogen is released for plant use. Light-colored soils contain little organic matter and will release little nitrogen. Dark-colored soils release about a half pound of nitrogen per 1000 sq ft and would require less nitrogen fertilizer than a light-colored soil.

If this is your first year in your present location, have your soil tested and follow those fertilizer recommendations. If you do not have your soil tested, follow the general recommendations in Table 3.11 or 3.12. Most nutrients in organic fertilizers and natural deposits are slowly available to the plant, and much more of these must be applied than chemical fertilizers. Soil nutrients can be present in unbalanced amounts by using organic fertilizers just as well as by using chemical fertilizers. A specific balance among elements is necessary for each vegetable to achieve maximum yield and quality. Few organic fertilizers contain a balanced amount of nitrogen, phosphorus and potassium, as chemical fertilizers do. Three different organic fertilizers (see Table 3.12) are needed to provide the proper amount of nitrogen, phosphorus, and potassium. Fertilizers high in nitrogen only promote vegetative growth, rather than fruits, seeds, or root and tuber crops. High nitrogen fertilizers should be used on lawns instead of on gardens.

If rock phosphate is being used, 10 lb of sulfur should be mixed with each 50 lb of finely ground rock phosphate. The sulfur will lower the pH of the soil immediately near the phosphorus and make it more available. The com-

TABLE 3.11. Chemical Fertilizer Recommendations for Vegetables.

Previous fertilizer treatment	Fertilizer analysis	Fertilizer to apply (lb/1000 sq ft)	Amount applied (lb/1000 sq ft)		
			N	P_2O_5	K_2O
None or little	3-12-12	40	1.2	4.8	4.8
	or				
	5-20-20	30	1.5	6.0	6.0
Some	5-10-10	25	1.2	2.5	2.5
Heavy (previous gardens have produced well)	10-10-10	12	1.2	1.2	1.2
	or				
	15-15-15	10	1.5	1.5	1.5

Source: Adapted from Vandemark *et al.* (1977).

mercial fertilizer superphosphate is made from rock phosphate treated with sulfuric acid. Adding elemental sulfur with the rock phosphate produces a similar although limited effect (Gerber 1980).

It is usually not harmful to add excess plant nutrients to a soil. However, large amounts may injure some plants. Large amounts of nitrogen induce vegetative growth rather than seed production and result in poor storage quality of vegetables. All plant nutrients will kill plants if added to the soil in large amounts.

When to Apply. Chemical fertilizer should generally be applied during the growing season of the plant. The fertilizer should be spread over the garden area in the fall or spring and the soil spaded or plowed. If organic

TABLE 3.12. Organic Fertilizer Recommendations for Vegetables.

Organic material	Previous fertilizer treatment	Main nutrient applied	Amount to apply	Amount of available nutrient applied
			(lb/1000 sq ft)	
Dried blood	All treatments	N	10	1.2
Fish meal	All treatments	N	15	1.2
Cottonseed or soybean meal	All treatments	N	20	1.2
Steamed bonemeal	Little or none	P	25	4.8
	Some	P	15	2.4
	Heavy	P	5	1.1
Rock phosphate[1]	Little or none	P	300	5.0
	Some	P	150	2.5
	Heavy	P	75	1.2
Greensand	Little or none	K	300	5
	Some	K	150	2.5
	Heavy	K	75	1.2
Wood ashes	Little or none	K	100	5
	Some	K	50	2.5
	Heavy	K	25	1.2
Manure	See later section on Manures			

[1] Nutrients in rock phosphate are not readily available the first year but become more available in subsequent seasons, resulting in high initial application rates.

fertilizers, slow-release chemical fertilizers, sulfur, or limestone is used, it is best applied in the fall to allow it to partially decompose and make the nutrients available for plant growth. Chemical fertilizers can be applied just before planting the crop, if they are worked into the soil. Phosphate fertilizers particularly give best results when worked into the soils, so plant roots can contact the fertilizer.

When transplants are used, a starter fertilizer should be applied to give faster plant growth. About 1 tablespoon of 10-52-17 or 10-50-10 per gallon of water should be made up and 1 cup of this solution applied around the roots of each plant. A solution of 1 cup on 0-45-0 or similar fertilizer in 12 qt of water may also be used and 1 cup of this solution used for each plant. As nutrients in organic fertilizers are not readily available, there are no good organic fertilizers that can be used as a transplant fertilizer.

Often the plants require more fertilizer during the growing season. Nitrogen particularly becomes limiting late in the growing season. Fertilizer can be applied in a band along one side of the row about 4 in. from the crop, a process called sidedressing. This fertilizer can be applied to leafy crops, sweet corn, greens, and root crops when they are half-grown; and to tomatoes, peppers, cucumbers, and squash when they begin to produce fruit. About 2½ lb of ammonium nitrate (33-0-0) or 2 lb of urea (42-0-0) per 1000 sq ft should be used, or about 8 lb of dried blood or fish meal. The chemical fertilizer should not come into contact with the plant leaves as it will burn them. All sidedressed fertilizers should be hoed or worked into the soil; and in dry weather, water should be applied to make the fertilizer more quickly available.

Effect on Soil Microorganisms and Earthworms. Microorganisms and earthworms live on organic matter in the soil. When fertilizer is used, more crop is produced and more plant residues are returned to the soil, providing more food. Microorganisms require relatively large amounts of nitrogen to degrade organic matter, and fertilizer stimulates their growth. When organic matter and insulative cover during the winter is given, earthworms prosper, even if the soil is spaded or plowed each year. Earthworm populations are limited most by a lack of fertilizer, water, and adequate insulation during cold temperatures.

Manure

Stable or barnyard manure was practically the only fertilizer available in the early days of vegetable production. The use of manure can greatly affect the amount of fertilizer needed, as it contains many plant nutrients and supplies an important source of nitrogen. Manure does more than add nutrients and sometimes produces better results than chemical fertilizers alone. Manure provides needed organic matter, which improves the physical condition of the soil (see Soil Amendments—Organic) and in this manner also improves plant growth. If animal manure can be obtained cheaply, it is the best material for maintaining the organic matter content of the soil and supplying a source of nutrients. Manure is best used for vegetables that require a large amount of nitrogen. Manure is usually low in phosphorus, and fresh manure should be reinforced with superphosphate at 50 lb per ton of cattle or horse manure. This phosphate will absorb ammonia and reduce the loss of nitrogen and increase the phosphorus content of the manure. Organic gardeners can use rock phosphate instead of superphosphate.

Continued heavy applications to the garden can create soil fertility and excessive salt problems. Some manures contain excessive salts, which must be leached from the soil by adequate rainfall or irrigation. Manures not reinforced with phosphate can lead to soils where the level of nitrogen and potassium are excessive or out of balance with other nutrients. This can result in excessive vegetative growth without fruit production, delayed maturity, and poor storage quality of vegetables. In addition, the large amounts of potassium found in some manures can "salt out" and reduce the amount of magnesium in the soil, creating a magnesium deficiency. If a soil test indicates that the soil contains an excessive amount of potassium, the application of manure should be avoided. If manure is used where carrots are grown, misshapen roots frequently result (Mansour and Baggett 1977). The approximate percentage of the primary nutrients in some manures is given in Table 3.13. Manure contains numerous micronutrients required for plant growth in small amounts. Soils low in these nutrients would greatly benefit from the use of manure (see Table 3.14). Manures vary

TABLE 3.13. Approximate Percentage of the Primary Nutrients in Animal Manures.

	Percentage		
Animal	N	P_2O_5	K_2O
Cattle, fresh	0.5	0.2	0.5
Cattle, dried	1.5	2.0	2.3
Goat, dried	1.4	1.0	3.0
Horse, fresh	0.7	0.3	0.5
Swine, fresh	0.7	0.6	0.7
Sheep, fresh	1.4	0.7	1.5
Sheep, dry	4.2	2.5	6.0
Chicken, fresh	1.5	1.0	0.5
Chicken, dry	4.5	3.5	2.0

TABLE 3.14. Micronutrients in Animal manure[1].

	lb nutrient/1000 lb manure								
Animal	B	Ca	Cu	Fe	Mg	Mn	Mo	S	Zn
Cattle	0.08	15.0	0.03	0.21	5.9	0.05	0.005	2.7	0.08
Horse	0.06	31.4	0.02	0.54	5.6	0.04	0.004	2.8	0.06
Sheep	0.04	25.4	0.02	0.69	8.0	0.04	0.004	3.9	0.11
Swine	0.20	28.5	0.03	1.40	4.0	0.10	0.005	6.8	0.30
Chicken	0.20	123.6	0.05	1.55	9.7	0.30	0.018	10.4	0.30

[1] Based upon dry (15% moisture) manure. B—Boron. Ca—Calcium. Cu—Copper. Fe—Iron. Mg—Magnesium. Mn—Manganese. Mo—Molybdenum. S—Sulfur. Zn—Zinc.

greatly in their nutrient content, depending upon the type of feed used, the amount of bedding or straw, the moisture content, losses in the liquid portion, and degree of rotting. About half the nutrient content is in the liquid portion, which is easily lost by leaching, runoff, and volatility. Cat, dog, bird, and other pet wastes should not be added as they may contain diseases which are readily transmitted to humans eating the plants. Fresh manure is best applied to the garden in the fall. The advantage of applying the material fresh is that (1) less loss of nutrients has occurred, (2) more microorganisms are added to the soil, and (3) more organic material is supplied. This provides food for microorganisms resulting in some additional release of nutrients. Manure may be composted and applied to the garden at any time as a fertilizer, soil amendment, or a mulch. It has the advantage of (1) containing a higher percentage of nutrients (but not total amount), (2) the nutrients are in a more readily available form, (3) it will not burn the crop, (4) it contains more phosphorous in relation to nitrogen and potassium, (5) it is easier to plow or spade into the soil, and (6) many weed seeds in the manure have been destroyed.

The availability of the nutrients in fresh manure depends partly upon the amount of bedding or straw. Mircroorganisms will degrade this organic material and compete with the plants for use of the nitrogen. The amount of manure to apply to the garden is shown in Table 3.15. These amounts are based upon the nitrogen requirements only and take into account the fact that much of the nitrogen will be used by microorganisms to degrade the organic matter. If the garden has previously received heavy applications of fertilizer, about 1 or 2 lb of P_2O_5 and K_2O are needed per 1000 sq ft. At the recommended rates to provide nitrogen, ample potassium is also supplied but many of the manures are short of phosphorus. The addition of superphosphate or rock phosphate would also be needed to provide an adequate balance of nutrients. The amount of manure applied should not exceed that recommended in Table 3.15. If excessive vine or foliage growth of vegetables occurs, chicken or sheep manure should not be used (Hinish and Jordan, undated). The nutrients in sheep and chicken manure are available the year the material is applied and may cause this excessive growth.

TABLE 3.15. Acceptable Rates of Manure for Vegetables.

Type of manure	Amount to apply[1] (lb/1000 sq ft)	Approximate amount of available nutrient applied (lb/1000 sq ft)		
		N	P_2O_5	K_2O
Cattle, fresh	500	1.5	0.5	1.3
Cattle, dry	180	1.2	1.6	1.8
Horse, fresh	500	1.2	0.5	0.9
Swine, fresh	500	1.2	1.1	1.2
Sheep, fresh	285	1.4	0.6	1.3
Sheep, dry	100	1.3	0.8	1.8
Chicken, fresh	280	1.4	0.8	0.4
Chicken, dry	100	1.4	1.1	1.6

[1] Rates are based upon the nitrogen requirement only. One bushel of manure weighs about 50 lb.

Manure should be composted or applied in the fall and spaded under as soon as possible after spreading. This will conserve the nutrients, minimize odors and flies, and speed up the decomposition of the organic matter. Manure can often be obtained from local riding stables, race tracks, and the county fairgrounds. Frequently, it can be obtained for the hauling, as manures are becoming a disposal problem for many livestock producers.

SOIL AMENDMENTS

The tilth or physical condition of the soil is important in relation to plant growth. If the soil is in poor physical condition, it will be hard and crusty when dry and sticky when wet. Vegetables will not grow and develop properly. The physical condition of a soil may be improved by the addition of a soil amendment. A soil amendment is a material that improves the chemical and/or physical condition of the soil and in this way indirectly improves plant growth. It may be organic or non-organic and does not have to supply nutrients.

Nonorganic Amendments

The most common amendments are limestone and sulfur, which are used to adjust the soil pH. By adjusting soil pH, many insoluble nutrients are made available for plant growth. Heavy clay soils can be improved by the addition of coarse sand, cinders, vermiculite, or perlite. About 2 in. of these materials can be worked into the soil to improve drainage, aeration, and workability. These non-organic soil amendments are of limited value compared to organic soil amendments.

Organic Amendments

Organic matter affects soil structure and fertility. The rate organic matter is decomposed depends upon adequate water, nitrogen and temperature. This decomposed organic matter improves the soil (Fletcher and Dutt, undated) in the following ways: (1) it serves as a source of food for microorganisms and earthworms, which help condition the soil; (2) it increases the water-holding capacity of sandy soils by filling in the excess spaces between particles; (3) it increases the amount of usable water in clay soils by keeping the tiny particles apart, allowing excess water to drain away; (4) it keeps the tiny particles in clay soils from cementing together, thereby reducing crusting and allowing easier root penetration; (5) it provides more pore space in clay soils, which allows more aeration for better root growth; (6) it keeps more even soil temperatures; (7) it releases plant nutrients when the organic matter decays; (8) the acids produced in the decay process help dissolve mineral nutrients from the soil and applied natural deposits; and (9) when decayed, it increases the cation exchange capacity, thereby allowing the soil to retain more nutrients.

Types of Organic Amendments

The organic matter in a garden soil can be increased by the addition of manure, sawdust, leaves, lawn clippings, compost, and similar materials. These materials vary greatly in nitrogen content. Nitrogen may have to be added to these materials when they are applied in the garden and again during the growing season, if a nitrogen deficiency develops. Walnut leaves contain toxic materials and should not be used. They may be burned and the ash used, however. Lawn clippings from lawns that have been treated with pesticide sprays (weed killers, insecticides, fungicides) should not be added to the garden. The amount of organic matter that can be added to the garden soil and the amount of available nitrogen required to be added to decompose the material without causing a nitrogen deficiency is given in Table 3.16. Either chemical or organic nitrogen sources may be used as the nitrogen source. For example, if leaves were used, which require about 1 lb of actual nitrogen (Table 3.16), and the fertilizer had an analysis of 10-0-0, then 10 lb of this fertilizer would be needed to supply 1 lb of actual nitrogen. As manure varies greatly in its nutrient composition, about 750 lb per 1000 sq ft of cattle or horse may be used, and about 300 lb of sheep or chicken manure. Organic matter is best added in the fall and spaded or plowed under with whatever additional nitrogen is required. This allows the material to partially decompose over the winter. Probably, the easiest obtainable source of organic matter is leaves. These can frequently be obtained from homeowners for the asking. Many cities collect and grind used Christmas trees, and these wood chips are also a good source.

In an undisturbed soil, the organic matter content reaches an equilibrium. The major losses of organic matter result from crop removal and oxidation as a result of cultivation. In large fields it is seldom possible to increase the percentage of organic matter to any degree. However, garden soils can be increased in organic matter content, particularly the sands and clays. The importance of organic matter should not be overlooked and gardeners should attempt to add organic matter yearly until their soil contains at least 5% organic matter (Bartelli *et al.* 1977). Most sands and clays contain 1% or less.

TABLE 3.16. Amount of Organic Material and the Nitrogen Required to Be Added to a Garden Soil.

Organic material[1]	Amount to add per 1000 sq ft	Actual nitrogen available in fertilizer to be added with the organic material[2]
Lawn clippings	40 bushels	none
Compost	750 lb (15 bushels)	none
Manure (cattle or horse)	750 lb (15 bu)	none
Manure (sheep, chicken)	300 lb (6 bu)	none
Hay	600 lb (10 bales)	none
Straw	600 lb (10 bales)	½ to 1 lb
Leaves	750 lb (40 bu)	½ to 1 lb
Sawdust, wood chips	500 lb (20 bu)	1 to 1½ lb
Corncobs	500 lb (20 bu)	1 to 1½ lb

[1] See earlier section on Manure for nutrients added with the various manures.
[2] See section on Fertilizers for nitrogen sources.

Compost

Compost is basically decomposed organic material. It is best described as a soil amendment since the amounts of nitrogen, phosphorus, and potassium are low. It can be added to the garden soil, used with soil to grow transplants or used as a mulch. Through the use of compost, organic materials are not wasted and higher quality and yields of vegetables can be achieved.

Decomposition of organic materials requires adequate nitrogen, air, and water. The microbes in the compost pile will decompose the carbon from the organic material into CO_2, and this CO_2 will escape into the air. This reduces the size of the compost pile and concentrates the remaining nutrients. However, some diseases and pesticides are not decomposed, and they will also be concentrated in the compost and be a problem when added to the garden. Some weed seeds and nonviral and nonspore disease organisms are killed by the temperatures produced in the decomposition process. The outside layers of the compost pile must be turned into the center so that weed seeds and disease organisms in these outer layers are destroyed by the heat. In the decomposition process, heat is released and temperatures may become 120° to 175°F. The intensity depends upon the amount of nitrogen material included. When the pile starts to cool, it can be considered finished. Gardeners can use the "look and touch" method. If the organic material is broken up and not recognizable as the original material and has developed a dark, rich color, the material is composted.

Carbon : Nitrogen Ratio. The microorganisms that decay organic material into compost need about 1 part of nitrogen for every 15–30 parts of carbon present in the material. This is called the carbon : nitrogen or C : N ratio. If this value is greater than 30 : 1, then nitrogen will be deficient. The organic matter will still decompose, but a long time will be required. Table 3.17 lists the C : N ratio of some common organic materials. If the organic material has a C : N ratio of greater than 30 : 1, fertilizer should be added to the compost material (Table 3.18).

TABLE 3.17. The C : N Ratio of Some Common Organic Materials.

Material	C:N ratio
Liquid manure	10:1
Alfalfa or sweet clover hay	12:1
Lawn clippings	20:1
Composted manure	20:1
Kitchen garbage	30:1
Green rye or oats	36:1
Cornstalks	60:1
Leaves	70:1
Straw	75:1
Strawy manure	80:1
Sawdust, wood chips	400:1
Corncobs	420:1

TABLE 3.18. Chemical and Organic Fertilizers for Use in Making Compost[1].

Chemical fertilizer	Organic fertilizer
25 lb 10-10-10 fertilizer	15 lb wood ashes
10 lb finely ground limestone	7 lb dried blood or fish meal
	4 lb steamed bonemeal

[1] The fertilizer is first mixed thoroughly together. Use 1 lb of the chemical or 3 lb of the organic fertilizer per 10 lb dry weight (40 lb green). Other organic fertilizers that will supply 0.07 lb each of N, P_2O_5, and K_2O per 10 lb dry weight of organic material to be composted may be substituted (see Table 3.8).

Materials for Compost. Virtually any organic material can be composted: kitchen wastes, eggshells, coffee grounds, nutshells, leaves, lawn clippings, garden residues that are not eaten, straw, hay, weeds, manure, newspaper, and so on. Material that should not be added to a compost pile includes metal and glass, as these will not decompose. In addition, there are a number of organic materials gardeners should not compost. Eggs, cheese, grease, fat, and meat scraps (protein foods) should be avoided as they decompose slowly and may develop offensive odors and attract rodents and flies. Walnut wastes (leaves, nuts, limbs) contain material that inhibits plant growth and should not be included. Cooked bones that are meat free and ground in a shredder may be included as they are similar to steamed bonemeal. Avoid lawn clippings treated with insecticides and herbicides and avoid obviously diseased plants and weed seeds (Erhardt and Littlefield 1973). Some diseases, weed seeds, and pesticides are not destroyed during composting, and they are then added back to the garden with the compost. Pet wastes may contain diseases that can be transmitted to humans and should not be included.

Time Required to Produce Compost. The time required to produce compost depends upon two major factors: the amount of nitrogen in the compost pile and the particle size of the organic material added. Ordinarily, 12 to 18 months are required to make coarse materials usable (Vandemark and Shurtleff 1970), but fertilizer and grinding of the organic material will greatly reduce this time. The microorganisms derive their energy from the organic material and require nitrogen to live and multiply rapidly. About 15–30 parts of carbon for every part of nitrogen is ideal. If the organic material has a C : N ratio of greater than 30 : 1 (see Table 3.17), fertilizer should be added. The decomposition process will make the compost slightly acid, and limestone should be added unless the compost is for use with acid-loving plants. The organic fertilizer (Table 3.18) uses wood ashes that contain 26% calcium, and do not need limestone. For each 40 lb of green material (10 lb dry material), about 1 lb of the chemical fertilizer or 3 lb of the organic fertilizer mixture shown in Table 3.18 should be mixed with the compost. A 1 or 3 lb coffee can is easily used as an approximate measure.

The size of the particles of organic matter also influences the speed of decomposition. The smaller the particle size, the greater is the surface area exposed to the microorganisms and fungi to decompose the material. Leaves and such may be chopped with a rotary grass mower. Limbs, cornstalks, and other coarse material should be ground to less than one-half inch in a chipper-shredder, if possible. If tree branches are ground they will decompose in less than a year, while whole tree limbs will take several years to decompose.

Crop residues naturally contain all the microbes needed for the decomposition process. Various types of compost activators do little to speed up this process (Fletcher *et al.* 1972). A small amount of soil, manure, or old compost can be added to increase the number of microbes in the compost pile. The addition of fertilizer and grinding the organic material will speed up the composting process more than anything else.

Making a Compost Pile. Before a compost pile is built, local ordinances should be checked to be sure that composting is permitted in the community. The pile should be in an area screened from the public and care taken to prevent the pile from becoming a home for various rodents. Compost piles are often 6 ft square and 5 ft deep. Compost piles of this size are often made by gardeners who have a lot of coarse material available, such as leaves, straw, or hay.

A compost pile can be built below ground as a pit if drainage is not a problem, or be made in a bin above ground. Aboveground bins are often made with snow fence, boards, block or stakes and chicken wire (Fig. 3.5). A division through the center is desirable, to separate the new from the usable compost. Materials are placed in a compost pile in layers, as this causes faster and more complete decomposition.

FIG. 3.5. Composting leaves in a wire bin (left) and a wooden box (right) for next year's garden. *From Anon. (1971).*

The first layers form the base of the pile and should be rather coarse, so that air can enter the pile. This layer is about 6 in. deep and as large as possible, as a small compost pile loses heat rapidly, slowing decomposition. The second layer should be 2 in. of animal manure. Purchased steer manure may be used, or dried blood and bonemeal (Table 3.18) may be substituted. Next, a few shovel-fulls of soil or, preferably, old compost are sprinkled over the pile. This is to provide additional microorganisms, which will spread up the composting process. The last layer is a light sprinkling of ground limestone or wood ashes. This is to neutralize the acidity naturally produced in the pile. The pile is now moistened with water. The pile should be damp, not soggy, and will need sprinkling periodically during the summer. Additional layers are added until all of the organic material has been added. The compost pile does not have to be made all at once, and additional layers can be added as organic material becomes available. The top of the pile should be flat and slant slightly to the center to assure even penetration of rainfall.

The compost pile will heat up to 120–175°C, and after about 3 weeks begins to cool down. The pile should then be turned, with the center going to the outer edges and the outer edges and top going into the center. This will speed up the process and make a more uniform compost. Additional water should be added as needed. In about a month, the compost should be ready for use. The material being composted affects the composting time. Very fibrous materials take longer than soft, leafy wastes. The outside temperature also affects the speed of composting. A compost pile started in late fall may not be completely composted by the next spring.

Some gardeners do not wish to turn the pile. A large sheet of black plastic can be placed completely over the pile, and sealed at the bottom with soil. This will prevent problems with insects, rodents, and odors. The composting will be somewhat faster and the pile does not have to be turned. Adequate water should be applied before sealing the pile. In about 3 months, composting should be complete.

Container Composting. Aerobic composting can take place in almost any container as long as air and water requirements are met. A compost container may be purchased (Fig. 3.6), or a standard garbage can can be used. About eight slots are made in the sides of the can to allow ventilation and five made in the bottom to allow drainage. Leaves, lawn cuttings, weeds, straw or other materials are run through a shredder-grinder or chopped with a rotary mower and collected with a grass catcher. This will increase the surface area for decomposition. These shredded materials are mixed with fertilizer (see Table 3.18) and placed in the garbage can. The can is placed on an elevated rack about one foot off the ground to allow drainage and ventilation. The material is watered until a slight dripping is observed through the bottom of the can. The lid is placed on the can, and in about three days the temperature inside the can should have reached about 138°F. In 10–14 days the material is composted and ready to use.

A plastic garbage container may also be used to compost fruit and vegetable peelings and various other kitchen wastes (Fig. 3.6). Dried materials such as

FIG. 3.6. A small amount of ground leaves or wood chips should be added to cover kitchen wastes to prevent odors during composting. *Courtesy of J.S. Vandermark II.*

leaves should be placed at the bottom of the container and the kitchen wastes added daily. During the summer, some ground wood chips or chopped leaves saved from the previous fall should be added periodically to cover the kitchen wastes to prevent flies and odors. The container may be covered and water applied to the compost weekly. Eggshells will help replace limestone, but should be crushed, as they compost slowly. In the fall, the compost can be added to the garden and spaded under.

Some gardeners prefer anaerobic decomposition, composting without air. A 32 gal. plastic garbage or trash bag can be filled with chopped leaves, garden debris, and fertilizer. When the bag is full, about a quart of water is added (more if dry material is used) and the bag tied tightly and set outside. The anaerobic method should be used outdoors and sufficient water added. The process involved is similar to the process involved in spontaneous combustion, which has resulted in fires in moldy hay and accumulated used paint rags.

Sheet Composting. Sheet composting grew out of the needs of large-scale gardeners. Instead of composting in piles, the organic material can be spread on the soil and will be composted at the soil surface. The basis for sheet composting is to provide the soil with the material to produce its own compost. The organic material can be spread between the vegetable rows 3 to 6 in. deep. Green material will decompose rapidly while dry material will act as a mulch and slowly decompose at the soil surface. Green material should not be placed next to the plants. The heat produced during decomposition and the microorganisms and insects may damage some of the vegetable plants. The undecomposed material can be spaded under at the end of the growing season. Sheet composting works best with shredded, dry material placed around trees, shrubs, and with perennial vegetables such as rhubarb, asparagus, and perennial herbs. In the fall, the unwanted vegetables may be chopped with a tiller or lawn mower and left on the surface as part of a no-till mulch system.

MULCHES

In a broad sense, a mulch is any substance applied to the soil surface that protects the roots of plants from extremes in temperature or drought or keeps fruit clean. Mulching vegetables can increase yields, promote early harvest, and reduce fruit defects when the plant is growing under less than ideal conditions (Courter *et al.* 1969). Mulches modify the soil and air microclimate in which a plant is growing, and specific mulches are used to create this favorable environment. Mulches do not always increase crop yields and may, under some conditions, reduce yields.

Temperature. Mulches may either increase or decrease soil temperature. Loose dry material such as straw or wood chips act as insulation and protect against high temperatures. Light reflective mulches such as white paper or aluminum foil can reflect sunlight and also decrease the soil temperature. Soil mulched with these materials is about 5°–10°F cooler than bare soil (Courter *et al.* 1969). During the spring growing season, this lower soil temperature may reduce plant growth and yields. If these mulches are applied when soil temperatures have increased, yields may increase; particularly with potatoes which require cooler temperatures for tuber formation or late summer plantings of zucchini squash and cool season crops such as Brussels sprouts and lettuce.

Black paper or black and transparent plastic mulches increase soil temperatures 10°–15°F compared with bare soil (Courter *et al.* 1969). This hastens early plant development, particularly for the production of warm season crops under cool conditions or in northern climates. Melons, cucumbers, pumpkins, squash, watermelons, and early crops of sweet corn and tomatoes respond well to soil-warming mulches (Fig. 3.7).

Moisture. Mulches reduce evaporation of water by 10–50% or more. Permeable organic mulches increase the rate at which the soil will absorb rainfall. These permeable mulches will also absorb irrigation water. Nonpermeable mulches such as plastic, aluminum, and paper will not absorb water. Adequate moisture should be available before applying the material or a trickle irrigation system installed. In rainy seasons, mulching may keep the soil too wet for adequate aeration. Mulches also save water for crop use by reducing compe-

FIG. 3.7. Yield of muskmelons grown on bare soil (left) and soil mulched with black plastic (right). *From Topoleski (1972).*

tition from weeds. Plastic mulches prevent water loss by evaporation, and frequently increased growth of these plants is noted. This increased plant growth on mulched plots results in greater water use, as the plants are larger and transpire more. Thus, on sandy, well-drained soils under dry conditions these crops may require more irrigation water than unmulched crops (see Fig. 3.8).

FIG. 3.8. Potatoes mulched with straw to conserve moisture and reduce soil temperature.

Weed Control. Mulches help control weeds but do not reduce weed infestations if the weeds are already present and established. If the material is weed-free and applied properly, weed seeds do not have a chance to germinate or the mulch layer is so deep that the germinated seedlings cannot push through it. Perennial weeds and grasses will grow through most organic mulches. Black plastic or paper mulches prevent light penetration which is necessary for the weeds to grow. Weeds will grow under a clear plastic mulch. Shredded organic material is more effective and provides longer weed control than loose, porous materials.

Soil Structure. Mulches help maintain good soil structure by preventing soil crusting and compaction. Mulched soil remains loose, providing good aeration for root growth. Many plants develop an extensive root system in the upper two inches of mulched soil while in unmulched soil these roots are reduced by drying, crusting, and cultivation. The mulch provides a physical barrier that prevents root pruning and injury due to cultivation and hoeing. An extensive root system allows more efficient use of nutrients. Organic mulches spaded under after the growing season improve soil tilth as they decompose and provide some nutrients.

Disease Control. Mulches do not eliminate plant diseases but keep the fruit from coming in contact with the soil where the disease is located. Tomato fruit rots and fruit defects in cucumber and melons can be reduced by mulching. The fruits are also cleaner than when in contact with the soil. Tomatoes should be either grown in wire cages or mulched to prevent fruit rots in areas where rainfall is plentiful during the summer.

Organic Mulches

Organic materials which have been used for mulching include (1) plant residues such as straw, hay, crushed corncobs, leaves, composts, dry lawn clippings, and peanut hulls; (2) peat; (3) wood products such as sawdust, wood chips, wood bark, old newspapers, and shavings, and (4) composted animal manure.

Many mulching materials require the addition of nitrogen fertilizer to reduce the chance of nitrogen deficiency in the plants that have been mulched (see Table 3.16). The fertilizer should be added with the mulch at a rate of about 2 lb of 5-10-5 per 100 sq ft. Additional fertilizer should be applied if the lower foliage of the plants becomes yellow or the plants appear stunted during the growing season.

Organic mulches should be placed on the soil after the plant is well established and preferably when the soil has warmed up sufficiently for active root growth. A mulch placed on cool soil will slow root development. A 3 to 6 in. layer of material can be spread between the rows and around the plants, making certain you do not cover the plants.

Plant Residues. Plant residues are commonly used as a mulch. Legume hay does not require additional nitrogen. Straw should be weed free and should not be used where a cigarette could carelessly create a fire hazard. Leaves are the least expensive mulch available where trees are abundant. Large leaves such as from sycamore should be chopped with a rotary mower to prevent them from matting down and preventing water penetration. Lawn clippings are best used when dry. If applied fresh, they mat down, produce heat during decomposition that damages the plants, and give off an offensive odor. Do not use lawn clippings for a mulch if they have been treated with a pesticide. Compost is probably the best mulch you can use.

Peat. Peat is one of the most commonly used mulches. The fine grade of peat has a tendency to blow away. When very dry, peat sheds water rather than allowing it to soak in.

Wood Products. Wood products decompose slowly and may create a nitrogen deficiency if additional nitrogen is not added. Wood chips can frequently be obtained from park districts, cities, and electric power companies, which remove and grind limbs overhanging streets or electric lines, and dispose of used Christmas trees as wood chips. Shredded newspapers are used to a limited extent. Unshredded paper mats down and prevents water penetration.

Manure. Well-rotted or composted manure can be used, but chicken manure should be excluded as a mulch since it burns plants too easily. Manure should not have been treated with odor-reducing chemicals as these substances sometimes injure the vegetable plants.

Synthetic Mulches

Synthetic mulches include paper, aluminum foil, plastic, and various combinations of these. Aluminum foil and some types of plastic are not biodegradable, and they must be removed at the end of the growing season. The plastic mulches are usually used to increase the soil temperature and obtain an early crop and are applied to the soil before the crop is planted. The edge of the plastic is covered with soil to prevent it from blowing away and holes then punched in the plastic. The seeds are planted in the soil beneath these holes. A number of plastic mulches are permeable, which allows air and water to reach the soil, and some come with pre-cut holes, large enough to allow water penetration. Some mulches are made of heavy-duty plastic that can be removed at the end of the season and reused for several years.

PLANT PROTECTORS

Nearly all gardeners have used some weather-modifying practice. In northern climates and at higher altitudes, these practices and devices are required to assure a successful garden. Mulches can modify excessively hot or cold temperatures, and were discussed earlier in this section. The use of transplants and early-maturing cultivars will help overcome a short growing season. Other practices are aimed at lengthening the growing season in both spring and fall; and in providing protection from the wind.

Frost Protection

The easiest frost protection is to apply water to the foliage. Dew forms on the foliage at night and the sun shines on the leaves causing the death of leaf cells. This can be prevented by spraying water on the foliage in the early morning. This applied water increases the plant temperature to the water temperature and prevents the damage that may be caused by sunlight on the plant leaves. In addition, the sprinkler system may be turned on the plants throughout the night. This increases the plant temperature and the water releases heat as it changes from water to ice. These methods are effective in the fall, but in the spring the warm-season crops and some transplants are too small and tender to withstand the frosts. Other methods may be used.

Various plastic materials may be placed over the plants to protect against frost damage. The plastic lets in the sun's rays, warming the plants, surrounding air, and soil. At night, the soil releases the sun's energy as heat, protecting the plants. In the fall, mulch should be removed from around pepper and tomato plants, and the bare soil will release enough heat to protect the plants from 30°F frosts. Lower temperatures will require some type of covering over the plants.

In the spring and fall, a movable plastic greenhouse, about the size of a camping tent, may be moved over the plants to protect against spring and fall frosts.

Hotcaps of paper or plastic may be placed over small plants for frost protection. Holes in the top allow for some ventilation, but excessive heat buildup may occur, and the hotcap should be removed during warm days. When the outside temperature is 60°F, temperatures under the hotcap have sometimes reached 100°F. As the plant grows, the top of the hotcap can be opened, leaving only the sides for protection.

Various water-filled plant protectors can be used for small plants (Fig. 3.9). They are filled with water and placed over the plant. The principle is the same as applying water to the plant foliage. The sun warms the water and the water releases this heat as temperatures decrease.

Garden blankets may be used. The blanket is left over the plants and provides some protection from spring and fall frosts. Care must be taken to prevent heat buildup under the blankets. It can be used in a fall garden to protect cool-season crops that tolerate cool temperatures. In some areas, it will extend the growing season by several months.

Row covers (Fig. 3.10) made of spunbonded polyester are designed for spring frost protection. The material is porous and allows water and air movement. It is best used for vine crops, and should be removed when the vine crops produce flowers. This allows bees to pollinate the flowers. With transplanted tomatoes

FIG. 3.9. A plastic, water-filled plant protector provides plant protection against early frosts. *Courtesy of Harris Seed Co.*

and peppers, the temperatures under these covers usually become so hot that the flowers and fruit are aborted. Covers can be used if temperatures are measured under the covers and heat units calculated (see Chapter 2—Germination requirements; temperature) using a germination temperature of 50°F. Tomato and pepper row covers should be removed after 4–7 weeks, depending on the season, when 650–675 heat units have accumulated. Clear polyethylene plastic accumulates too much heat under the covers to be used without rigid maintenance. Row covers may also be placed over a row of cool-season vegetables grown as a fall garden, to extend the growing season. In windy areas, a wide hoop should be placed over the row to keep the cover off the plants, which may cause considerable damage (Splittstoesser and Gerber, 1989).

FIG. 3.10. Row covers, placed over vegetables, will provide spring frost protection.

Wind Protectors

The best wind protection is a properly located fence or hedge. Any of the frost-protection devices also protect against wind damage. Cardboard boxes, gallon plastic milk jugs with the bottom removed, boards, shingles, or shakes may also be conveniently placed around the plants to protect against wind damage.

SELECTED REFERENCES

Anon. 1968. Optimum pH range for vegetable crops. Ill. Agric. Exp. Stn. Bull. *H-420*.

Anon. 1988. World production of phosphate rock. Better Crops *Fall*:6–8.

Bartelli, L.J., Slusher, D.F., and Anderson, K.L. 1977. Know your soil and how to manage it. *In* Growing Your Own Vegetables. U.S. Dep. Agric. Bull. *409*.

Courter, J.W., Hopen, H.J., and Vandemark, J.S. 1969. Muching vegetables. Ill. Agric. Exp. Stn. Circ. *1009*.

Delwiche, C.C. 1970. The nitrogen cycle. Sci. Am. *223*, 126–146.

Erhardt, W.H. and Littlefield, L.E. 1973. Natural gardening. Maine Agric. Exp. Stn. Bull. *567*.

Fletcher, R.F. and Dutt, J.O. No Date. Garden soil management. Pa. Agric. Exp. Stn. Circ. *540*.

Fletcher, R.F. and Ferretti, P.A. No Date. Soil testing for the organic gardener. Pa. State Agric. Exp. Stn. Hortic. Ser. *1*.

Fletcher, R.F., Ferretti, P.A., Hepler, R.W., MacNab, A.A., and Gesell, S.G. 1972. Extension agent's guide to organic gardening, culture, and soil management. Pa. Agric. Exp. Stn.

Gerber, J.M. 1980. Organic gardening and soil fertility. Dep. Hortic., Univ. Ill. Hortic. Facts *VC-5-80*.
Gerber, J.M. and Swiader, J.M. 1981. Fertilizing your vegetable garden. Dep. Hortic., Univ. Ill. Hortic. Facts *VC-9-80*.
Gerber, J.M., Swiader, J.M., and Peck, T.R. 1981. Sewage sludge on vegetables—A mixed blessing. Ill. Res. *23* (2) 12–13.
Gowans, K.D. and Rauschkolb, R.S. 1971. Organic and inorganic fertilizers and soil amendments. Calif. Agric. Exp. Stn. *AXT-357*.
Hindawi, I.J. 1970. Air Pollution Injury to Vegetation. U.S. Dep. Health, Educ. and Welfare, Washington, D.C.
Hinish, W.W. and Jordan, H.C. No Date. Profitable use of poultry manure. Pa. Agric. Exp. Stn. Spec. Circ. *146*.
Jacobson, J.S. and Hill, A.C. 1970. Recognition of Air Pollution Injury to Vegetation: A Pictorial Atlas. Air Pollution Control Association, Pittsburgh.
Janssen, K.A. 1976. Snap bean nitrogen needs. Va. Truck Organ. Exp. Stn. *30* (10).
Lerner, B.R. and Utzinger, J.D. 1986. Wood ash as a soil liming material. HortScience **21**:76–78.
Luckhardt, R.L. 1976. What happens to applied nitrogen? Agrichemical Age (Sept.–Oct.), 27.
MacGillivray, J.H., Sims, W.L., and Johnson, H., Jr. 1968. Home vegetable gardening. Calif. Agric. Exp. Stn. Circ. *499*.
Mansour, N.S. and Baggett, J.R. 1977. Root crops more or less trouble free, produce lots of food in a small space. *In* Growing Your Own Vegetables. U.S. Dep. Agric. Bull. *409*.
Shanmugam, K.T. and Valentine, R.C. 1976. Solar protein. Calif Agric. *30* (11) 4–7.
Splittstoesser, W.E. and Gerber, J.M. 1989. Vegetative growth, fruit set, and quality of bell pepper as influenced by row tunnel removal date. Proc. Nat. Agr. Plastics Cong. *21*:283–287.
Topoleski, L.D. 1972. Growing vegetables organically. N.Y. Agric. Exp. Stn. Bull. *39*.
Utzinger, J.D., Trierweiler, J., Janson, B., Miller, R.L., Saddam, A., and Crean, D.E. 1973. Organic gardening. Ohio Agric. Exp. Stn. Bull. *555*.
Vandemark, J.S. 1973. Home garden recommendations for soil test interpretations. Ill. Agric. Exp. Stn. Veg. Growing *25*.
Vandemark, J.S., Jacobsen, B.J., and Randell, R. 1977. Illinois vegetable garden guide. Ill. Agric. Exp. Stn. Circ. *1091*.
Vandemark, J.S. and Shurtleff, M.C. 1970. The garden compost pile. Ill. Agric. Exp. Stn. Veg. Growing *1*.
Vandemark, J.S., Splittstoesser, W.E., and Randell, R. 1973. Organic gardening can be successful if you follow sound principles. Ill. Res. *15* (2) 5.

4

Pest Control

Very few gardens will escape attack from various pests. Weeds and insect pests will usually be the major problem. Some vegetables, however, can be grown with little or no danger from insect or disease pests. Other vegetables will require pest control measures to reduce the pest population to a tolerable level. Home gardeners use mechanical, cultural and biological methods of pest control. If a chemical or botanical pesticide is used, it must be used with care. Pesticides are generally sold as dusts, emulsible concentrates, wettable powders, and granulars. Dusts are used in the form they are purchased. Emulsible concentrates and wettable powders are diluted with water and sprayed on the area. Granulars are small clay pellets that are impregnated with pesticide and usually applied to the soil to reduce soil-borne pests.

The use of pesticides should be limited to those crops which would otherwise be seriously damaged by insects or diseases. Sprays and dusts can drift to other crops in the garden or garden area and damage these plants. Gardeners are limited in the number of pesticides that can be economically used and some vegetable crops do not have any pesticides which have been cleared for use by the U.S. Government. If pesticides are handled or applied improperly or if unused pesticides are disposed of improperly, they may be injurious to humans, domestic animals, desirable plants, pollinating insects, and fish or other wildlife, and they may contaminate water supplies. Use pesticides only when needed and handle them with care. Follow the directions and heed all precautions on the container labels.

Store all pesticides in a cool, dry, locked storage area so that they are not accessible to children, irresponsible persons, and animals. Do not dispose of pesticides through sewage systems. Haul them or have them hauled to a sanitary landfill for burial. Various pressurized pesticide cans should not be placed on a stove or heater or near a source of heat that might exceed 120°F as they may explode. They should also be stored in a cool place, and empty cans should be disposed of in a sanitary landfill. Pesticide containers should never be used for anything except to contain the original pesticide, and pesticides should not be stored in anything but the original container. By the time any publication on insect or disease control reaches the gardener, it is usually already outdated by more recent research. Thus, the chemical pesticides listed here are general control measures rather than specific ones.

Due to changing laws and regulations, no liability is assumed for the following recommendations. The recommendations for using pesticides that are included in this handbook are incomplete; therefore, they should be used *only* as guidelines. Complete instructions for the use of a specific pesticide are on the pesticide label. Read and follow the label directions and precautions before applying any pesticide. The pesticide user is responsible for applying pesticides according to label directions, as well as for problems that may arise through misapplications or misuse of the pesticide. Trade names have been used for clarity, but their use does not constitute an endorsement nor imply discrimination against other products. Label changes, product cancellations, and changes in recommendations may have occurred since the publication of this handbook. Check with the county advisor or Agricultural Experiment Station for current recommendations.

TOXICITY AND HAZARDS OF PESTICIDES

Pesticides are poisonous. They have to be poisonous to kill undesirable plants, insects, diseases, or other pests. Safe and proper use of pesticides depends upon a knowledge of their use.

Toxicity is the inherent capacity of a material to produce injury or death (Bever *et al.* 1981). If you know the toxicity of a pesticide, you will know what precautions to take.

Tests are performed with each pesticide to determine the toxicity to rats, rabbits, guinea pigs, or other animals. These tests are helpful in determining how hazardous the pesticide probably would be to humans.

In oral tests, the animal is given quantities of the pesticide by mouth according to the animals' body weight. The dose is increased until the dose which will kill 50% of the test animals is found. This lethal dose is called "Oral LD_{50}." The dose is expressed in milligrams of pesticide per kilogram of body weight (mg/kg). [There are 1000 mg in a gram and 454 grams in a pound. A kilogram (kg) is 2.2 pounds.] If a pesticide has an Oral LD_{50} of 100, then 100 mg of the pesticide are required to kill 5 out of 10 test animals each weighing a kilogram. A compound with an Oral LD_{50} of 100 is dangerous because 1/90 of a pound (one or two teaspoons) could kill a human (Bever *et al.* 1981). The lower the LD_{50} number, the more toxic the pesticide.

In dermal tests, the pesticide is placed on the skin of the test animal and covered with a bandage so that it will remain on the skin for 24 hr. If 500 mg of the pesticide are required to kill 5 out of 10 test animals weighing 1 kg, the Dermal LD_{50} is 500. Most materials are absorbed more rapidly by the body when taken by mouth than when placed on the skin. Thus the Oral LD_{50} is almost always lower than the Dermal LD_{50}.

In inhalation tests, the test animals are placed in an airtight container with specific quantities of the pesticide. The animals remain in the container for 1 hr. Inhalation values called LC_{50}'s are measured in micrograms per liter (μg/liter). The LC_{50} is the lethal concentration that will kill 50% of the test animals. [*NOTE:* There are 1000 micrograms (μg) in 1 milligram (mg). One liter is equal to 1.06 quarts.] An LC_{50} of 100 means that 100 μg per 1 qt of air are required to kill 5 out of 10 test animals when they are exposed for 1 hr. LC_{50} also refers to the toxicity of a material to fish in water and then is expressed in parts per million (ppm). Table 4.1 gives the Oral and Dermal LD_{50} and the inhalation LC_{50} for pesticides with high, moderate, low, and slight toxicity. Highly toxic materials contain a drawing of a skull and crossbones and the words "Danger—Poison" on the label; moderately toxic pesticides contain the word "Warning" on the label; and pesticides with low and slight toxicity contain the word "Caution" on the label. All pesticide labels carry a "Keep Out of Reach of Children" warning.

Oral exposure can occur because of an accident, but it is more likely to be a result of carelessness. Blowing out a plugged spray nozzle by mouth and smoking or eating without washing contaminated hands can result in oral exposure. Dermal exposure is skin contamination and can occur any time a pesticide is handled, mixed, or applied. Inhalation exposure results from breathing in pesticide vapors, dust, or spray particles. This can occur from smoking, breathing smoke from burning pesticide containers, or inhaling the pesticide immediately after applying it.

The lethal dose is dependent upon the body weight of the person. A person who weighs 150 lb can tolerate 50% more of the material than a 100 lb person, while a 50 lb child can only tolerate one-third as much as the 150 lb person. This is another reason for keeping all toxic materials out of reach of children.

It should be emphasized that *all* materials are toxic at some concentration. Remember that toxicity is the inherent capacity of a material to produce injury or death, but that does not mean that it will if handled properly. Some common household materials are as toxic or moreso than some pesticides. In Illinois, there were 11 deaths due to aspirin in 1977 while 35 persons have died in the past 18 years due to all agricultural chemicals (Gentry 1978). Table 4.2 lists the various insecticides and fungicides that are discussed in this chapter on Pest Control. It is obvious from Table 4.2 that the materials derived from plant sources are also toxic to humans, and some of them are more toxic than manmade chemical prod-

TABLE 4.1. Oral, Dermal, and Inhalation Toxicity Ratings of Pesticides.

Toxicity rating	Signal words on label	Oral LD_{50} (mg/kg)	Dermal LD_{50} (mg/kg)	Inhalation LD_{50} (μg/liter or ppm)	Lethal oral dose, 150 lb person
High	Danger—Poison	0–50	0–200	0–2000	few drops to 1 tsp
Moderate	Warning	50–500	200–2000	2000–20,000	1 tsp to 1 oz
Low	Caution	500–5000	2000–20,000	20,000 +	1 oz to 1 pt + or 1 lb
Slight	Caution	5000 +	20,000 +	—	1 pt + or 1 lb +

Source: Bever *et al.* (1981).

TABLE 4.2. Relative Toxicity of Some Pesticides and Common Household Items.

Material (trade name)	Oral LD_{50}[1] (mg/kg)	Chemical class
High toxicity		
Nicotine (tobacco)	10	Botanical insecticide
Kerosene	50	Mechanical insecticide
Moderate toxicity		
Diazinon	80	Chemical insecticide
Nicotine sulfate	83	Botanical insecticide
Gasoline	150	(For comparison only)
Rotenone	80–400	Botanical insecticide
Caffeine (coffee, tea, cola)	200	(For comparison only)
Low toxicity		
Carbaryl (Sevin)	700	Chemical insecticide
Pyrethrum	1000–1300	Botanical insecticide
Malathion	1100–1300	Chemical insecticide
Ryania	1200	Botanical insecticide
Aspirin (various brands)	1300	(For comparison only)
Ziram (Zerlate, Karlate, Karbam White Z-C spray dust)	1400	Chemical fungicide
Table salt (various brands)	3300	(For comparison only)
Ethyl alcohol (various brands)	4500	(For comparison only)
Slight toxicity		
Zineb (Dithane 78, Zineb 75)	5200	Chemical fungicide
Maneb (Manzate D, Dithane M22)	6750–7500	Chemical fungicide
Captan (Captan, Orthocide 50)	9000–15,000	Chemical fungicide
Bacillus thuringiensis (Dipel, Bactur, Thuricide)	15,000	Biological control insecticide

[1] (mg/kg) means milligrams of actual material taken by mouth (orally) per kilogram of body weight; 1 kilogram = 2.2 pounds; 454 grams = 1 pound; 1 milligram = 0.001 of a gram.

ucts. This point needs to be re-emphasized as many organic gardeners think that if the material is a botanical insecticide (or "organic"), it is not poisonous.

The toxicity values alone are not a measure of the hazards of a material to humans. Hazard and toxicity are not the same. Hazard is a combination of toxicity and exposure. It is the potential threat that injury will result from the use of a material in a particular formulation or quantity. Some hazards do not involve toxicity to humans or other animals. For example, oils and numerous other chemicals are considered safe or relatively safe to animals but may cause considerable injury to some plants.

A compound may be extremely toxic but present little hazard when used (1) in a very dilute formulation, such as 2 or 5% dusts; (2) in a formulation that is not inhaled or absorbed through the skin; or (3) only occasionally and under conditions in which people are protected. Conversely, a chemical may be relatively nontoxic but present a hazard because it is used in a concentrated form or carelessly applied. A concentrated material may be toxic if taken orally but may not be a hazard when diluted with water. Chemicals are sold in different forms so that highly toxic materials are less of a hazard. Rotenone, nicotine sulfate, and Sevin are sold as 2 or 5% dusts. Diazinon is impregnated on a granule and sold in this form. Dichlorvos (Vapona or DDVP) is a highly toxic material but is impregnated on a resin strip ("no-pest strip") and enclosed in a container to reduce the hazard to humans. Thus hazard and toxicity are *not* the same. Table salt is listed in Table 4.2 as having a low toxicity. Although humans require some salt in their diet, they cannot add 2% salt to their liquid intake for two days and live. Thousands of people have died from drinking seawater (2% salt) as castaways in lifeboats. This again emphasizes that all chemicals have a toxicity rating, but that toxicity and hazard are not the same.

WEEDS AND THEIR CONTROL

Weeds have been defined by some ecologists as "a plant growing out of place." If this definition is used, then the growing of a "weed" cannot be done intentionally; the growing of a weed is something that could be done only by accident. A better definition of a weed is a plant that is unsightly (crabgrass in your lawn), causes disease (ragweed for hay fever or poison ivy), reduces yield (sunflower with sweet corn), or grows where it is not wanted (volunteer pumpkin seeds from the added compost among your cucumbers). Weeds compete with vegetables for water, nutrients, and sunlight. Weeds also harbor insects and diseases that may attack plants around the house and garden.

Herbicides. There has been a substantial amount of research conducted on the use of weed-killers (herbicides) to control weeds without harming the vegetable plant. These materials do an excellent job of controlling weeds on commercial farms. However, they are not recommended for the homeowner (with the possible exception of an herbicide on a large planting of asparagus).

The problems associated with herbicides used by the homeowner in the garden are as follows:

(1) Herbicides are manufactured for specific crops to control specific weeds, and some vegetable crops do not have any herbicide that has been cleared for use by the U.S. Government. No one specific herbicide can be used on all the crops growing in the garden. The gardener would need to apply a herbicide to vegetables in one or two rows, then apply a different herbicide to the different vegetables in the next row and so on. In addition, some herbicides are applied before the seeds are planted and many gardeners forget which area was treated and which was not. The gardener would need to purchase,

safely use, and store several different materials. This would be expensive and create a storage problem as some herbicides cannot be frozen.

(2) Gardens will contain many different types of vegetables at different stages of growth, and drift from sprays can damage the vegetable plants. Similarly, spray drift from herbicides used on the lawn or yard can damage vegetable plants in the garden. The damage created by herbicide drift has frequently resulted in legal action against the herbicide user.

(3) Herbicides must be applied at the proper time and at the proper rate. There is a tendency to apply high rates if the quantity measured out "looks" as if it is not enough. These excessive rates not only damage plants this season but also may damage plants the next growing season. These high rates prevent the gardener from growing two or more crops of different vegetables in the same area. If rates lower than recommended are used, weeds are not killed.

(4) Some weeds cannot be controlled easily with herbicides, particularly perennials. Excessive rates would be required which would also damage the vegetable plants (Utzinger *et al.* 1973).

FIG. 4.1. Hoeing is the most common method of weed control in a garden.

Mechanical Methods. Cultivation and mechanical removal are the most common methods for controlling weeds in a garden (Hopen 1972). The best time to remove weeds is just as they appear on the soil surface. If the weeds are allowed to grow too large, they will shade the vegetables, causing them to grow poorly. Only those weeds that are present can be controlled and repeated cultivations are necessary.

Weeds should be cut off just below the surface with a sharp hoe (Fig. 4.1). On some soils, cultivation early in the season may loosen the soil, break up the soil crust, and aerate the roots better. Shallow hoeing is best as roots of many vegetables are near the soil surface and can easily be damaged. In large gardens, rototillers or garden-type tractors can be used, but they should cultivate shallowly. Deep tillage brings buried weed seed to the surface where it can germinate and require additional cultivations.

Mulching. The easiest and best method of weed control in the garden is to mechanically remove the weeds and then apply an organic mulch. An organic mulch 6 in. deep or an opaque synthetic mulch will control weeds by preventing light from reaching the seedlings (Binning 1975). This will control annual weeds which germinate from seeds each year but not perennial or established annual weeds. These must be removed before the mulch is applied and perennial weeds mechanically removed if they come through the mulch (see Mulches in Chapter 3 for types of mulch to use and Cropping Systems in Chapter 1 for no-till mulch systems).

INSECTS AND THEIR CONTROL

There are now 1100 species of alien insects in the United States. Most of these were introduced by the immigrants who brought fruits and vegetables with them when they settled here. Such pests as Hessian fly on grains, codling moth on apples, San Jose scale, gypsy moth, European corn borer, and Oriental fruit moth are only a few of the imported pests (Sailer 1972). There are 615 species of imported insects that have some adverse effect on our plants, and 217 are pests of major importance. When the settlers came, they brought few beneficial insects and there are few native beneficial insects that attack the imported destructive ones. The insect ecosystem in North America is not as well developed as many European countries. Since 1888 (Sailer 1972), the U.S. Department of Agriculture (USDA) has been combing the world for beneficial insects and so far has introduced 128 species of beneficial insects. Gardeners recognize, however, that not all destructive insects are under control.

If you decide to garden without the use of any insecticide, either chemical or botanical, the following recommendations (Utzinger *et al.* 1973) should be considered:

(1) Grow few or no vegetables highly susceptible to insect damage. Vegetables that have few or no insect problems include beets, chard,

Chinese cabbage, chives, leeks, lettuce, mustard, onions, parsnips, peas, salsify, spinach, an early crop of sweet corn in northern areas, sweet potatoes, tomatoes, turnips, and most herbs (Vandemark *et al.* 1973).

(2) Expect a certain amount of insect damage to occur.

(3) Be prepared to spend considerably more time and effort using mechanical methods of control.

(4) Plant and take care of a larger garden than required for your needs. This will allow you to obtain the necessary yields to offset losses due to insects.

(5) Expect considerable damage from soil insects if they are a problem in your garden. Frequently these insects do little damage, but areas of lawn which are converted into a garden are often infested with various grubs.

Destructive Insects

Many destructive insects can be found in the garden, but not all of these are important. Figures 4.2–4.11 and Table 4.3 show some vegetable insects that are important garden pests.

Cabbageworms. Cabbage looper (Fig. 4.2) and imported cabbageworms are serious pests of cabbage, cauliflower, broccoli, and sometimes lettuce. They overwinter as pupae attached to the stems of last year's plants. In the spring, white moths emerge and lay white to yellow eggs on the lower side of leaves. The eggs hatch and the larvae eat the leaves and stems. Frequently, they can be found in the heads of cabbage and cauliflower, causing them to rot.

Cabbageworms generally increase during hot, dry periods. Cabbage loopers are almost never a serious problem during wet seasons, primarily because of the

FIG. 4.2. Cabbage looper on a bean leaf. *Courtesy of Illinois Natural History Survey.*

TABLE 4.3. Common Garden Insects and Their Control.

Insect	Crop	Dust formula	Spray formula	Remarks
Aphid	Cabbage Cucumbers Melons Peas Potatoes Tomatoes	5% malathion	2 tsp 50–57% emulsifiable malathion	Apply on foliage when aphids appear. Repeat weekly as needed.
Blister beetle	Potatoes Corn Tomatoes Beans	5% Sevin	2 tbsp wettable Sevin in 1 gal. water	
Cabbageworms	Broccoli Cabbage Cauliflower Greens		*Bacillus thuringiensis*; follow label directions	Thorough treatment is necessary. Repeat weekly as needed. Begin treatment when worms are small.
Corn earworm (⅔ nat. size)	Sweet corn Tomatoes	5% Sevin; *Bacillus thuringiensis* (Thuricide, Dipel, Bactur) on tomatoes	Inject ½ medicine dropperful of mineral oil into silk channel as silks start to dry *or* 2 tbsp wettable Sevin in 1 gal. water	Dust or spray silks with Sevin every other day for 10 days. Dust or spray tomatoes with Sevin 3 to 4 times at 10-day intervals; begin when first fruits are small.
European corn borer	Sweet corn	5% Sevin *or* 5% Sevin granules	2 tbsp wettable Sevin in 1 gal. water *or* 2 tbsp 25% diazinon in 1 gal. water	Apply insecticide four times at 5-day intervals beginning with egg hatching near mid-June. Avoid early spring plantings. On late corn, dust as for corn earworm.
Striped cucumber beetle	Cucumbers Melons Squash	5% Sevin	2 tbsp wettable Sevin in 1 gal. water	Treat as soon as beetles appear. Repeat when necessary.
Cutworm	Most garden crops		2 tbsp 25% diazinon in 1 gal. water	At transplanting, wrap stems of seedling cabbage, pepper, and tomato plants with newspaper or foil to prevent damage by cutworms.

(Continued)

TABLE 4.3. *(Continued)*

Insect	Crop	Dust formula	Spray formula	Remarks
Flea beetle	Most garden crops	5% Sevin	2 tbsp wettable Sevin in 1 gal. water	Apply as soon as injury is first noticed. Thorough application is necessary.
Grasshopper	Most garden crops	5% Sevin	2 tbsp wettable Sevin in 1 gal. water	Treat infested areas while grasshoppers are still small.
Hornworm (½ nat. size)	Tomatoes	5% Sevin *or* *Bacillus thuringiensis* (Thuricide, Dipel, Bactur)	2 tbsp wettable Sevin in 1 gal. water	Ordinarily, hand-picking is more practical in the home garden.
Leafhopper	Beans Carrots Potatoes Cucumbers Muskmelons	Use Sevin dust or 5% methoxychlor dust	2 tbsp wettable Sevin in 1 gal. water	Spray or dust once a week for 3 to 4 weeks, beginning when plants are small. Apply to underside of foliage.
Mexican bean beetle	Beans	5% Sevin	2 tbsp wettable Sevin in 1 gal. water	Apply insecticide to underside of foliage. Also effective against leafhoppers on beans.
Potato beetle	Potatoes Eggplant Tomatoes	5% Sevin	2 tbsp wettable Sevin in 1 gal. water	Apply when beetles or grubs first appear and repeat as necessary.
Squash bug	Squash	5% Sevin	2 tbsp wettable Sevin in 1 gal. water	Adults and brown egg masses can be hand-picked. Trap adults under shingles beneath plants. Kill young bugs soon after they hatch.
Squash vine borer	Squash	5% Sevin	2 tbsp wettable Sevin in 1 gal. water	Dust or spray once a week for 3 to 4 weeks beginning in late June when first eggs hatch. Treat crowns of plants and runners thoroughly.

Source: Illinois Cooperative Extension Service.
Insects are approximately natural size except where otherwise indicated. Where two drawings are shown, the smaller one is natural size. One pound of dust or 3 gallons of spray should be sufficient to treat 350 feet of row. tbsp = tablespoon. tsp = teaspoon.

activities of a naturally occurring nuclear polyhedrosis virus, which is spread more readily during wet periods (Hofmaster 1981).

Cabbage loopers are pale-green measuring worms with light stripes down their backs. They are up to 1½ in. long. They double up or loop when they crawl. The larva of the imported cabbage worm is velvety green, and up to 1¼ in. long. The adult is a white moth with 3–4 black spots on its wings.

Aphids. Several species of aphids may damage vegetables. They overwinter as an egg. Nymphs hatch out in the spring, mature, and give birth to young aphids (Fig. 4.3). Many generations occur over a summer, and large populations can be built up rapidly. Aphids suck the plant juices from leaves, pods, blossoms, and stems; and dwarfed plants may result.

There are many species of aphids. They are a major problem on fava beans. They are tiny insects, green or powdery blue to black in color, and are soft bodied.

Hornworms. Hornworms overwinter as a dark-brown pupa in the soil. In the late spring, a hawk or sphinx moth emerges. It lays a single greenish-yellow egg on the lower surface of leaves of tomato, pepper, eggplant, and potato plants. The larvae that hatch are bright green with diagonal white stripes on the side. They contain a slender horn at the rear end of the worm, hence the name hornworm. These worms eat large amounts of plant foliage for 3 or 4 weeks and grow quite large. Frequently, a parasitic wasp will inject her eggs

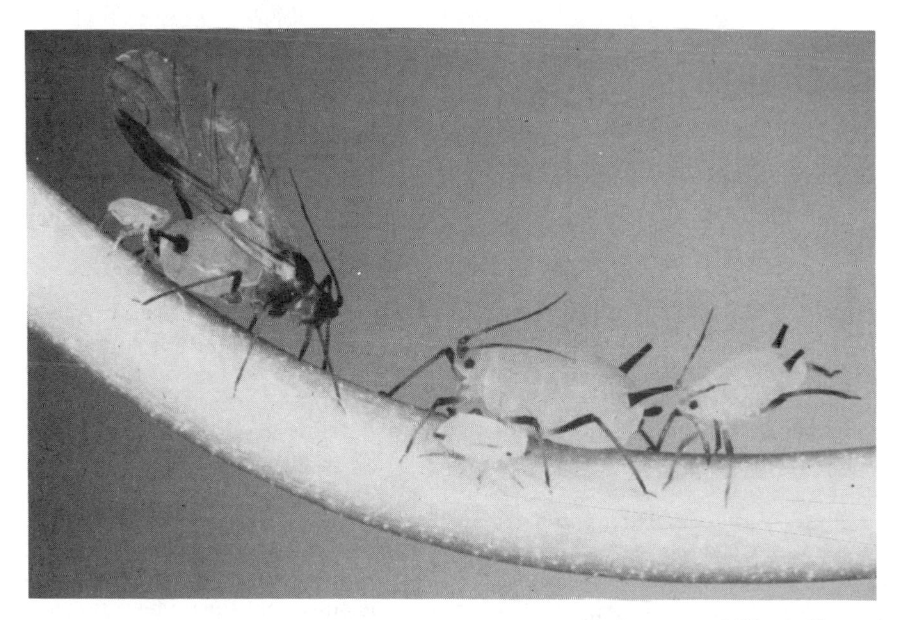

FIG. 4.3. Young aphids (right) and a winged aphid (left). *Courtesy of Illinois Natural History Survey.*

FIG. 4.4. A parasitized tomato hornworm with the cocoons of the wasp parasite on its back. *Courtesy of Illinois Natural History Survey.*

into a hornworm (Fig. 4.4). The eggs hatch, and the wasp larvae eat the inside of the hornworm. When the larvae are fully grown, they emerge from the hornworm. Hornworms with cocoons attached should not be destroyed as any damage to the vegetable plant has already occurred and the wasps should be encouraged.

Mites. Mites are not true insects but very small 8-legged animals. They appear as tiny specks, frequently under a webbing. Mites suck the juices from the underside of leaves, causing them to turn yellow, then brown, and finally to fall off. Many generations can occur in a single season, and large populations can occur rapidly. Mite buildup is encouraged by dry conditions.

Bean Beetles. The adults of the bean leaf beetle overwinter in plant debris near where beans were grown the preceding year. In the spring, these adults emerge and eat the new bean plants. They then lay eggs at the base of the bean plants. The eggs hatch into slender white larvae which eat the roots, nodules, and stems of the plant just below ground. These larvae develop into adults that emerge a month or so later and eat irregular holes in the leaves of the plant. These adults also eat bean blossoms and pods.

Adults of the Mexican bean bettle overwinter in plant debris in the garden, fencerows, wooded areas, and roadsides. The adults move to snap beans and lima beans in the spring. Here they lay eggs on the lower surface of the leaves (Fig. 4.5). The larvae hatch, and both larvae and adults eat the underside of the leaf. This gives the leaf a lacelike appearance. Larvae also eat the bean blossoms and pods.

FIG. 4.5. Mexican bean beetles feeding on the underside of a leaf (damage, upper right). Full-grown larva (left), adult (right). *Courtesy of Illinois Natural History Survey.*

The adult bean leaf beetle is reddish to yellowish with black spots on the back, and up to 1 in. long. The larva is a slender, white grub which is brown at each end. The Mexican bean bettle adult is copper-colored, oval shaped, about ¼ in. long, and has 16 black spots on its back. The larva is orange to yellow, fuzzy or spiny, and up to ⅓ in. long.

Thrips. Thrips are small insects that eat the leaf surface by rasping away small areas. This causes white streaks to appear on the leaves. The leaves wither, turn brown, and fall to the ground. Thrips damage beans, onions, and vine crops and are particularly a problem under drought conditions.

Root Maggots. Several different species of root maggots eat the roots of different vegetable plants. They overwinter as pupae in the garden soil. In the spring, the adult fly emerges and lays eggs near the stems of the vegetable plants. The eggs hatch into small maggots that eat the roots and may tunnel into root crops. Different species of maggots can be a problem in beans, broccoli, carrots, cabbages, cauliflower, onions, radishes, sweet corn, and turnips.

Cucumber Beetles. The striped cucumber beetle (Fig. 4.6) overwinters as an adult in sheltered areas. In the spring, they move to the vine crops (cucumbers,

FIG. 4.6. Two types of cucumber beetles. The striped cucumber beetle (left) and the spotted cucumber beetle (right). *Courtesy of Illinois Natural History Survey.*

gourds, pumpkins, squash) and eat the leaves. They lay eggs near the plants that hatch into larvae. The larva eats the roots of the plants. The spotted cucumber beetle is really the adult southern corn rootworm. In life cycle and damage it is similar to the striped cucumber beetle. Cucumber beetles do little damage by eating the leaves of mature vine crops. However, they carry bacterial wilt and mosaic virus and infect the plants with these diseases. The plants then die from these diseases.

The adult striped cucumber beetle is yellow to black with 3 black stripes down its back, and is about ⅕ in. long. The adult spotted cucumber beetle is yellowish green with 12 black spots on its back, and is ¼ in. long. As the adult often emerges in late fall, it may do little damage to vine crops; but it also eats pods of beans and leaves of many garden plants.

Colorado Potato Beetle. The beetles overwinter as adults in the soil. In the spring they emerge and lay orange-yellow eggs on the lower surface of the leaves. The eggs hatch into larvae, and both larvae and adults eat the potato leaves. The beetles may eat enough leaves to prevent the potato tubers from being formed.

The adult Colorado Potato Beetle is yellow with 8 black strips on the wing covers and black dots just behind the head. It is hard shelled and about ⅓ in. long (Fig. 4.7).

Leafhoppers. Leafhoppers suck plant juices from the leaves (Fig. 4.8). They secrete a toxin into the plant that causes browning of the leaf tips and edges of the leaf (called "hopperburn"). Leafhoppers migrate from the South to the

FIG. 4.7. Colorado potato beetle larva (left), adult, and eggs on a potato leaf. *From: Cooper and Lindquist (1983).*

FIG. 4.8. Leafhoppers suck plant juices from the leaves. *Courtesy of Illinois Natural History Survey.*

North and may suddenly appear in the summer in large numbers. Other years they may not be found at all. Leafhoppers can completely destroy potato plants.

There are several species of leafhoppers. The adults are green and wedge shaped about ⅛ in. long. They fly rapidly when disturbed. The young numphs resemble adults, but are smaller and crawl sideways like crabs.

Flea Beetles. Flea beetles are a group of small beetles with enlarged hind legs. They jump readily when disturbed. The adults overwinter and appear on plants in the late spring. The beetles feed on the leaf surface. Potatoes and eggplant are quite susceptible to flea beetle damage.

There are many species of flea beetles, which infest nearly every type of vegetable. They are black and shiny with curved yellow or white stripes (Table 4.3), and about $\frac{1}{10}$ in. long.

Squash Vine Borers. These insects overwinter as pupae in the soil. In late spring, the moths emerge and lay single brown eggs on the stems of vine crops. The eggs hatch into larvae which bore into the stem and eat the inside (Fig. 4.9). They must be controlled before the larvae tunnel into the stem. Fortunately, they are seldom a problem in most home gardens.

Squash Bugs. The adults overwinter under old squash vines and move to new plantings in the spring. Here they lay brownish-bronze eggs on the lower

FIG. 4.9. Squash vine borer and damage. *Courtesy of Illinois Natural History Survey.*

FIG. 4.10. Squash bug nymphs and an adult. *Courtesy of Illinois Natural History Survey.*

surface of the leaves. The eggs hatch into nymphs (Fig. 4.10). Both nymphs and adults suck the plant juices from the leaves, often killing the plants.

The adult squash bug is dark brown to black with orange or brown on the perimeter abdomen, and about ½ in. long. They emit a peculiar odor when crushed, and are sometimes called "stink bugs." The nymph varies from bright green with a red head and legs to dark greenish-gray with a black head and legs.

Cutworms. There are many different species of cutworms. Cutworms may overwinter in the soil as larvae, pupae, or adults. The larvae begin feeding in the spring and change in the soil into pupae and then adult moths. The adults lay their eggs on the stems of many weeds and grasses, and different species may have several generations during the growing season. The larvae are plump, soft-skinned worms that attack nearly all garden vegetables. They eat the stem of the plants just at the soil surface at night (Fig. 4.11). The plants can be found cut off and wilting on the soil. The worms are usually located in shallow holes in the soil near the stem of the plants. When the worms are disturbed they roll up into a tightly coiled spiral.

Slugs. Slugs are not insects; they resemble snails without shells (Fig. 4.12). Slugs frequently cause damage in mulched gardens, which provide excellent living conditions for them (Judkins 1977). They eat most vegetables and can be a major problem on lettuce leaves and radish roots. Slugs

FIG. 4.11. A cutworm eating a plant stem at the soil surface. *Courtesy of Illinois Natural History Survey.*

range in size up to 2 in. long. Slugs are seldom seen in the daytime but leave shiny mucous trails.

Integrated Pest Management

Pest control should be conducted in as safe a manner as possible. Reducing the use of pesticides through integrated pest management (IPM) is one way to accomplish this goal. The two main components of IPM are the scouting for pests and utilization of a variety of pest population control methods. These include mechanical, cultural, biological, and botanical or chemical tools; the use of resistant cultivars and prevention. Often, pesticides are used for cosmetic reasons to produce blemish-free produce. If you peel the carrots, potatoes, and cucumbers, some damage can easily be tolerated.

Scouting. The first step is to learn to recognize both destructive and beneficial insects. This point must be reemphasized, as too many gardeners do not recognize the larval stage of the lady beetle, for example. The garden should be

FIG. 4.12. Spotted garden slug. *From Judkins (1977).*

inspected three or four times a week for pests, and those pests identified. Their numbers should be estimated to determine whether the pest is present in large enough numbers to justify control.

Indirect evidence may or may not indicate the presence of a pest. For example: (1) holes in leaves may be caused by late frost damage, not by chewing insects; (2) white flecks on onion leaves may be caused by ozone damage, not thrips; but (3) holes in tomato leaves in midsummer almost always indicates the presence of tomato hornworms. It is important to find the insect responsible for the damage observed. Do not assume guilt through association with the damage.

Identification. Once found, the suspect pest must be identified. In some situations, an insect that is present in great numbers may not be the cause of damage. For example, lady beetle larvae and adults and other aphid-destroying insects are often present in large numbers in the midst of damage caused by aphids. Identification is also important because some types of insects are more damaging than others. Once you know what types of insects are present, you can better judge whether or not the potential damage justifies control measures. Knowing the identity of a pest also helps you learn about the insect's biology, enabling you to use other IPM tools to control the insect.

Insects or diseased plants that the gardener cannot identify can be taken to the local county extension advisor or sent to agricultural specialists at the agricultural experiment stations (see Table 1.3). Insect pests should be wrapped in facial tissue, placed in a strong container and sent by first class mail to the Extension Entomologist. Diseased plants should be sent to the Extension Plant Pathologist. The material should be sent so it arrives on a weekday, and be accompanied by a letter describing the crop, location in the state, any pesticides used, and type of damage noted. The more information available, the easier it will be for the specialist to identify the problem and recommend a solution.

Pest Population Size. Knowing the number of pests present will help you estimate their impact and whether it is sufficient cause to spend time and money on control. Different numbers of pests may be damaging in different situations. For example, (1) a small number of white grubs may cause heavy damage to root crops (potatoes, carrots, onions), but cause no yield loss in non-root crops (beans, cabbage, tomato); (2) a small infestation of aphids will probably be controlled biologically on all except fava beans; but (3) if you see three or four cabbage moths, you can be sure you will have cabbageworms. Some years, destructive insects fail to appear in large enough numbers to cause damage. Other years, considerable damage or complete loss may occur. Rainfall, humidity, temperature, and beneficial insects all influence the level of destructive insects.

IPM Tools. Before various control measures are used on a pest population, make sure that the pest is in large enough numbers and that major damage will occur. In the spring, when the plants are small, small amounts of damage will greatly reduce yields. But, when an eggplant begins to form fruit, for example, it can withstand 30% leaf damage from flea beetles with little yield loss. A good IPM program will usually combine two or more tools to control the pest. The following sections give various insect control methods.

Beneficial Insects and Biological Controls

Biological control is the direct or indirect use of living organisms to reduce the number of damaging insects below a level of economic importance (Fletcher *et al.* 1972). It has been investigated by the U.S. Department of Agriculture since 1888 (Sailer 1972). A number of weeds and insects are presently controlled biologically, but once the pest is no longer economically important, it is soon forgotten as a problem. Gardeners, however, may wish to control present insect populations biologically. The insect population may be controlled with parasites, predators, and pathogens.

Parasites normally complete their life cycle on or in a single host and are smaller than the host. Frequently, a number of parasites live on the same host and may or may not result in the death of the host. Parasites cannot be depended upon to control insects in the garden, but if they are present they should be encouraged. A good example is a parasitized hornworm.

Predators are usually larger than their prey and require several hosts to complete their development. They eat the pest and cannot be depended upon to control garden insects. Examples are lady beetles and praying mantids.

Pathogens are bacterial agents, fungi, and viruses that are used for biological control. One pathogen that is effective as a biological control agent and is effective when needed is the bacterium *Bacillus thuringiensis*, a microbial preparation sold in several commercial formulations.

Parasites and predators are most effective when the insect population is relatively low. When the insect population is expanding rapidly, these forms of biological control have little effect. Pathogens are most effective when pest populations are high.

There are several advantages in favor of biological control (Flectcher *et al.* 1972): (1) once established, it is a relatively permanent method of control; (2) there are few undesirable side effects as compared with chemical and botanical insecticides; and (3) once established, there is no additional cost.

There are also several disadvantages in the use of biological control methods: (1) many parasites and predators are not present in early spring when insects can easily destroy young plants; (2) many predators are poor searchers for food and eat beneficial insects instead of destructive ones; (3) many destructive pests have no known biological control agents; (4) destructive insects must be present to support the biological control methods, meaning some insect damage to plants will occur; and (5) most control agents are not easily handled, for example, lady beetles shipped in bottles usually die from lack of food.

Biological control agents should be encouraged. Home gardeners should learn to recognize the various beneficial agents. Some of these follow.

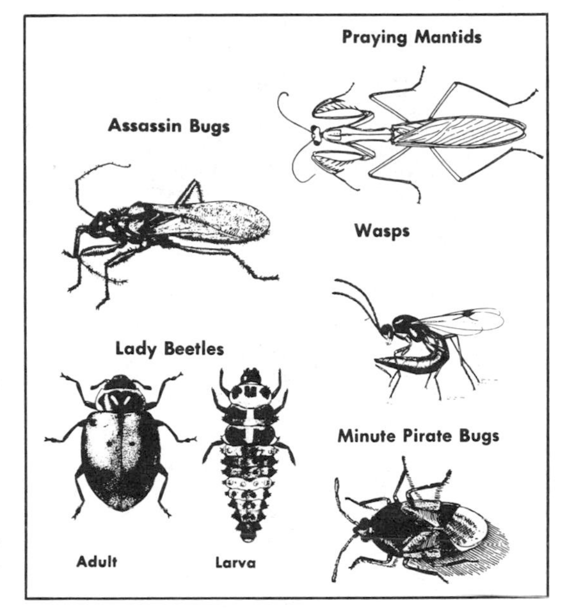

FIG. 4.13. Some common beneficial insects in the home garden. *From Erhardt and Littlefield (1973) (USDA figure).*

Assassin Bug. They are sometimes called the "kissing bug" as they bite painfully when handled or when they fly against a person's face at night. Assassin bugs (Fig. 4.13) eat the eggs, larvae, and adults of many destructive insects.

Assassin bugs are brown or black, with a flat, sculptured body with a groove that holds the stout beak when at rest. The abdomen may flare outward beneath the wings. They often have a peculiar hooded structure behind the head. They are about 1½ in. long, and walk on leaves with a slow, clumsy manner. Their forelegs are usually in a prayful position and are used to capture and hold other insects.

Bacillus thuringiensis. *Bacillus thuringiensis kurstaki,* or B.T., is a naturally occurring microbial preparation. It acts as a pathogen and is an extremely effective biological control agent. As any material that kills insects is an insecticide, B.T. is an insecticide. It is sold as Thuricide, Dipel, and Bactur, among others. B.T. is effective on various cabbageworms, hornworms, fruitworms, and bagworms, regardless of the worm's size (Fig. 4.14). When eaten by the worm, B.T. disrupts and paralyzes the worm's gut. Then it infects the worm with a highly specific disease bacterium (Anon. 1978). The worms stop eating within minutes, but it takes 3 or 4 days before the worms die. Therefore, expect to see cabbageworms for several days after spraying. B.T. is a biological control agent and as such has no tolerance requirements and is not restricted in its uses. It can be used on tender plants and is not harmful to bees, beneficial insects, or fish.

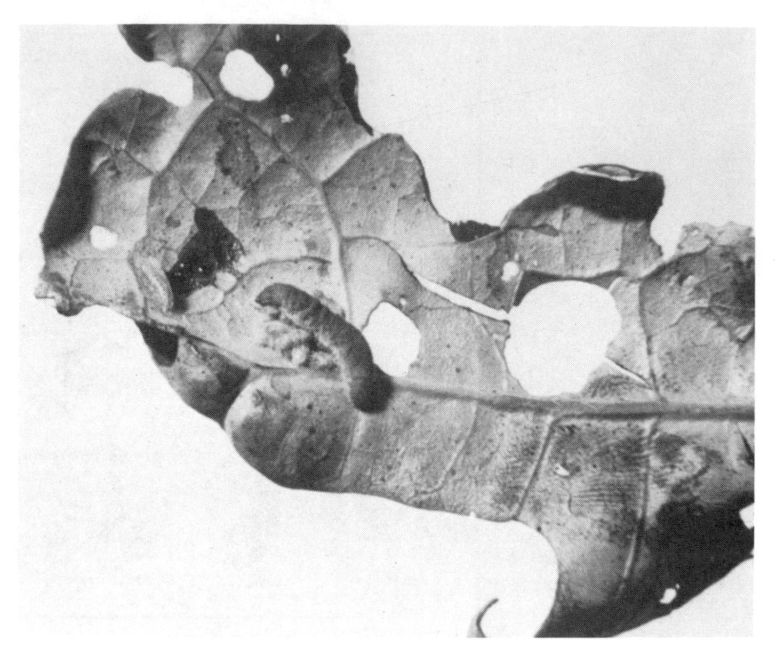

FIG. 4.14. A cabbageworm paralyzed by *Bacillus thuringiensis*. *Courtesy of Illinois Natural History Survey.*

Birds. Many species of birds are helpful in controlling insect pests. They are more important in preventing insect problems than in controlling the insect once it is established. To use birds as a biological control agent, those species that feed largely on insects must be encouraged. If all species are encouraged, damage to fruit trees, sweet corn, and peas frequently result. Insect-feeding birds can be encouraged by (1) providing cover to hide and nest; (2) by supplementary feed, including the growing of sunflowers for bird use; and (3) preventing various predators, particularly cats, from entering the area. Such insect-feeding birds can be attracted to your home by growing the following ornamentals, which provide food and cover (Taber *et al.* 1974): bittersweet, cherry, cotoneaster, crabapple, dogwood, elderberry, firethorn, hawthorn, highbush cranberry, holly, mountain ash, red cedar, Russian olive, sumac, and wild plum.

Brown creepers, kinglets, nuthatches, and titmice (Erhardt and Littlefield 1973) will eat ants and ant eggs. Juncos, ruby-crowned kinglets, swallows, and sparrows will eat scale insects. Barn swallows, flycatchers, gnatcatchers, phoebes, red-eyed vireos, and scarlet tanagers will help control all types of moths. Gnatcatchers and many warblers will control leafhoppers. Bluebirds, brown thrashers, catbirds, flycatchers, meadowlarks, and mockingbirds will eat grasshoppers. Juncos and towhees eat various insects that live in the soil. Downy woodpeckers eat snails.

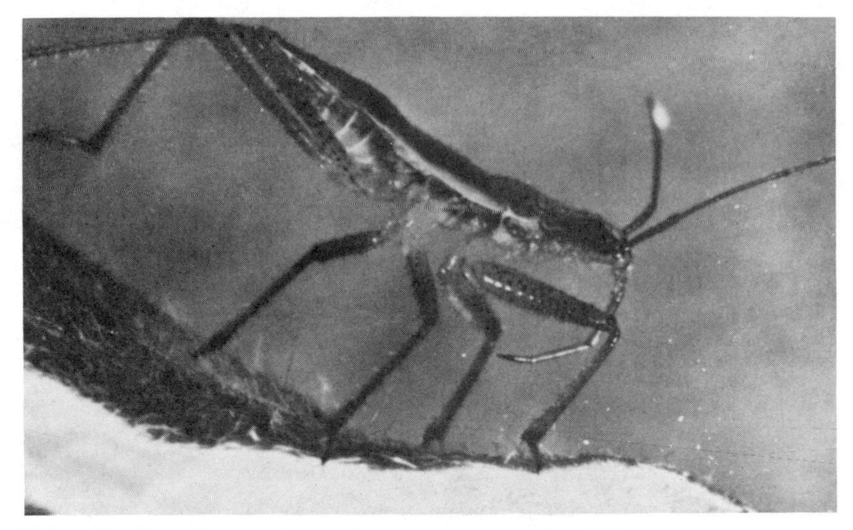

FIG. 4.15. Nymph damsel bug. *From Johansen et al. (1972).*

Damsel Bugs. Damsel bugs somewhat resemble assassin bugs. They are pale gray and about 1 in. long (Fig. 4.15). Their long forelegs are used for capturing prey. Both nymphs and adults eat eggs, larvae, and adults of many destructive insects including aphids, caterpillars, leafhoppers, and mites. The adults are very fast and aggressive insects with enlarged front legs used to catch and hold their prey. They are one of the few predators that can catch lygusbugs.

Lacewings. There are many types of lacewings found throughout North America (Fig. 4.16). These insects are one of the most beneficial insect types in the garden. The adult brown lacewings are dark brown with yellow just behind the head, and are about ⅕ in. long. The adult green lacewings are pale green with a slender body, and are about ¾ in. long. Both have long hairlike antennae (Johansen *et al.* 1972). The name lacewing comes from their large membranous wings, which have a netlike venation. The adults feed on honeydew and nectar and attach their eggs with a long silken stalk to stems and leaves of plants. The eggs hatch into larvae that are about one-half inch long when fully grown. The larva is called an "aphis lion" and looks somewhat like an alligator (Fig. 4.17). These larvae eat large amounts of aphids, mealybugs, scale, insect eggs, spider mites, and some small insects. They grab their prey with the two hollow mandibles and completely suck the body dry.

Lady Beetle. Lady beetles are the best-known predator insect (Fig. 4.13). All lady beetles and their larvae eat aphids (Figs. 4.18 and 4.19), insect eggs, scale insects, mites, and mealybugs. One convergent lady beetle can eat 2400 pea aphids and can lay 1700 eggs in her lifetime (Johansen *et al.* 1972). Lady

FIG. 4.16. Adult green lacewing (top) and brown lacewing (bottom). *From Hoy et al. (1978) (Univ. Calif. Div. Agric. Sci.).*

beetles eat about 100 aphids before they are ready to lay eggs and need a minimum of 2 aphids per day for each egg produced. Lady beetles do not kill grubs, Japanese beetles, caterpillars, or other large insects. If there is not an ample supply of live aphids on the plants, the beetles will eat each other or leave the area. Lady beetles should be encouraged but are undependable as an effective biological control measure to quickly reduce a pest population (Vandemark *et al.* 1973).

The most common lady beetle in North America is the convergent lady beetle (Fig. 4.18). The adult is orange or red, usually with 12 black spots; has a black area (rimmed in white) behind the head, which contains two white stripes, and

FIG. 4.17. Larva stage of a lacewing (aphis lion) (left) eating a worm. *Courtesy of Illinois Natural History Survey.*

FIG. 4.18. Convergent lady bettle feeding on an aphid. *From Johansen et al. (1972).*

is ¼ in. long. The larva are black with orange spots (Fig. 4.19). The second most common lady beetle adult is gray to pale yellow with black spots (often 16 of them), and ¼ in. long. Their larva are black with yellow markings on the head. Another species is found in the Western part of North America and the South and West. The adult has orange to bright red wing covers with few, if any, markings. It has black and white marks just behind the head, and is ¼ in. long. The larva are grayish, flat bodied, and wrinkled. Another species is usually only found in California and Florida. The adult female is red with irregular black marks while the male is predominantly black. They are round and about ¼ in. long. The larva are pinkish with black markings.

Lady beetles require an overwintering site. Allow fallen leaves and other debris to accumulate in areas such as at the base of trees and shrubs to provide a safe overwintering site.

FIG. 4.19. Lady beetle larva feeding on an aphid. *From Johansen et al. (1972).*

Pirate Bugs. Pirate bugs are small predator insects. The minute pirate bug (Fig. 4.13) is about one-eighth of an inch long. They are black and white or brown and white as adults. Pirate bugs eat eggs, larvae, and adults of small insects such as aphids and mites.

Praying Mantids. Praying mantids (Fig. 4.13 and 4.20) are large predators. They commonly rest with the front part of the body upraised, and their enormous front legs are in an attitude of prayer (hence their common name). Mantids are exclusively carnivorous, eating only other insects.

FIG. 4.20. A resting praying mantis. *Courtesy of Illinois Natural History Survey.*

Their wings are fully developed, but mantids are poor searchers for food and wait for their prey to come to them. Praying mantids prefer grasshoppers, crickets, bees, wasps, and flies and may destroy more beneficial insects than destructive insects. In addition to destroying few garden insects, most mantids hatch in the middle or late part of the growing season. Thus, they are seldom present in sufficient numbers to suppress a pest population, particularly in the early part of the growing season. The eggs are laid in large masses about 1 in. long. They are contained in a frothy, gummy substance, which hardens and fastens them to twigs of trees. These egg masses, containing about 200 individual eggs, are frequently available for sale. The first mantids hatched usually eat the subsequent mantids as they hatch from the egg mass and few nymphs survive the first week of life (Fletcher *et al.* 1972).

There are about 20 different species of praying mantids in North America. The bodies of most species are green with green wings, with brown on the front wing margins. The bodies may sometimes be brown. The abdomens of the females are larger than that of the males. They range in size from 2½ to 5 in. in length.

Syrphid Flies. Syrphid flies are also known as flower, or hover, flies. There are many species. The adult fly is about ½ in. long, and its abdomen contains bright yellow and black bands. They are often confused with wasps and bees. The adults hover over flowers and dart away quickly. The adults feed on pollen and nectar (Fig. 4.21) and pollinate many flowers. The adults lay their white eggs among aphid colonies. These eggs hatch into maggots as the larval stage. The larval stage is the predator stage and these larvae equal lady beetles and aphis lions(green lacewing larvae) as important biological control agents. Learn to recognize these brown, green, or bright orange larvae about one-fourth inch long. Aphid colonies usually have at least one syrphid fly larva preying upon it. These larvae eat aphids at the rate of one a minute. They grasp the aphid with their pointed jaws (Fig. 4.21), raise it in the air, suck and pick out the body contents, and then toss the shell aside.

FIG. 4.21. Adult syrphid fly on a flower (left). Larva stage feeding on an aphid (right). *From Johansen et al. (1972).*

Wasps and Flies. The two most important parasite insects are ichneumon wasps (Fig. 4.13) and tachinid flies. Ichneumon wasps vary in size from one-eighth inch to 2 in. in length. They usually have a long egg-laying structure at the tip of their abdomen (Johansen *et al.* 1972). These wasps parasitize various cutworms and caterpillars. Tachinid flies resemble a large housefly. They can be differentiated from houseflies by their entirely bare bristle on the antenna. The adults eat leaves and flowers. Tachinid fly adults lay eggs that are glued to the skin of the host, or lay eggs on the leaves where the host insect will swallow them whole while feeding, or deposit already hatched larvae directly into the host insect. Usually only one fly develops per host insect. The larvae eat nonessential fat and muscle of the host, and the host caterpillar usually forms a cocoon before it dies. However, a tachinid fly, not a moth or butterfly, emerges from the cocoon. Tachinid flies attack caterpillars, cutworms, and some beetles. Chalcid wasps include species that are both beneficial and harmful to humans. One species parasitizes the imported cabbageworm. Braconid wasps parasitize various caterpillars and beetles. One species attacks the tomato hornworm (Fig. 4.4). This species injects the eggs into the hornworm, where the eggs hatch. The larvae eat the inside of the hornworm and, when fully grown, eat a hole through the back of the hornworm. Here they spin small white cocoons and attach these to the host. A very small braconid wasp parasitizes aphids and may completely control an aphid population (Metcalf *et al.* 1951).

Other Predators. Gardeners usually think of predators of destructive insects as other insects. However, geese, garden snakes, and box turtles help control insects and slugs. Insects move from the plants to the ground at night and can be eaten by these predators. If the garden area is fenced to keep out dogs, box turtles are quite beneficial and seldom seen.

MECHANICAL AND CULTURAL CONTROL MEASURES

Mechanical and cultural control measures are the oldest methods used to control insects (Erhardt and Littlefield 1973). These methods use direct or indirect measures to (1) destroy the insect directly, (2) modify the environment or planting conditions to make it undesirable for the insect, or (3) prevent or disrupt the normal life processes of the insect. These methods are used in many parts of the world, and some of them are particularly effective in gardens. However, some of them require considerable time and effort.

Crop Rotation. Rotate the garden plot if possible. If this is not possible, the sequence of plants grown in the garden can be changed (Taber *et al.* 1974). Do not grow the same vegetables in the same place year after year. Soil insects build up into large populations. Do not plant on recently plowed sod or lawns, as these usually contain various grubs. If soil insects such as wireworm, maggots, or grubs are a problem, avoid root crops such as carrots, potatoes, radishes and turnips and avoid broccoli, cabbage, and cauliflower, as these are also susceptible to soil insects. Beans, Swiss chard, peas, and spinach are more tolerant of these insects.

Sanitation. Many garden insects overwinter on old plant debris and weeds near the garden. The weeds and plant debris should be removed and composted or spaded under as soon as the harvest of annual vegetables is completed (Reed and Webb 1975). Unused or unwanted vegetables should also be removed, as many insects continue to use them to reproduce. Manure and compost should be spaded under or used sparingly to prevent a buildup of grubs, slugs, snails, pillbugs and millipedes. If the garden is in an area infested with Japanese beetles, roses, raspberries, and grapes should not be grown as these plants harbor the beetles.

Planting Practices. The garden should be well maintained. Adequate fertilizer and water should be used to provide good plant growth. Healthy plants can withstand some insect damage. Plant crops that are adapted to your climate and best suited for your soil.

By paying attention to the planting dates, an early crop of sweet corn can be grown in northern areas without damage by corn earworms (Irwin and Hendrie, 1985). Corn earworm moths are blown from the south to the north by the prevailing winds, and the moth must lay the eggs on the fresh corn silk. If sweet corn is planted by the middle of May, the silks will be produced and have dried up before the moths arrive. Late season corn frequently has an earworm problem.

Grow your own or use purchased transplants. The longer the time that a plant is grown in the garden, the greater the time available for the plant to be destroyed by insects. A small plant just developing from a planted seed is less able to withstand as much insect damage as a larger transplanted plant. This is particularly true for cool season crops that are planted from seed early in the growing season, such as cabbage.

Plant warm season crops when the soil is warm. The seeds will germinate quickly, and the plants will grow rapidly. This will reduce maggot damage in beans and sweet corn.

With the exception of sweet corn, do not plant solid plantings of a given vegetable. If the plants are isolated from each other, there will be less insect damage. Use interplantings if possible.

Mechanically cultivate the garden to kill weeds and to work the soil. When the soil is worked, some insects will be injured or killed and some others will be exposed for birds to feed on. Plowing or spading in the fall in northern climates exposes many buried insects to cold temperatures. Many insects cannot withstand low temperatures or freezing and thawing winter conditions.

Resistant Cultivars. All gardeners should use resistant cultivars or, at least, cultivars that are tolerant to certain insects. Sometimes a resistant cultivar will be susceptible to some other condition or disease and offer little advantage. Butternut squash, for example, is more resistant to squash vine borer than are acorn or blue hubbard squashes. Zucchini squash appears to be the most susceptible summer squash cultivar.

Through biotechnology, various microorganisms which are not toxic to humans are being incorporated into plants. Cabbage plants, which carry the *Bacillus thuringiensis* gene for the toxin responsible for killing cabbage worms, are being tested. These plants are resistant to cabbageworms. Various bacteria and fungi are being incorporated into seed pellets and seed tapes that prevent the seeds from being attacked by root insects. Various fungi are being applied to tomato transplants that prevent insects from feeding on the foliage. More insect-resistant cultivars will be available in the future, and gardeners should look for them.

Physical Barriers. Hot-caps, paper collars, light roofing paper, or used milk cartons can be placed around plants for some types of insect control. For cutworm control, the collar should extend 1 in. into the soil and at least 2 in. above the soil about 1 in. from the plant stem. For root maggot control on cabbage, broccoli, and cauliflower, the collar should fit snugly around the base of the plant. Hot caps can be placed over cucumber, squash, and pumpkin seedlings in the early part of the growing season to prevent cucumber beetle damage.

Reflectors. Reflective mulches (aluminum foil, aluminum-coated paper, and white polyethylene plastic) will repel migrating aphids (Toscano *et al.* 1979) and reduce Mexican bean beetle damage to garden beans (Topoleski 1972). The mulches reflect the sun's ultraviolet rays, which the aphids "see" instead of the blue-green light (color) of plants. In effect, aphids receive a signal to "keep flying" instead of landing. In summer squash plantings, aluminum foil reduced aphid damage over the entire season by 85%, while white plastic reduced damage by 63% (Toscano *et al.* 1979).

Traps. Slugs can be trapped in a shallow pan filled with beer and sunk to ground level. The pans need to be checked frequently to remove the dead slugs and replenish the beer. Slugs and other insects can be trapped under boards placed on the ground. The slugs and insects hide under them at night and can be collected and destroyed the next morning.

Some insects are attracted to the color yellow. Traps made of yellow plastic and coated with adhesive can be suspended throughout the garden to trap aphids and other insects. Adhesive materials can be applied to yellow cardboard sheets or screens that will trap the insects, similar to "fly-paper." The adhesive recommended by Erhardt and Littlefield (1973) is a mixture of hydrogenated castor oil, natural gum resins, and vegetable wax. The sheets or screens can be placed on the ground or on the plants to trap leafhoppers, flea beetles, and other insects.

Light traps have been recommended to trap destructive insects. Black light traps will collect a wide variety of insects, some of which will be destructive insects. However, some important insect pests, such as Colorado potato beetle, Mexican bean beetle, and flea beetles, are not attracted to light. Wingless aphids and other immature sucking insects are relatively nonmobile and will not be caught (Hofmaster 1981). Do not expect light traps to collect soil insects, daytime-feeding insects, and various worms. Light traps are a good way to determine which insects are in the area, but they have little effect upon the garden insect population (Fletcher *et al.* 1972). High light trap counts do not necessarily mean a correspondingly severe garden problem since other factors, such as host plant condition, are also involved. Light traps will also attract insects into the area that were not present originally. These insects frequently remain and cause plant damage.

Pheromones. Various biological attractants are effective in luring a specific insect species. Other insects are unaffected. This material may be a sex attractant, which was obtained from similar insects and lures those types of insects into a trap.

Plants give off various chemicals that act as pheromones and attract insects. Tomatoes give off chemicals that stimulate some moths to lay their eggs on the leaves (Coates et al. 1988). These pheromones can be used to attract and trap those moths, or be used in a poison bait. In the future, tomato cultivars may be available that do not contain these chemicals.

Wild cucurbits contain chemicals called cucurbitacins that attract cucumber beetles, which then feed on the leaves (Rhodes et al. 1980). The leaves and fruit taste bitter, and only small amounts of these chemicals are found in domesticated cultivars. These pheromones are present in greater quantities in the wild plants, which are often found in desert areas of the southwestern U.S., and in some domesticated cultivars, such as bitter melon. The fruits from these plants have been cut in half, insecticide applied to the fruit, and used to attract and kill cucumber beetles.

Trap Crops. Plants may be planted around the garden to attract insects before they enter the garden. The weed, black nightshade, and eggplant are very attractive to egg-laying for the Colorado potato beetle. On nightshade, neither the larva nor adult eat the leaves and they soon die. On eggplant, they may eat all of the leaves. Trap crops similar to those in the garden may also be used to help reduce the number of insects moving into the garden. Once the insects have infested these trap plants, they should be destroyed. An insecticide can be used, as these plants and fruit will not be eaten.

Repellent Plants. This is based upon the theory that certain kinds of plants will give off an odor that will specifically prevent insect damage to certain other plants when they are grown nearby. Usually herbs or flowers with aromatic foliage or blossoms are used. For this technique to be successful, the major garden pests, such as cabbageworms, flea beetles, bean beetles, and cucumber beetles must be repelled. The repellent plants must be large enough to ensure that any protection or repellent qualities are fully present when they and the vegetable transplant are placed into the garden. This technique has been researched in numerous places in the USA and Canada with limited success.

In California tests (Koehler et al. 1983), the four repellent plants were planted in a square configuration with the vegetable seedlings in the center. With anise, the number of cabbageworm eggs deposited on the cabbage were less (5.9) when compared to cabbage protected by basil (7.3), thyme (8.5), sage (8.8), or the controls (10.4) with no repellent plants. When only two repellent plants were used, cabbageworm egg numbers were similar between the treatments. However, even a few cabbageworms can cause considerable damage. Cabbage yields were greatly reduced due to cabbageworms, and competition for sunlight, water, and nutrients. Little repellent effect was noted with other vegetables that had different insect problems.

Handpicking. Tomato hornworms are usually few in number, and handpicking is the recommended control. Colorado potato beetles and bean beetles may be shaken off the plants into a pail containing kerosene, diesel fuel, or cooking oil.

Handpicking requires the gardener to recognize the egg masses and injurious stages of the pest. It is impractical to handpick many small insects such as mites and aphids.

Plant Part Sprays. Some gardeners suggest boiling or grinding up plant parts in water to use as an insect spray. Boiling rhubarb leaves or soaking tobacco stems in water, for example, releases toxic materials into solution. These toxic materials are more toxic to humans than most chemical insecticides. Therefore, homemade sprays made from plant parts cannot be recommended.

CHEMICAL CONTROLS

Despite all precautions against insects, pest epidemics may become a threat, and an insecticide is then needed to reduce the pest population to a tolerable level. The gardener should check the crops every 2 or 3 days and examine the plants for signs of an insect infestation such as egg masses, insect droppings, and damaged leaves or fruit. In this way, most insect problems can be discovered and control measures taken before the problem becomes serious. The gardener should identify the insect and be sure it is causing damage before applying an insecticide. Sometimes lady beetle larvae (see Fig. 4.19) are not recognized and are intentionally killed with an insecticide, often resulting in an aphid problem. On the other hand, if the gardener recognizes two or three cabbage moths in the garden, broccoli, cabbages, cauliflower, and some greens will soon be infested. Table 4.3 lists two chemical or manmade insecticides (Sevin and malathion), one biological control agent *(Bacillus thuringiensis)*, and one mechanical method. Although these latter two methods of insect control are not chemical controls, they are the best control method for gardeners.

Bacillus thuringiensis. This microbial preparation is not a chemical control but a biological control agent (see earlier section on Beneficial Insects and Biological Controls). It is repeated as many gardeners do not recognize it as a nonchemical method. For those uses listed in Table 4.3, it is the safest and most effective control for gardeners.

Diazinon. Diazinon, sold as Spectracide, is a moderately toxic insecticide. It is listed in Table 4.3 as it is used to control soil insects of all types on all vegetable crops. The dust formulation is really a granule, and label directions should be followed. For a spray formula, 2 tbsp of 25% diazinon should be added per gallon of water. Both formulations should be applied to the soil at the beginning of the season and raked or rototilled into the soil.

Malathion. Malathion, sold as Cythion, is a widely used manmade insecticide that exhibits activity on a wide range of insects. It degrades very rapidly and disappears completely in about a week. Many parasites and predators can tolerate its use without major disruption. It has a low toxicity on birds but is highly toxic to fish and bees. It has been approved by the Environmental Protection Agency (Fletcher *et al.* 1972) for use on most vegetables. To give the reader some indication of the safety of malathion, 1 oz of the 1% dust formulation per adult is recommended as the control for human lice (Anon. 1988).

Carbaryl or Sevin. Carbaryl is the common name and Sevin is a trademark of this widely used manmade insecticide. It exhibits activity on a wide variety of insect pests. Carbaryl has a low toxicity and degrades quite rapidly. About 50% disappears within 3 days with nearly complete disappearance in 2 weeks. It has a low toxicity to animals and birds, a slight toxicity to fish, and high toxicity to bees. Sprays should be used when bees are not in the garden. The dust formulation presents less of a problem to bees than the spray formula. The Environmental Protection Agency has approved (Fletcher *et al.* 1972) carbaryl for use on most vegetables. To give the reader some indication of the safety of carbaryl, 1 oz of the 5% dust formulation per adult is recommended as the control for human lice (Anon. 1988).

Time to Wait Between Application and Harvest. Gardeners must ensure that the insecticide is not applied and the vegetable then immediately harvested and eaten. There is an established time the gardener must wait between application of the chemical and harvest of the vegetable (Table 4.4). These waiting times are designed to protect the consumer. The federal government and most states have laws that require pesticide containers to

TABLE 4.4. Days to Wait Between Application of the Chemical and Harvest of the Vegetables[1].

Vegetable	Chemical	
	Carbaryl	Malathion
Beans	0	1
Cabbage and related crops	3	7
Collards, kale, other leafy crops	14	7
Eggplant	0	3
Lettuce	14	14
Onions	—	3
Peas	0	3
Potatoes	0	0
Pumpkin[2]	0	3
Other vine crops[2]	0	1
Sweet corn	0	5
Tomato	0	1

Adapted from: Anon. (1988).
[1] *Bacillus thuringiensis* has no time limit.
[2] Apply insecticide late in the day after blossoms have closed to avoid bee kill.

state the directions for rate and time of application of the material. By following these directions, the gardener is assured of producing vegetables that are safe and meet the tolerances established by the Food and Drug Administration.

BOTANICAL CONTROLS

Despite all other methods of natural pest control, pest epidemics will sometimes occur. Most organic gardening magazines recommend, and organic gardeners use, botanical insecticides as a last resort. Botanical insecticides are derived from plant sources and are not synthethic or manmade. Contrary to popular opinion, botanical insecticides are toxic to humans. Some are more toxic (Table 4.2) to humans than their synthetic counterpart. Thus, many insect control recommendations do not list these botanical insecticides. Table 4.5 give some common garden pests and their botanical control. The table lists three botanical insecticides (nicotine sulfate, pyrethrum, and rotenone), one biological control agent *(Bacillus thuringiensis),* and one naturally occurring deposit (sulfur). The latter two methods of control are not botanical methods but accepted organic methods and are included to make the table more complete.

Botanical insecticides such as rotenone are sold as a 1 or 2% dust. Nicotine sulfate is sold as 40% liquid, which is diluted and applied as a spray. Pyrethrum is normally combined with rotenone and sold in a pressurized container. If you read the label on many insecticide containers for homeowners, you will note that many of them are botanical insecticides.

Nicotine Sulfate. Nicotine sulfate is not the same as nicotine extracted by soaking tobacco in water. Nicotine is eight times more toxic to humans

TABLE 4.5. Some Botanical Insecticides and the Vegetable Insects They Control.

Pest	Insecticide[1]				
	B.T.	Nicotine sulfate	Pyrethrum	Rotenone	Sulfur
Ants	0	0	X	0	0
Aphids	0	X	X	X	0
Bean beetles	0	0	X	X	0
Beetles and true bugs	0	0	X	X	0
Cabbageworms	X	0	X	X	X
Caterpillars	X	0	X	X	X
Cucumber beetle	0	0	X	X	0
Leafhoppers	0	0	X	X	X
Mites	0	X	0	0	X
Moths	X	0	0	0	0
Potato beetle	0	0	X	X	0
Thrips	0	X	X	X	0
White fly	0	0	X	X	0

[1] B.T.—*Bacillus thuringiensis*. X—Indicates the insecticide will control that particular pest. 0—Indicates little or no control.

than nicotine sulfate (Table 4.2). Only the commercial formulation should be used.

Nicotine sulfate is a moderately toxic insecticide that has largely been replace by malathion, which has a lower toxicity rating. Nicotine sulfate controls aphids, thrips, and some soft-bodied sucking insects. It is usually sold as a 40% liquid concentrate such as Black Leaf 40. The material is then diluted with water and used as a spray. Dusts are not used because of their toxicity to the applicator. Nicotine sulfate is most effective when applied during warm weather. Nicotine sulfate degrades rapidly and has a short residual. It can be used within several days of harvesting of the vegetables. The Environmental Protection Agency has approved registration of nicotine sulfate for use on a large number of vegetables.

Pyrethrum. These insecticides are obtained from dried flowers of *Chrysanthemum cinerariaefolium* from Ecuador and Kenya. Pyrethrum has been used for centuries as an insect control agent. They provide a rapid knockdown of a wide range of insects but will not control mites. Pyrethrum has a very short residual and may be used the day of harvest. It is frequently sold in combination with another insecticide, such as rotenone, and with an activator or synergist, such as piperonyl butoxide. It is also sold as Pyrenone in which both pyrethrum and activator are botanically derived. Because of its relatively low toxicity to humans (Table 4.2), its short residual, its crop tolerance, its broad-spectrum of insect control, and no residue, pyrethrum may be the best botanical insecticide to use. Several synthetic analogs of the natural pyrethrum are also available.

Pyrethrum degrades extremely rapidly and can be used around parasites and predators, although it is toxic to them while being used. Pyrethrum is a contact insecticide and kills those insects it touches. Pyrethrum is not hazardous to birds and wildlife but is slightly toxic to fish. Pyrethrum is registered for use on most vegetables.

Rotenone. Rotenone is a general-purpose botanical insecticide available for home gardens. It is derived from *Derris* roots from East Asia or *Cube* roots from South America. Rotenone is effective against bean and cucumber beetles and controls many sucking and chewing insects but will not control mites. It acts as both a contact and a stomach poison. It will control some sucking insects such as aphids that feed on the underside of the leaf and escape sprays of pyrethrum. Rotenone is a slow-acting insecticide and frequently does not control insects that are rapid feeders, such as caterpillars and cabbageworms.

Rotenone usually degrades within a week. After using rotenone, the gardener must wait one day before harvesting the vegetables. The potency of rotenone declines with time, and new material should be purchased each growing season.

Rotenone has a low toxicity to birds and wildlife but is extremely toxic to fish. In South America, the natives use this material to stun and kill fish, and it is still used in North America to kill unwanted fish in ponds and lakes. Rotenone is registered for use on most vegetable crops.

Ryania. Ryania is derived from stems and roots of *Ryania speciosa*, which is native to the Amazon River area and to Trinidad. Ryania acts both as a contact and stomach poison. It is slow acting, is effective against some caterpillars, and does not seem to affect some predator insects. It is not labelled for many garden vegetables.

Sabadilla. Sabadilla is derived from the dried, ripe seeds of a Central and South American lily, *Schoencavion officinale*. With the increasing interest in organic gardening, it is again available in limited supply. It is effective against chewing insects, and acts as both a contact and stomach poison. It has little effect on aphids and can be used on most vegetables. It has a low toxicity to humans and bees.

Milky Spore Disease. This material contains *Bacillus popullae* or a similar bacteria. It controls various grubs, including the grubs (larva stage) of Japanese Beetles. It is not harmful to other insects or humans, and is often applied on lawns. The spores remain active in the soil for about 10 years. It is sold as "Doom," "Safer Grub Killer," and others. This material is most effective on young larva and better control is often noted the second year after the initial application. A large grub population is seldom completely controlled by milky spore disease, and limited success has been noted in these areas, particularly in the northeastern USA.

Soap. The use of soap sprays for insect control has been used since 1787 (Taber et al. 1974). Soft-bodied insects, such as aphids, cabbageworms, and whiteflies are most susceptible to soap sprays. Sprays must be applied frequently, and can be used where 100% insect control is not necessary. In California tests (Koehler et al. 1983), commercial insecticide formulations (Safer or Acco soap) were not any better than Ivory Liquid Detergent (at 1–2%). Soaps were not as effective as chemical or botanical insecticides, and caused injury to some crop plants when used frequently. Unsightly soap deposits often remain on the edible plant parts. Beneficial insects may be killed if the soaps are sprayed on them, but there is no residual effect.

Sulfur. Sulfur is a natural deposit rather than a botanical insecticide. The finely ground yellow powder can be used as a dust or spray, and elemental sulfur should be used. Sulfur controls spider mites but should not be used on vegetables that will be canned. Sulfur creates an off-flavor in the canned vegetable and is converted to sulfur dioxide. This production of sulfur dioxide will cause the container to explode. Sulfur can be used on vegetables that are eaten fresh, frozen, or dried. Sulfur is very safe and is an essential plant nutrient. Applications of large quantities of sulfur to the garden soil will cause the soil to become acid.

DISEASES

Diseases often cause serious vegetable losses. Some diseases cause the seed to decay or the seedling to die before or at emergence. Others attack roots, stems, and fruits. In contrast to insect control, vegetable disease control measures must begin before the disease is noted in the garden. There is little control once the disease is present. Nearly all materials which will control the disease on an infected plant will injure or kill the plant as well. Vegetable diseases are caused by many different types of (1) fungi, (2) bacteria, (3) nematodes, (4) viruses, and (5) mycoplasma. Diseases develop only when (1) a susceptible plant is present, (2) the disease organism is present, and (3) a suitable environment occurs to allow development of the disease. Without all three requirements present, the disease will not develop. The gardener can use control measures to eliminate one of the three requirements and prevent many vegetable diseases without fungicides.

Most plant diseases are too small to be seen. Some virus particles must be magnified 100,000 times before they can be seen easily.

Nematodes are small wormlike animals that attack plant roots. Rootknot nematodes burrow into the roots of plants and cause small, knotlike or rounded swellings, or galls, in older, larger roots. They are easily seen in root crops such as carrots. Plants with severe root galling grow slowly, look unhealthy, and tend to wilt in hot, dry weather and may die prematurely. Galls are swellings within the root in contrast to beneficial bacterial nodules, which are attached loosely to the roots of beans and peas. Nematodes are most often found in areas that have mild winters, where the soil is not subject to freezing and thawing conditions. Gardeners who suspect nematodes are damaging their vegetables should contact their local county agent or Agricultural Experiment Station (see Table 1.3).

Mechanical and Cultural Controls. Disease control measures are aimed at preventing the disease from occurring. In general, mechanical and cultural control measures are sanitation practices and good gardening methods. If these methods are used, vegetable diseases should not be a major problem in the garden. Most good gardeners do not experience disease problems and chemical controls are rarely needed.

Crop Rotation. Always practice a rotation of vegetables within the garden area. Many soil-borne diseases become a major problem when the same or related crops are grown in the same area each year. The cabbage group (broccoli, cabbage, cauliflower, and others) are susceptible to clubroot, yellows, blackleg, and black rot and should not be grown in the same area each year. Eggplant, peppers, potatoes, and tomato are in the same family and have similar diseases. The cucumber family includes cucumber, gourds, melons, and squash and should be rotated in the garden area.

Sanitation. Sanitation is most important in preventing vegetable diseases. Most diseases survive on infected plants between growing seasons. Crop debris should be spaded under after harvest or composted. Diseased plants should not be composted and should be removed and discarded as soon as they are observed. If diseased plants are added to the soil, the disease is added also. Some diseases live a long time in the soil, even though susceptible plants are not grown. Clubroot disease of the cabbage family survives at least 7 years and verticillium wilt of the tomato family lives in the soil at least 10 years.

Gardeners should not smoke around tomato plants. The smoke may contain unkilled tobacco virus, which infects tomato plants.

Weeds often serve as the initial source of several vegetable diseases in the spring and should be destroyed. Dense weeds in the garden create an ideal microclimate for the growth of fungal and bacterial diseases.

Planting Practices. Locate the garden in a sunny area with good air and water drainage. Some diseases are encouraged by wet soils, and a sunny location will hasten the drying of dew and rain. Many diseases are most serious when plants remain wet for long periods of time. Cultivating or touching the plants when they are wet will spread some plant diseases, particularly bacterial wilt of beans. Avoid watering plants in the evening. If the soil surface stays damp all night, disease organisms can thrive.

Some diseases are carried in the seed, such as many cabbage, tomato, and bean diseases. Hot water-treated seed should be purchased as this treatment kills the disease but not the seed. Bean seed should be certified, western-grown seed that is free of bean leaf diseases. Do not save bean seeds from eastern or midwestern locations. Under their hot humid conditions, virtually all bean seed is infected. Potatoes should be certified seed potatoes to prevent scab disease on the tubers.

Vegetables should be planted at the proper time in a garden with optimum fertility and pH. Excessively weak or vigorous plants are more susceptible to some diseases than similar plants grown under optimum conditions. Warm-season crops such as supersweet corn, beans, gourds, cucumbers, melons, and squash should be planted when soil temperatures are warm. These crops require high germination temperatures; if they are planted into cold soil, the seeds rot. Under warm conditions the seeds germinate and the seedlings grow rapidly, preventing many diseases and damping-off fungi.

Damping-off fungi occur in most soil worldwide and attack the seed when planted. The fungi are stimulated to grow by nutrients released from a germinating seed, and the fungi infect the seed or seedling. Frequently, germinating seeds are killed before they emerge from the ground, accounting for poor stands in many crops. Older plants are usually not killed by damping-off fungi because the stem tissue has developed a protective barrier that limits fungal entry. Many purchased seeds will be treated with a fungicide such as Thiram or Captan to prevent damping-off fungi. However, many seed suppliers sell untreated seed. Watering in the evening produces cool, wet conditions that favor damping-off fungi.

Vegetable plants should not be planted closely together. Dense stands of plants increase the humidity around the plants and encourage plant diseases. Various mildews, blights, and white mold of beans are encouraged by high humidity. Applying water to the leaves of some plants also encourages mildews. Correct plant spacings allow for adequate air circulation and drying of the plants and soil.

Tomato fruit rots can be controlled by preventing the fruit from coming into contact with the soil. This can be achieved by using a 3 to 6 in. layer of mulch around the plants or growing the tomatoes in wire cages (Fig. 4.22). Cages can be made of 10-gauge 6 × 6 in. mesh concrete reinforcing wire. The wire can be cut into 4 ft lengths, 2½ ft wide. The center wires can be cut so the ends of the horizontal wires can be used as hooks to fasten the cage into a cylinder; and the ends of vertical wires can be pushed into the soil to support the cage. This forms a cylinder about 2½ ft high and 15 in. in diameter. These cages can be placed over the plants 2 weeks after transplanting and left for the growing season. The fruits are kept off the ground, and fruit rots are controlled.

Mulches can be used to help control potato scab. If 8 in. of straw mulch are applied when the plants are 6 in. high, the potato tubers will form in the straw. This will keep them out of the soil where scab diseases are present.

Fall plowing or spading to bury diseases is also helpful. Plant debris will be partially decomposed during the growing season and eliminate food sources for disease-carrying insects.

Resistant Cultivars and Disease-free Transplants. All gardeners should use disease-resistant cultivars. Cultivars are available for nearly all vegetable crops that are resistant to one or more diseases. Table 4.6 gives instructions to "plant resistant cultivars" as the control measure for a number of diseases that infect specific vegetables.

Grow your own or purchase disease-free transplants. Transplants should be grown from disease-free seed and grown in a disease-free soil. Many diseases are brought into the garden in diseased plants and soil. This is particularly a problem with transplants of the cabbage and tomato families.

FIG. 4.22. Tomato plant growing in a wire cage to prevent fruit rots.

Disease-carrying Insects. Several different kinds of insects carry virus, mycoplasma, and bacterial diseases in their body and transmit the disease from infected plants to disease-free plants. It is easier to control the insect than to control the disease. These insects carry the disease overwinter, from wild to cultivated plants and from diseased to healthy plants. Diligent insect control will reduce such damage. This is particularly a problem with cucumber beetles, which spread bacterial wilt; leafhoppers, which spread yellows; and various other insects that spread virus diseases.

Natural Controls. Roots of several plants contain chemicals that are leached out or released upon decay and are toxic to some diseases. Asparagus roots contain a material toxic to several types of nematodes (Erhardt and Littlefield 1973). French and African marigolds reduce the populations of lesion nematodes. Various mints destroy the disease organism that causes clubroot of the cabbage family. These plants have this effect regardless of which vegetable plant they are growing near.

Elemental sulfur has been used to control fungus diseases. The material can be applied as a dust or spray to control powdery mildew, rust, and leaf spot. Sulfur should not be used on vegetables for canning. Sulfur forms sulfur dioxide in the container, causing it to explode. Sulfur is very safe to humans.

Chemical Controls. About 90% of the plant diseases can and should be controlled by mechanical and cultural practices (Utzinger *et al.* 1973). However, some plants are attacked by diseases that are not easily controlled by these methods.

Most fungicides have a low level of toxicity to humans (Table 4.2). To be effective, they must be applied at the right time (Table 4.6). Applications should start before the disease appears, all aboveground parts of the plant must be sprayed or dusted, and the applications must be repeated at 10-day intervals during humid or wet weather.

Sometimes cucumbers, onions, potatoes, and tomatoes need a mild fungicide during humid weather to prevent leaf blights. Asparagus, beans, beets, carrots, lettuce, peas, peppers, and radishes do not need one (Topoleski 1972).

BIRD AND ANIMAL CONTROL

Animals such as toads and snakes and most birds eat large numbers of insect pests. However, at times and particularly in new housing developments and rural areas, animals and birds can become a nuisance.

Most chemical repellents are not suggested for use on food crops, although Thiram fungicide (1 tbsp per gal.), applied to the area, repels rabbits and a number of other animals. Some repellents are not recommended for use on

TABLE 4.6. Common Diseases of Vegetables and Their Control.

Crop	Disease	Control measure
Asparagus	Rust	Apply fungicide containing Zineb after harvest. Make 5 applications at 10-day intervals.
Beans	Mosaic	Plant resistant cultivars.
	Leaf and pod diseases	No fungicide recommended. Do not cultivate, weed, or harvest beans when plants are wet. Plant certified, western-grown seed.
Beets, Swiss chard, spinach	Leaf diseases	Apply fungicide containing Maneb or Zineb at 10-day intervals. Start when plants are 6 to 8 in. high.
Cabbage broccoli, Brussels sprouts, cauliflower, Chinese cabbage, kale, collards, kohlrabi, mustards, rutabaga, radish, turnip	Yellows	Plant resistant cabbage cultivars.
	Blackleg	Buy only hot water-treated seed.
	Black rot	Buy only hot water-treated seed.
	Clubroot	Apply 1 cup of transplanting solution containing pentachloronitrobenzene (Terraclor, PCNB) around the roots of each plant. The solution is made by mixing 3 level tablespoonsful of 50% wettable Terraclor in 1 gal. of water.
Carrots and parsnips	Leaf diseases	Apply fungicide containing Maneb or Zineb when spots first appear.
	Yellows	Control leafhoppers, which transmit the mycoplasm. Destroy infected plants.
Cucumbers, pumpkins, squash, gourds	Bacterial wilt	Control cucumber beetles, which spread the bacteria from plant to plant. Remove infected plants.
	Scab	Plant resistant cucumber cultivars. Buy hot water-treated seed.
	Mosaic	Plant resistant cultivars.
	Leaf and fruit diseases	Apply fungicide containing Zineb or Maneb at 7- to 10-day intervals. Begin after vines start to spread. If control is needed before vines start to spread, use Ziram or Captan.
Eggplants	Fruit rot	Apply fungicide containing Maneb, Zineb, or Ziram at 7- to 10-day intervals. Begin when the first fruits are 2 in. in diameter.
	Verticillium	Plant resistant cultivars.
Muskmelons (cantaloupes), honeydew melons, and watermelons	Fusarium wilt	Plant resistant cultivars.
	Bacterial wilt	See Cucumbers.
	Leaf and fruit diseases	See Cucumbers.
Onions, garlic, and chives	Leaf diseases	Apply fungicide containing Maneb or Zineb at weekly intervals. Begin when leaf spots are first noticed. Add 1 tablespoonful of powdered household detergent or 1 teaspoonful of liquid detergent to each gallon of spray solution.
	Smut	Plant disease-free onion sets. Smut only attacks onions grown from seed. Treat seed with Thiram before planting.

(Continued)

TABLE 4.6. *(Continued).*

Crop	Disease	Control measure
Peas	Fusarium wilt	Plant resistant cultivars.
	Root rots	Plant early and use a seed treatment.
Potatoes	Tuber diseases	Buy certified seed potatoes. Plant uncut tubers. Grow cultivars resistant to scab and late blight.
	Leaf diseases	Apply fungicide containing Maneb or Zineb at 5- to 10-day intervals. Start when plants are 10 in. high.
Sweet potatoes	Black rot, scurf, foot rot	Buy certified plants. Use 3- or 4-year rotation.
	Wilt, root-knot, soil rot	Plant resistant cultivars.
Tomatoes, peppers	Fusarium wilt, verticillium wilt	Plant immune or resistant cultivars.
	Leaf and fruit diseases	Apply fungicide containing Maneb or Zineb at 5- to 10-day intervals. Begin when the first fruits are 1 in. in diameter.

Source: Adapted from Anon. (1988).
Zineb fungicides such as Dithane Z-78, Chipman Zineb, New Dragon Tomato Dust, Science Zineb Fungicide, etc., contain zinc ethylenebis-dithiocarbamate.
Maneb fungicides such Maneb Spray, Science Maneb Garden Fungicide, etc., contains manganese ethylenebisdiothio-carbamate.
Ziram fungicides such Karbam White, Allied Ziram, Corozate, etc., contain zinc dimethyldiothiocarbamate.
Captan fungicides such Captan Garden Spray, Captan Seed Protectant, Captan 50 WP, etc., contain N-[(trichloromethyl)thio]-4-cyclohexane-1,2-dicrboximide.

edible plants. The most reliable, mobile, and sometimes vigilant repellent is the family dog.

Rabbits, squirrels, and other animals are game animals, and District Game Wardens should be consulted before trapping. Live trapping can be an effective means of removing the individual animals causing damage to the garden. The period of greatest activity for rabbits is just before sunrise and just after sunset; for squirrels it is just after sunrise and during late afternoon. Traps, other than live traps, must be used with caution, particularly around children or pets whose curiosity might result in serious injury.

Birds. In many areas, some precautions are necessary to protect young seedlings and transplants from bird damage. A favorite method is to hang reflectors or fluttering objects over the area to discourage the birds. This method is only partially effective. Plastic bird netting, aluminum fly screen, nylon net, or cheesecloth can be used to make a shield that will completely protect a new planting of corn, cucumbers, peas, and beans. These shields can be made of scrap lumber and need only be 6 to 8 in. high. Once the plants are established, birds do little damage and the shields may be removed.

Gophers. Gophers make elaborate tunnels 6–18 in. underground. Many gardeners who live in areas where gophers are abundant place a fence made of 1-in. galvanized wire about 2 ft in the ground around the garden area to prevent gopher damage.

Gophers frequently follow a person's footprints down a row of newly seeded corn, beans, squash, and pumpkins, eating the seeds on the way. Damage can be prevented by raking the area lightly to remove the footprints or by the use of shields similar to those used to control bird damage.

Moles. Moles make mounds of dirt from their main tunnel and make ridges or raised and cracked soil just above their shallow feeder tunnels. The bases of children's pinwheels can be placed into moles' tunnels to frighten them away by the vibrations produced when the pinwheels turn in the wind. Good insect control measures will eliminate the grubs that the moles feed on, and with no food available the moles will leave the area.

Rabbits. Rabbit-proof fences 2 ft high made of 1½ in. galvanized wire will prevent rabbit damage. The bottom of the fence should be buried in the ground 6 in. to prevent rabbits from digging or crawling under. Dried blood sprinkled around the edge of the garden also discourages rabbits. Many avid gardeners use mothballs. A mothball is tied with a piece of string and the string is run through a small inverted cup (such as a disposable, bathroom drinking cup). The string is then attached to a wire loop which holds the mothball off the ground 5 to 6 in. The paper cup prevents the mothball from being dissolved by water and the rabbits are repelled both by the movement of the cup swaying in the breeze and the odor of the mothball. These will frequently last for 1 or 2 months.

Raccoons. Raccoons are a serious problem in sweet corn as they eat the ears about two days before they should be harvested for optimum quality. The ears can be covered with a paper bag to prevent damage. Pumpkins, gourds, or winter squash can also be planted among the corn plants. The spiny vines hurt the raccoons' feet and discourage them from entering the area and eating the corn. If there are no trees that a raccoon can climb and enter the garden, a low electric fence about 1 ft high with two electric wires about 5 in. apart should keep them out.

Deer. Deer can cause severe damage in one night in the garden. Commercial deer repellents are available, but rains soon render them ineffective (Nelms, 1980). Various combinations of blood and bonemeal in cloth bags were similarly ineffective. If deer are a major problem, a special electric fence (Fig. 4.23) should be erected (Nelms, 1980). Deer can easily jump over standard fences, electric or otherwise.

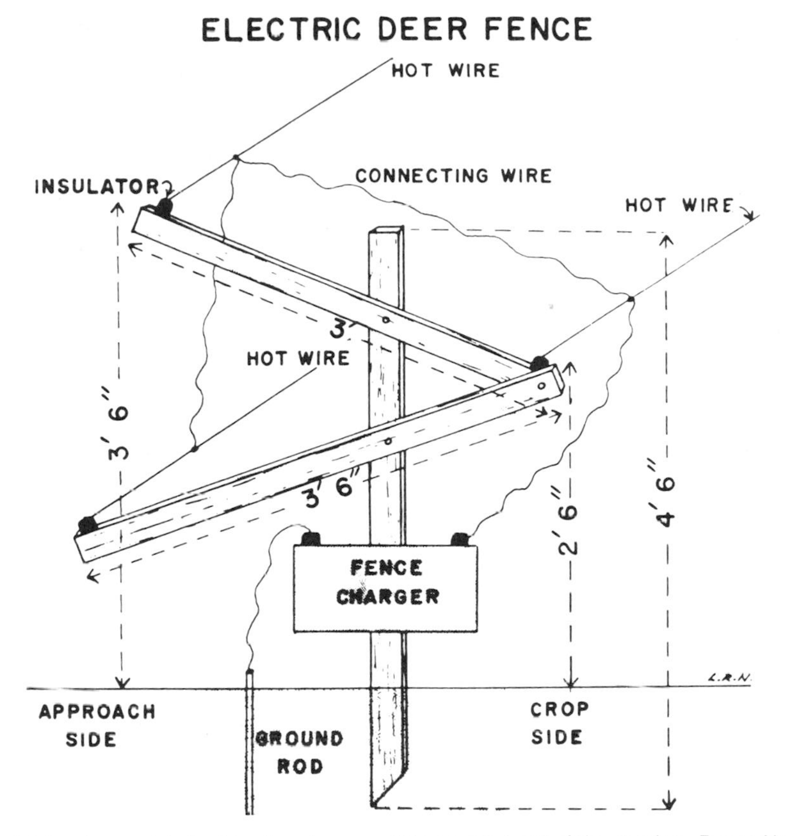

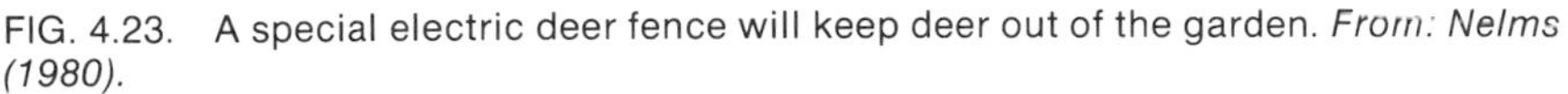
FIG. 4.23. A special electric deer fence will keep deer out of the garden. *From: Nelms (1980).*

SELECTED REFERENCES

Anon. 1978. One bite and the worm's a goner. Agrichemical Age (May).

Anon. 1988. Illinois pest control handbook. Illinois Cooperative Extension Service.

Bever, W., Bode, L., Jacobsen, B.J., McGlamery, M.D., and Moore, S., III. 1981. Illinois pesticide applicator study guide. Ill. Agric. Exp. Stn. Spec. Publ. *39*.

Binning, L.K. 1975. Vegetable gardens without weeds. *In* Weed Control in the Home Garden. C.R. Miller and F. McGourty, Jr. (Editors). Brooklyn Bot. Gard. *31* (2) 36–38.

Coates, R.M., Denissen, J.F., Juvik, J.A., and Babka, B.A. 1988. Identification of α-santalenoic and endo-β-bergamotenoic acid as moth oviposition stimulants from wild tomato leaves. J. Org. Chem. *53*:2816–2192.

Cooper, R.M. and Lindquist, P.L. 1983. Seek solution for problem with Colorado potato beetle. Ohio Rep. May:45–47.

Courter, J.W. and Vandemark, J.S. 1973. Growing tomatoes in wire cages. Ill. Res. *15* (1) 14–15.

Erhardt, W.H. and Littlefield, L.E. 1973. Natural gardening. Maine Agric. Exp. Stn. Bull. *567*.

Fletcher, R.F., Ferretti, P.A., Hepler, R.W., MacNab, A.A., and Gesell, S.G. 1972. Extension agents guide to organic gardening, culture and soil management. Pa. Agric. Exp. Stn.

Gentry, D. 1978. Proper storage methods curb pesticide poisoning. *Cited in* The Morning Courier, Champaign-Urbana, IL, June 24, 19.

Hofmaster, R.N. 1981. Light trap collection of insects and its value to agriculture. Va. Polytech. Inst. Veg. Grow. News *38* (2) 1–4.

Hopen, H.J. 1972. Controlling weeds in the home garden. Ill. Agric. Exp. Stn. Circ. *1051*.

Hoy, M.A., Ross, N.W., and Rough, D. 1978. Impact of NOW insecticides on mites in northern California almonds. Calif. Agric. *32* (5) 10–12.

Irwin, M.E. and Hendrie, L.K. 1985. The aphids are coming, Ill. Res. *27*(1):11.

Johansen, C., Eves, J., and Retan, A. 1972. Beneficial predators and parasites found on Washington crops. Wash. Agric. Exp. Stn. Bull. *646*.

Judkins, W.P. 1977. Organic gardening—Think mulch. *In* Gardening for Food and Fun. J Hays (Editor). U.S. Dep. Agric. Yearbook of Agriculture. Washington, DC.

Koehler, C.S., Barclay, L.W., and Kretchum, T.M. 1983. Pests in the home garden. Calif. Agric. *37*(9,10):11–15.

Metcalf, C.L., Flint, W.P., and Metcalf, R.L. 1951. Destructive and Useful Insects. McGraw-Hill Book Co., New York.

Nelms, L.R. 1980. Electric deer fence. Illinois Vegetable Growers Schools *18*:76–77.

Reed, L.B. and Webb, R.E. 1975. Insects and diseases of vegetables in the home garden. U.S. Dep. Agric. Bull. *380*.

Rhodes, A.M., Metcalf, R.L., and Metcalf, E.R. 1980. Diabrotocite beetle responses to cucurbitacin kairomones in cucurbita hybrids. J. Amer. Soc. Hort. Sci. *105*:838–842.

Sailer, R.I. 1972. A look at USDA's biological control of insect pests. 1888 to present. Agric. Sci. Rev. Coop. State Res. Serv. U.S. Dep. Agric. *Fourth Quarter*, 15–27.

Taber, H.G., Davison, A.D., and Trelford, H.S. 1974. Organic gardening. Wash. Agric. Exp. Stn. *EB648*.

Topoleski, L.D. 1972. Growing vegetables organically. N.Y. Agric. Exp. Stn. Bull. *39*.

Toscano, N.C., Wyman, J., Kido, K., Johnson, H., and Mayberry, K. 1979. Reflective mulches foil insects. Calif. Agric. *33* (7) 17–19.

Utzinger, J.D., Trierweiler, J., Janson, B., Miller, R.L., Saddam, A., and Crean, D.E. 1973. Let's take a look at organic gardening. Ohio Agric. Exp. Stn. Bull. *555*.

Vandemark, J.S., Splittstoesser, W.E., and Randell, R. 1973. Organic gardening can be successful if you follow sound practices. Ill. Res. *15* (2) 5.

5

Harvest and Storage of Vegetables

Vegetables should be harvested at peak quality and are often eaten fresh from the garden. However, most gardeners have a surplus of some vegetables during the season. The different vegetables differ widely in how long they can be stored fresh. Potatoes can be stored for months while eggplants can be stored only about a week. Some of these vegetables may easily be frozen or canned and provide a year-round supply of vegetables.

HARVEST

Vegetables should be harvested at the proper maturity to obtain maximum quality for immediate use or for use as a stored product. Vegetables picked at the peak of maturity and used promptly are almost always superior in nutritional content, flavor, and appearance. The difference in vegetable quality is due mainly to freshness of the produce.

Vegetables should be harvested all through the growing season rather than all at once. By harvesting the most mature vegetables first, the specific crop can be used as a fresh vegetable for an additional two or three weeks. To maintain quality, the vegetables should not be cut or bruised when harvested. After harvest, vegetables should be handled carefully, for injury will encourage decay. Unless used immediately, vegetables such as sweet corn, peas, asparagus, and leafy crops should be cooled to 40°F as soon as possible.

FREEZING VEGETABLES

Freezing is a most satisfactory method of preserving many vegetables. More of the original flavor, color, texture, and nutritive value is usually retained during freezing than when foods are preserved in any other way. Freezing, however, does not add anything to the original quality of the vegetable. Vegetables for freezing should be harvested at peak flavor and texture, the kind you would choose for immediate use. When possible, they should be harvested in the cool part of the morning and processed as soon as possible. If processing is delayed, the vegetables should be cooled in ice water or crushed ice and stored in the refrigerator to preserve flavor and quality and to prevent vitamin loss. Table 5.1 gives some suggestions on how to prepare vegetables for freezing.

Not all cultivars of vegetables freeze equally well. In general, cultivars that excel as fresh products have been found to be suitable for freezer storage.

Vegetables should be washed thoroughly in clean, cold water. All injured, bruised or substandard material should be discarded. This will remove dirt particles and reduce the number of bacteria.

Vegetables that are to be frozen must first be blanched (scalded) to inactivate enzymes, otherwise flavor and vitamins will be lost during storage. Rhubarb is satisfactory when it is frozen without blanching. Beets and winter squash are easier to handle if they are precooked rather than blanched before freezing. For tomato juice, the fruits are simmered before the juice is extracted. For tomato paste, the fruits are ground then simmered until most of the water has evaporated and the product is paste-like in consistency.

One gallon of water for each pound of vegetable is needed for blanching, except for leafy greens which need 2 gal. per pound. The water should be boiling and the blanching time counted from the moment the vegetable is placed in the boiling water. See Table 5.1 for blanching times for various vegetables. As long as the blanching water is clean and not too discolored, it may be used more than once.

After blanching, the vegetable should be cooled in cold running water or ice water for the same time used for blanching. The vegetable should be drained, packed into containers, and immediately frozen.

In two-year objective and subjective tests, microwave-blanched vegetables were not as acceptable as water-blanched vegetables. Microwave-blanched vegetables lose flavor, nutritional content, color, and texture; overall, they compare unfavorably to water-blanched vegetables.

Containers that are easy to fill and empty and occupy little freezer space are best. The more nearly moisture-proof and vapor-proof a container is, the better will be the frozen product.

Vegetables should be kept frozen until ready to use. Once frozen foods have thawed, the bacteria in them multiply and the food deteriorates in flavor, texture, and nutritive value. All vegetables may be cooked from the frozen condition, except corn on the cob, which should be partially defrosted. The frozen vegetables should be cooked in about ½ cup of boiling, salted water. They should be cooked until tender, which is about half as long as if the same vegetable were fresh. Vegetables frozen in a solid mass such as greens and spinach should be broken into smaller pieces as they are boiled.

TABLE 5.1. How to Prepare Vegetables for Freezing.

Vegetable	Preparation
Asparagus	Wash and sort medium and large stalks. Leave whole or cut in 1- or 2-in. lengths. Blanch medium stalks 3 min., large stalks (½ to ¾-in. diameter) 4 min. Cool.
Beans (green-podded)	Wash, snip off tips, and sort for size. Cut or break into suitable pieces or freeze small beans whole. Blanch 3½ min. Cool.
Beans (yellow-podded)	Process same as green-podded beans.
Beans, lima	Wash, shell, and sort. Blanch 3 min. Cool.
Beans, snap (Italian)	Wash, snap off ends, and cut or break into 1- or 1½-in. lengths. Blanch 3½ min. Cool.
Beets	Select small- or medium-sized beets. Remove tops and wash. Cook until tender. Chill. Remove skins. Slice or dice large beets.
Beet greens	Use tender, young leaves. Wash thoroughly and blanch 2 min. Cool.
Broccoli	Discard off-color heads or any that have begun to blossom. Remove tough leaves and woody butt ends. Cut stalks to fit container. Cut through stalks lengthwise, leaving heads 1 in. in diameter. Soak ½ hr. in salt brine (½ cup to 1 qt water) to drive out small insects. Rinse and drain. Blanch 4 min. in water. Steam-blanch 5 min. Cool. Pack heads and stalk ends alternately in container.
Brussels sprouts	Wash and trim. Soak ½ hr in salt brine (see broccoli). Rinse and drain. Blanch medium sprouts 4 min, large sprouts 5 min. Cool.
Carrots	Use tender carrots harvested in cool weather. Top, wash, and scrape. Dice or slice ¼ in. thick. Blanch 3½ min. Cool.
Cauliflower	Trim and wash. Split heads into individual pieces 1 in. in diameter. Soak ½ hr in salt brine (see broccoli). Rinse and drain. Blanch 4 min. Cool.
Sweet corn	Husk, remove silks, and trim ends. Use a large kettle (12 to 15 qt capacity). Blanch whole kernel corn to be cut from the cob 4½ min. For corn on the cob, 14 small ears 1¼ to 1½ in. in diameter should be blanched 8 min, and cooled 16 min; 10 ears over 1½ in. should be blanched 11 min and cooled 22 min.
Dandelion greens	Use only tender, young leaves. Wash and blanch 3 min. Cool.
Eggplant	Precooked eggplant is usually more satisfactory for freezing than blanched eggplant. Peel, cut into ¼ to ⅓ in. slices, or dice pieces immediately into cold water containing 4 tbsp salt per gallon. Blanch 4½ min in the same proportion salted water. Cool and package in layers separated by sheets of locker paper.

(Continued)

TABLE 5.1. *(Continued).*

Vegetable	Preparation
Garden herbs	Wrap sprigs or leaves in foil or seal in plastic bags and store in a carton or glass jar. Wash, but do not scald leaves.
Onions	Peel onions, wash, and cut into quarter sections. Chop. Blanch 1½ min. Cool. (They will keep 3–6 months.)
Peas, English Peas, Southern	Shell small amount at a time. Blanch 1½ to 2 min. Blanch Southern peas 2 min. Cool.
Peas (edible-podded)	Wash. Remove stems, blossom ends, and any string. Leave whole. Blanch 2½ to 3 min. Cool.
Peppers, green	Wash, cut out stem, and remove seeds. Halve, slice, or dice. Blanch halved peppers 3 min, sliced or diced ones 2 min. Cool. You can freeze chopped peppers without blanching them.
Peppers, pimiento	Oven roast at 400°F for 3 to 4 min. Cool, skin, and pack dry without additional heating.
Pumpkin	Cut or break into fairly uniform pieces. Remove seeds. Bake at 350°F, or steam until tender. Cool, scoop pulp from rind, and mash or put through dicer. You can prepare pie mix for freezing, but omit cloves.
Potatoes	Wash, peel, remove deep eyes, bruises, and green surface coloring. Cut in ¼- to ½-in. cubes. Blanch 5 min. Cool. For hash brown: Cook in jackets until almost done. Peel and grate. Form into desired shapes. Freeze. For French fries, peel and cut in thin strips. Fry in deep fat until very light golden brown. Drain and cool.
Rhubarb	Remove leaves and woody ends, wash, and cut in 1-in. lengths. Do not blanch. For sauce, pack in sugar syrup using 2½ cups sugar to 1 qt water. For pies, pack in dry sugar using 1 cup sugar to 4 cups rhubarb, or pack without sugar for a few months storge.
Soybeans	Blanch 5 min in pods, cool, and then hull.
Spinach and other greens	Sort and remove tough stems. Wash. Blanch most leafy greens 2 min. Blanch collards and stem portions of Swiss chard 3 to 4 min. Blanch very tender spinach 1½ min. Cool.
Summer squash	Wash, peel and cut in pieces. Blanch ¼ in. slices 3 min, 1½ in. slices 6 min. Cool.
Winter squash	Prepare same as pumpkin. You can blend two or more cultivars or blend squash with pumpkin.
Tomato juice	Wash, quarter, and simmer for 10 min. Strain off juice and add ½ tsp salt for each pint.
Tomato paste	Wash, grind, or blend entire fruit and simmer until most of water has evaporated. Roma types produce about ⅔ more paste.

Source: Adapted from Brill and Munson (1975) and Van Duyne (1970).
[1] Blanching times given are counted from the time the vegetable was placed in boiling water.

CANNING VEGETABLES

As in freezing, vegetables for canning must be picked at their peak flavor and color. Vegetables should be washed thoroughly in cool running water and substandard vegetables discarded. Successful home canning depends on the destruction of food spoilage agents by heat and exclusion of air from the jar. Enzymes, yeasts, molds, and bacteria cause spoilage in canned foods. These agents are inactivated by temperatures of 212°F (100°C) (boiling water) or higher. Some heat-resistant bacteria and bacterial spores of *Clostridium botulinum*, which causes food poisoning, are destroyed only at temperatures of 240°F (115.6°C) or higher. A pressure canner is needed to reach these temperatures.

For home canning purposes, tomatoes and pickles are high-acid foods, even the so-called low-acid tomato cultivars. The higher levels of acid inhibit the growth of botulism-producing bacteria but do not inhibit the growth of yeasts, molds, and some bacteria. These high-acid foods must still be canned in a boiling water bath at 212°F (100°C) (or a pressure canner) to destroy these other spoilage agents. All other vegetables must be processed in a pressure canner (Anon. 1972). Process directions are included with canning equipment or may be obtained from a State Agricultural Experiment Station (see Table 1.3 for addresses).

After canning, the seal on the jars should be checked after the containers have cooled. The self-seal lids are sealed if the lid is down and does not move when pressed with a finger. A ringing metal sound should be heard when the lid is gently tapped with a knife, indicating that there is a vacuum inside the jars.

Proper heating and correct sealing are absolutely essential for successful canning. Canned foods should be examined before using. The food should not be tasted if the container is leaky, the lid bulges, the seal is faulty, liquid spurts when the container is opened, the food has a peculiar odor, or mold is present. The food should be discarded. As a safeguard against botulism, all home-canned, low-acid vegetables should be boiled at least 10 minutes (spinach and corn 20 minutes) before tasting. If the material has an unusual odor or becomes foamy, it should be discarded.

FRESH STORAGE OF VEGETABLES

Vegetables may be stored fresh in a number of different storage facilities. All storage facilities should be above freezing temperatures during winter storage, and most vegetables are stored under cool conditions. Cool temperatures retard respiration of the vegetables; slow down aging due to ripening which results in softening and color changes; prevent moisture loss and wilting or shrivelling; retard spoilage due to various fungi, bacteria, and molds; and prevent sprouting of potato tubers, onion bulbs, and their relatives.

A well-ventilated basement under a house with central heat may be used for short-term storage of potatoes, sweet potatoes, and onions and for ripening tomatoes. For long-term storage, an unheated basement is preferable. If the basement is heated, a well-ventilated area may be partitioned off and insulated to keep it cool.

Outdoor storage cellars can be constructed partly or entirely below ground. Cellars constructed belowground are preferable because they maintain a more uniform and desirable temperature longer (Fig. 5.1). Storage of vegetables in aboveground buildings is practical only where the climate is consistently cold. A thermostatically controlled heater is necessary on cold nights to prevent the vegetables from freezing.

Many vegetables will keep for a short time in cold storage, but others require warm temperatures as shown in Table 5.2. Gardeners can harvest small amounts of vegetables and keep many of them in the refrigerator. Most refrigerators are set at 35°–40°F (1.7–4.4°C). If a high humidity is required, the vegetables can be placed in plastic bags or containers and sealed to prevent water loss and wilting. Cucumbers in particular shrivel badly if kept under dry conditions. If a large amount of vegetables is to be stored under cold or cool and moist conditions, an older model refrigerator without an automatic defrosting device may be used. Automatic defrosting devices remove water from the refrigerator, resulting in lowered humidities and wilted vegetables. Small amounts of many vegetables can be stored fresh in these refrigerators for several weeks; but as many gardeners will harvest more vegetables than the refrigerator will hold, they may wish to store them for several months. In this case a basement room, outdoor pit, or aboveground building may be constructed. Various plans are available from the U.S. Dept. of Agriculture and the Cooperative Extension Service.

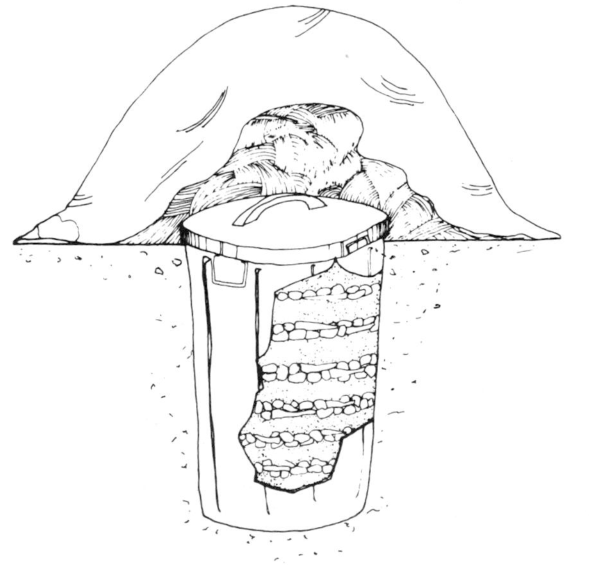

FIG. 5.1. A large garbage container packed with alternating layers of straw and vegetables for outdoor storage. *From Herner (1975).*

TABLE 5.2. Storage Recommendations and the Approximate Length of the Storage Period.

Vegetable	Temperature (°F)	Humidity (%)	Approximate length of storage period
Cold moist storage			
Asparagus	32–35	85–90	2–3 weeks
Beets, topped	32	95	3–5 months
Broccoli	32–35	90–95	10–14 days
Brussels sprouts	32–35	90–95	3–5 weeks
Cabbage, late	32	90–95	3–4 months
Cabbage, Chinese	32	90–95	1–2 months
Carrots, mature and topped	32–35	90–95	4–5 months
Cauliflower	32–35	85–90	2–4 weeks
Celeriac	32	90–95	3–4 months
Celery	32–35	90–95	2–3 months
Collards	32–35	90–95	10–14 days
Corn, sweet	32–35	85–90	4–8 days
Endive, escarole	32	90–95	2–3 weeks
Greens, leafy	32	90–95	10–14 days
Horseradish	30–33	90–95	10–12 months
Kale	32	90–95	10–14 days
Kohlrabi	32	90–95	2–4 weeks
Leeks, green	32	90–95	1–3 months
Lettuce	32–35	90–95	2–3 weeks
Onions, green	32–35	90–95	3–4 weeks
Parsnips	32–35	90–95	2–6 months
Peas	32–35	85–90	1–3 weeks
Potatoes, late crop	35–40	85–90	4–9 months
Radish	32–35	90–95	3–4 weeks
Rhubarb	32–35	90–95	2–4 weeks
Rutabagas	32–35	90–95	2–4 months
Spinach	32–35	90–95	10–14 days
Turnips	32	90–95	4–5 months
Cool moist storage			
Beans, snap	40–45	90–95	7–10 days
Beans, lima	32–40	90	1–2 weeks
Cantaloupe	40	85–90	15 days
Cucumbers	40–50	85–90	10–14 days
Eggplant	40–50	85–90	1 week
Okra	45	90–95	7–10 days
Peppers, sweet	40–50	85–90	2–3 weeks
Potatoes, early	50	90	1–3 weeks
Potatoes, late	40	90	4–9 months
Squash, summer	40–50	90–95	5–14 days
Tomatoes, ripe	40–50	85–90	4–7 days
Tomatoes, unripe	60–70	85–90	1–3 weeks
Watermelon	40–50	80–85	2–3 weeks
Cool dry storage			
Beans, dry	32–40	40	Over 1 year
Garlic, dry	32	65–70	6–7 months
Onions, dry	32	65–70	1–8 months
Peas, dry	32–40	40	Over 1 year
Peppers, chili, dry	32–50	60–70	6 months
Shallots, dry	32	60–70	6–7 months
Warm dry storage			
Pumpkins	55–65	40–70	2–4 months
Squash, winter	55–65	40–70	3–6 months
Sweet potato	55–60	70–85	4–6 months
Tomatoes, unripe	55–70	85–90	1–3 weeks

Source: Adapted from Wright *et al.* (1954).

SELECTED REFERENCES

Anon. 1972. Ball Bluebook. Ball Corp., Muncie, IN.

Brill, G.D. and Munson, S.T. 1975. Freezing fruits and vegetables. Minn. Agric. Exp. Stn. Ext. Folder. *156*.

Herner, R. 1975 Research on ethylene yields storage tips. *In* Michigan Science in Action. Mich. Agric. Exp. Stn. Sci. Home Gard. *28*.

Van Duyne, F.O. 1970. How to prepare fruits and vegetables for freezing. Ill. Agric. Exp. Stn. Circ. *602*.

Wright, R.C., Rose, D.H., and Whiteman, T.M. 1954. The commercial storage of fruits, vegetables and florist and nursery stock. U.S. Dep. Agric. Handb. *66*.

Nutritional Value of Vegetables

Vegetables play an important role in meeting the needs of humans for vitamins and minerals. Some vegetables, such as potatoes and sweet potatoes, may also make a significant contribution as a source of calories. Peas and beans are an important source of vegetable protein. Some vegetables also possess laxative qualities. People consume vegetables as fresh (45%), canned (45%) and frozen (10%) vegetables and their use in diets have been increasing (Table 6.1).

The major food category consumed (Table 6.1) is dairy products, including milk (25 gal), cheese (22.3 lbs), ice cream (17 lbs), butter (4 lbs), and binders in processed meats. As potatoes are one of the major foods used worldwide (Splittstoesser 1977), the USDA lists it separately and potatoes are not included in the vegetable category. The major fruits consumed are bananas (22 lbs), oranges (21 lbs), and apples (20 lbs). Sweeteners include sugar (65 lbs), corn syrup (45 lbs), candy (18 lbs), and artificial sweeteners (7.1 lbs). Oils include soybean oil (42 lbs), margarine (11 lbs), and lard. Included in potatoes are potato chips (17 lbs). Liquids include coffee (565 cups), soft drinks (612 cans), and others.

Although over 3000 plant species are utilized by humans as food, only about 150 are commercially important. The only vegetables consumed in the top ten foods worldwide are the potato (4th), various beans (5th), and the sweet potato (9th) (Splittstoesser 1977). Vegetables are an excellent source of vitamins and minerals. Vegetables have a high water content, and the energy content is usually low. Thus, vegetables have a relatively high content of nutrients per unit of energy (Table 6.2). Humans require about 2500 calories per day. For those with a low energy requirement, due to a limited physical activity, vegetables are essential to balance the diet and to prevent an overweight condition. Vegetables such as tomatoes are low in nutrient content and rank fourteenth, but they are consumed in such large quantities that tomatoes rank number one in their nutrient contribution to the diet. The nutrient content of our diets can easily increased by the proper selection of vegetables in our diets (Table 6.2).

QUALITY

Vegetable quality is not restricted to a single characteristic. The consumer looks at external quality (size, shape, color, cosmetic—free from blemishes, etc.). "Quality is the composite of attributes that differentiate among units of a

TABLE 6.1. Average U.S. Per Capita Consumption of Twelve Major Food Categories[1].

Food	Rank	Lbs/person
Dairy products	1	264.9
Vegetables, except potatoes	2	172.7
Cereal, flour	3	151.4
Sweetners	4	138.3
Poultry	5	82.8
Oil	6	73.6
Beef	7	72.8
Potatoes	8	70.6
Fruit	9	66.0
Pork	10	63.4
Fish	11	13.0
Various liquids	12	330.5
Total		1500

Adapted from Anon. (1989).
[1] Values include juices and are retail weights. They do not take into account home trimming, peeling, shrinkage during cooking, bones, or table waste.

product and have significance in determining the degree of acceptability of the unit by the buyer" (Nilsson 1978). The emphasis on the cosmetic value of a vegetable does not always equate with the body's need for a high nutritive value and an acceptably low level of harmful or unhealthy substances. Potatoes may have some scab on the skin, for example, if they are to be peeled anyway.

TABLE 6.2. Sixteen Common Vegetables Ranked According to Their Total Nutrient Concentration and Their Contribution of Nutrients to the Diet, Based on Consumption.

Nutrient content		Contribution of nutrients to diet	
Vegetable	Rank	Vegetable	Rank
Broccoli	1	Tomato	1
Spinach	2	Potato	2
Brussels sprouts	3	Lettuce	3
Lima beans	4	Sweet corn	4
Peas	5	Carrots	5
Asparagus	6	Cabbage	6
Artichoke, globe	7	Onion	7
Cauliflower	8	Sweet potato	8
Sweet potato	9	Peas	9
Carrots	10	Spinach	10
Sweet corn	11	Broccoli	11
Potato	12	Lima beans	12
Cabbage	13	Asparagus	13
Tomato	14	Cauliflower	14
Lettuce	15	Brussels sprouts	15
Onion	16	Artichoke, globe	16

External quality also implies that a vegetable is at the normal stage of maturity and that it contains the normal amount of nutrients.

Utility quality is the suitability of vegetables for commercial use. About 55% of the vegetable consumed are in some processed product—canned, frozen, individual meals, etc. The color of peas is important in the frozen product and the water content of tomatoes influences the amount of tomato paste produced. The size and shape of onions and tomatoes influences their slicing utility and determines whether they will be used in fast food chains. Utility quality does not equate with nutritional quality (Nilsson 1978), but processors are placing more emphasis on nutrition, as food containers now show the ingredients and nutritional value of the product.

Nutritional quality includes not only fat, protein, carbohydrates, calories (see Table 6.3), and vitamins (see Table 6.4), but also the ability of the product to maintain human health. For example, the balance of individual amino acids within a protein is more important than the total amount of protein. Gelatin is mostly protein, but contains amino acids which are not required to be ingested for human health. Nutritional quality is a complex term that cannot be defined as the presence of a few chemical substances in a vegetable (Nilsson 1978).

Nutritional quality is influenced by many factors. Soil fertility affects the quality of most vegetables. Generally, the proper soil fertility increases the size, yield, and quality of vegetables. Fertilizers tend to correct soil nutrient deficiencies that would otherwise reduce vegetable quality. Moisture stress induces fiber formation, resulting in tough root crops, celery stalks, and leaf tissue of salad vegetables. Erratic irrigation can lead to tomato fruit rots, cracking in melons and cabbage heads, leaf burn in lettuce, and regrowth of onion bulbs and potatoes. Harvest time influences quality, and spring-grown beets contain more sugar, but less red color, than fall-grown ones. Crops harvested in the fall store best. Pesticides may influence quality and taste (Courter 1979). Soap residues may be left upon the leaves from soap insecticide sprays. Generally, the application of insecticide sprays at the wrong time or at excessive rates reduces quality. Storage and processing affect the nutritional value, and vitamins are affected most. Each vitamin is affected differently. Niacin is relatively unaffected by most treatments. Cold storage usually improves the retention of vitamins. Iced asparagus lost 15% of its vitamin C after 30 hr, while that kept at room temperature lost 50%. Potatoes stored at room temperature have starch converted to sugar, reducing quality. Blanching may leach out vitamins and solids (Quebedeaux and Bliss 1988). Food preparation also affects the nutritional value. Boiling in salted water removes vitamins and minerals. The salt draws out the plant juices and the vitamins with it. Water should be boiling before the vegetable is added, and it can be salted later, if needed. The use of a microwave with minimal amounts of water is better. Baking is best. In many root vegetables, the minerals are concentrated under the skin and peeling the vegetables removes them. Many vegetables should be washed and then eaten, unpeeled and raw, to preserve the most nutritional value. Almost everything that the gardener does during the growth, storage, and food preparation of a vegetable influences nutritional quality.

TABLE 6.3. Nutritional Value of the Edible Part of Vegetables[1].

Vegetable	Weight (g)	Calories	Protein (g)	Fat (g)	Carbohydrate (mg)	Calcium (mg)	Iron (mg)
Asparagus	145	94	30	3	5	30	0.9
Beans, dry							
Great Northern	180	210	14	1	38	90	4.9
navy or pea	190	225	15	1	40	95	5.1
lima	190	260	16	1	49	55	5.9
Beans, fresh							
lima	170	190	13	1	34	80	4.3
snap, green	125	30	2	T	7	63	0.8
snap, yellow	125	30	2	T	6	63	0.8
Bean, mung, sprouts	125	35	4	T	7	21	1.1
Beet	170	55	2	T	12	24	0.9
Beet, greens	145	25	3	T	5	144	2.8
Broccoli	155	40	5	1	7	136	1.2
Brussels sprouts	155	55	7	1	10	50	1.7
Cabbage, raw							
common cultivars	70	15	1	T	4	34	0.3
red	70	20	1	T	5	29	0.6
savoy	70	15	2	T	3	47	0.6
celery or Chinese	75	10	1	T	2	32	0.5
Cabbage, cooked							
common cultivars	145	30	2	T	6	64	0.4
Chinese	170	25	2	T	4	252	1.0
Carrot							
raw	110	45	1	T	11	41	0.8
cooked	145	45	1	T	10	48	0.9
Cauliflower	120	25	3	T	5	25	0.8
Celery, raw	100	15	1	T	4	39	0.3
Collard	190	55	5	1	9	289	1.1
Corn, sweet	140	92	3	1	20	2	0.5
Cucumber, raw, pared							
6⅛ in. center slices	50	5	T	T	2	8	0.2
Dandelion greens	180	60	4	1	12	252	3.2
Endive, escarole							
2 oz	57	10	1	T	2	46	1.0
Kale	110	30	4	1	4	147	1.3

Lettuce, raw							
butterhead or Boston							
1 head (4 in.)	220	30	3	T	6	77	4.4
crisphead or iceberg							
1 head (4¾ in.)	454	60	4	T	13	91	2.3
looseleaf—2 large leaves	50	10	1	T	2	34	0.7
Mustard greens	140	35	3	1	6	193	2.5
Okra, 8-3 × ⅝ in. pods	85	25	2	T	5	78	0.4
Onion, cooked	210	60	3	T	14	50	0.8
Onion, raw							
6 young, green—no tops	50	20	1	T	5	20	0.3
1 mature 2½ in.	110	40	2	T	10	30	0.6
Parsnip	155	100	2	1	23	70	0.9
Pea, English	160	115	9	1	19	37	2.9
Pea, Southern	160	175	13	1	29	38	3.4
Pepper, sweet							
1 raw, green pod—no stem or seeds	74	15	1	T	4	7	0.5
1 cooked pod	73	15	1	T	3	7	0.4
Potato, medium 5 oz							
baked, then peeled	99	90	3	T	21	9	0.7
boiled, then peeled	136	105	3	T	23	10	0.8
peeled, then boiled	122	80	2	T	18	7	0.6
mashed, with milk, 1 cup	195	125	4	1	25	47	0.8
Pumpkin	228	75	2	1	18	57	0.9
Radish, raw, 4 small	40	5	T	T	1	12	0.4
Spinach	180	40	5	1	6	167	4.0
Squash							
summer	210	30	2	T	7	52	0.8
winter, baked	205	130	4	1	32	57	1.6
Sweet potato, medium 6 oz							
baked, then peeled	110	155	2	1	36	44	1.0
boiled, then peeled	147	170	2	1	39	47	1.0
canned	218	235	4	T	54	54	1.7
Tomato							
raw, 1 medium, 7 oz	200	40	2	T	9	24	0.9
juice	243	45	2	T	10	17	2.2
Turnip	155	35	1	T	8	54	0.6
Turnip greens	145	30	3	T	5	252	1.5

Source: Anon (1971).

[1] Unless otherwise stated, the vegetable was cooked, drained, and 1 cup used. T—Trace found. 1000 mg = 1 g. 1 ounce (oz) = 28.35 g.

TABLE 6.4. Vitamin Content of the Edible Part of Vegetables[1].

Vegetable	Vitamin A (IU)	Thiamin (mg)	Riboflavin (mg)	Niacin (mg)	Vitamin C (mg)
Asparagus	1,310	0.23	0.26	2.0	38
Beans, dry					
Great Northern	0	0.25	0.13	1.3	0
navy or pea	0	0.27	0.13	1.3	0
lima	0	0.25	0.11	1.3	2
Beans, fresh					
lima	480	0.31	0.17	2.2	29
snap, green	680	0.09	0.11	0.6	15
snap, yellow	290	0.09	0.11	0.6	16
Bean, mung, sprouts	30	0.11	0.13	0.9	8
Beet	30	0.05	0.07	0.5	10
Beet, greens	7,400	0.10	0.22	0.4	22
Broccoli	3,880	0.14	0.31	1.2	140
Brussels sprouts	810	0.12	0.22	1.2	135
Cabbage, raw					
Common cultivars	90	0.04	0.04	0.2	33
red	30	0.06	0.04	0.3	43
savoy	140	0.04	0.06	0.2	39
celery or Chinese	110	0.04	0.03	0.5	19
Cabbage, cooked					
common cultivars	190	0.06	0.06	0.4	48
Chinese	5,270	0.07	0.14	1.2	26
Carrots					
raw	12,100	0.06	0.06	0.7	9
cooked	15,220	0.08	0.07	0.7	9
Cauliflower	70	0.11	0.10	0.7	66
Celery, raw	240	0.03	0.03	0.3	9
Collard	10,260	0.27	0.37	2.4	87
Corn, sweet					
yellow cultivar	310	0.09	0.08	1.0	7
white cultivar	T	0.09	0.08	1.0	7
Cucumbers, raw, pared					
6⅛ in. center slices	T	0.02	0.02	0.1	6
Dandelion greens	21,060	0.24	0.29	—	32
Endive, escarole					
2 oz	1,870	0.04	0.08	0.3	6
Kale	8,140	—	—	—	68
Lettuce, raw					
butterhead or Boston					
1 head (4 in.)	2,130	0.14	0.13	0.6	18
crisphead or iceberg					
1 head (4¾ in.)	1,500	0.29	0.27	1.3	29
looseleaf, 2 large leaves	950	0.03	0.04	0.2	9
Mustard greens	8,120	0.11	0.19	0.9	68
Okra, 8-3 × ⅝ in. pods	420	0.11	0.15	0.8	17
Onion, cooked	80	0.06	0.06	0.4	14
Onion, raw					
6 young, green—no tops	T	0.02	0.02	0.2	12
1 mature 2½ in.	40	0.04	0.04	0.2	11
Parsnip	50	0.11	0.12	0.2	16
Pea, English	860	0.44	0.17	3.7	33
Pea, Southern	560	0.49	0.18	2.3	28
Pepper, sweet					
1 raw, green pod—no stem or seeds	310	0.06	0.06	0.4	94
1 cooked pod	310	0.05	0.05	0.4	70

TABLE 6.4. *(Continued)*.

Vegetable	Vitamin A (IU)	Thiamin (mg)	Riboflavin (mg)	Niacin (mg)	Vitamin C (mg)
Potato, medium 5 oz					
baked, then peeled	T	0.10	0.04	1.7	20
boiled, then peeled	T	0.13	0.05	2.0	22
peeled, then boiled	T	0.11	0.04	1.4	20
mashed with milk, 1 cup	50	0.16	0.10	2.0	19
Pumpkin	14,590	0.07	0.12	1.3	12
Radish, raw, 4 small	T	0.01	0.01	0.1	10
Spinach	14,580	0.13	0.25	1.0	50
Squash					
summer	820	0.10	0.16	1.6	21
winter, baked	8,610	0.10	0.27	1.4	27
Sweet potato, medium, 6 oz					
baked, then peeled	8,910	0.10	0.07	0.7	24
boiled, then peeled	11,610	0.13	0.09	0.9	25
canned	17,000	0.10	0.10	1.4	30
Tomato					
raw, 1 medium, 7 oz	1,640	0.11	0.07	1.3	42
juice	1,940	0.12	0.07	1.9	39
Turnip	T	0.06	0.08	0.5	34
Turnip greens	8,270	0.15	0.33	0.7	68

Source: Anon. (1971).

[1] Unless otherwise stated, the vegetable was cooked, drained, and 1 cup used. T—Trace found. Dashes in the columns show that no suitable value could be found although a measurable amount of the vitamin may be present. 1000 mg = 1 g. 1 ounce (oz) = 28.35 g.

WATER

Water is not considered a nutrient, but it is required to maintain the proper osmotic concentration in cells and to remove metabolic waste products from the body. Humans should drink about 2½ pints of liquid per day. In addition, about ½ pints of liquid are eaten as part of food products. The body also releases water (½ pint) during the metabolism of starches, fats, and protein.

This 5½ pints of water are removed daily by waste products (3 pints) and evaporation from the skin and lungs (2½ pints). The body releases about one pint of water from the lungs while sleeping.

CARBOHYDRATES

The carbohydrate content per fresh weight is generally low in most vegetables, except potatoes and some root crops, such as sweet potatoes and beets. Beets contain about 15% sugar, which is similar to sugar beets. Most vegetables have low amounts of calories (Table 6.3).

Sweetness is dependent upon the amount of glucose, fructose, and sucrose present. In many vegetables, such as sweet corn, these three sugars comprise about half of the dry matter content. The proper ratio of these sugars and citric acid and malic acid is required for the acceptable taste of tomatoes. These sugars are more important in the taste of vegetables than for their calories.

Some vegetables contain the sugars starchyose and raffinose, which are not digested by humans. Instead they reach the large intestines, where they are fermented by anaerobic bacteria, causing intestinal disturbances and flatulence. This is a problem with various legume seeds such as the mung bean and soybean. These seeds may be germinated as sprouts, as these two sugars are metabolized by the seed and disappear by six days (Adjei-Twum et al. 1976). Beans are usually cooked to destroy trypsin inhibitors, an antinutritional factor.

Much of the sugar from carbohydrates that is absorbed by the body is converted to fat, which the muscles use as fuel for mechanical energy. The nervous system also uses considerable amounts of energy to maintain itself in working order, and 20% of the basal metabolism takes place in the brain. The nervous system appears to get its energy entirely from carbohydrates; as it seems to have no important source of stored energy. It must depend entirely upon the carbohydrates supplied by the bloodstream. Vegetables can provide these carbohydrates.

FIBER

Plant materials that are indigestible by humans are loosely defined as dietary fiber. They consist mostly of nondigestible carbohydrates such as celluloses, hemicelluloses, pectins, polysaccharides, and pentosans; and substances such as lignin. These substances have a high water-binding capacity. Some fiber types bind cholesterol and lower blood cholesterol levels. Fiber enhances the reabsorption of bile and helps prevent constipation. Fiber may also reduce the incidence of colon cancer (Quebedeaux and Bliss 1988) and cardiovascular (heart) disease, and reduces the rate of glucose absorption (Hoff 1985); it has been used clinically to treat diabetes.

Not all fiber acts alike. Bread made with purified cellulose (wood pulp) doesn't have the beneficial effect upon cholesterol and glucose absorption, and it reduces mineral absorption. High quantities of fiber in the diet reduces the absorption of calcium, magnesium, fat, and nitrogen needed by the body. High fiber levels induce gas and bloating, and too much fiber may be harmful (Hoff 1985). The National Cancer Institute has recommended the consumption of 30–40 grams of fiber per day from a variety of sources. Some vegetables (Table 6.5) are good sources of fiber, but seemingly minor amounts in vegetables should not be ignored (Hoff 1985).

FAT

Fats provide a convenient, concentrated source of calories, and diets should have about 15% of the total calories derived from fats. Fats may be solid or liquid (oils) at room temperature. Fats are made up of glycerol and fatty acids;

TABLE 6.5. Fiber Content in Selected Vegetables.

Vegetable	Serving	Fiber (grams)
Beans, baked	1 cup	18.6
Beans, green	1 cup	3.5
Broccoli	1 cup	5.6
Cabbage, as slaw	1 cup	1.9
Cantaloupe	$\frac{1}{4}$ melon	2.5
Carrots	1 cup	3.2
Cauliflower	1 cup	2.5
Celery	1 stalk	0.7
Cucumber	6 slices	0.2
Lettuce	1 cup	0.8
Onion	1 small	1.4
Peas	1 cup	11.3
Potato, boiled	1 medium	1.4
Spinach	1 cup	3.5
Tomato	1 medium	3.0

Source: Hoff (1985).

vegetable oils have a large diversity in their fatty acid composition, both in the length of the fatty acid and in the degree of unsaturation of the fatty acid. Corn and sunflower oils contain a high percentage of unsaturated fatty acids. Most vegetables do not contain large amounts of fats (Table 6.3).

Cholesterol and other sterols comprise a small, but important part of the total fat in food. Cholesterol occurs only in products of animal origin and does not occur in vegetables. Cholesterol is essential for humans and it is synthesized within the body in normal metabolic processes. Thus, cholesterol supplied by the diet is sometimes referred to as exogenous cholesterol.

PROTEINS AND AMINO ACIDS

All living matter contains proteins. They serve as the building blocks of living cells, and are found in enzymes, many hormones, antibodies, in association with genetic material, and have many other functions. The common component of proteins are the amino acids. Some amino acids can be produced by the human body; others cannot and are called "essential" amino acids. Humans must obtain the essential amino acids from animal or vegetable proteins. The separation between essential and nonessential amino acids is not always clear. Arginine and histidine, nonessential amino acids in adults, are not formed fast enough in infants, and must be supplied in the diet. Tyrosine may be nonessential, if enough phenylalanine is supplied; and cystine may substitute for about 30% of the methionine requirement. The amount of the essential amino acids required in adults is given in Table 6.6

Phenylalanine and tyrosine contain a benzene ring, which is used to make the hormones adrenaline and thyroxine. They are also used to make melanin, a pigment found in hair, eye color, and skin. About one in ten thousand infants have a genetic disorder known as phenylketonuria. These infants cannot metabolize phenylalanine normally and dietary levels must be controlled.

TABLE 6.6. Approximate Amounts of Essential Amino Acids Found in the Edible Part of Vegetables, Compared to that Required[1].

	Amino acids (grams)									
	Try	Thre	Isol	Ieu	Lys	Meth	Cys	Phen	Tyr	Val
	(Required/day)									
Male adult	0.25	0.50	0.70	1.10	0.80	0.60	0.55	1.10		0.80
Female adult	0.16	0.31	0.45	0.62	0.50	0.29	0.26	0.22	0.90	0.65
	(Found in Vegetables)									
Asparagus	0.02	0.06	0.08	0.09	0.10	0.02	0.02	0.05	0.03	0.08
Beans,										
Kidney	0.04	0.16	0.17	0.28	0.22	0.04	0.04	0.20	0.13	0.20
Lima	0.70	0.24	0.36	0.44	0.37	0.06	0.07	0.28	0.18	0.35
Snap	0.01	0.04	0.03	0.06	0.05	0.01	0.01	0.03	0.02	0.05
Beets	0.01	0.03	0.03	0.04	0.03	0.01	0.01	0.03	0.02	0.03
Broccoli	0.02	0.08	0.09	0.11	0.12	0.03	0.02	0.07	0.05	0.11
Brussels sprouts	0.01	0.02	0.02	0.02	0.02	0.01	T	0.02	T	0.03
Cabbage, green	0.01	0.03	0.04	0.04	0.03	0.01	0.01	0.04	0.03	0.06
Chinese	0.01	0.04	0.08	0.08	0.08	0.01	0.01	0.04	0.03	0.06
Carrots	0.01	0.03	0.03	0.04	0.03	0.01	0.01	0.03	0.02	0.04
Cauliflower	0.02	0.04	0.04	0.07	0.06	0.02	0.01	0.04	0.03	0.06
Celery	0.01	0.01	0.01	0.02	0.02	T	T	0.01	0.01	0.01
Collard	0.01	0.04	0.04	0.07	0.05	0.01	0.01	0.04	0.03	0.05
Corn, sweet	0.02	0.11	0.11	0.29	0.12	0.06	0.02	0.13	0.10	0.16
Cucumber (raw)	T	T	T	0.01	0.01	T	T	0.01	0.01	0.01
Eggplant	T	0.01	0.02	0.03	0.02	T	T	0.02	0.01	0.02
Endive (raw)	T	0.01	0.02	0.03	0.02	T	T	0.01	0.01	0.02
Kale	0.02	0.06	0.07	0.09	0.07	0.01	0.02	0.06	0.04	0.07
Kohlrabi	0.01	0.04	0.07	0.06	0.05	0.01	0.01	0.03	—	0.04
Leek (raw)	0.01	0.03	0.03	0.05	0.04	0.01	0.01	0.02	0.02	0.03
Lettuce (raw)										
Cos//Romaine	T	0.02	0.03	0.03	0.03	0.01	0.01	0.02	0.01	0.02
Mustard greens	0.02	0.04	0.06	0.05	0.07	0.02	0.02	0.04	0.08	0.06
Okra	0.01	0.05	0.05	0.08	0.06	0.02	0.01	0.05	0.07	0.07
Onion	0.01	0.02	0.03	0.03	0.05	0.01	0.02	0.02	0.02	0.02
Pea										
Edible-Podded	0.03	0.09	0.15	0.21	0.19	0.01	0.03	0.08	0.09	0.26
English	0.03	0.16	0.15	0.26	0.26	0.07	0.03	0.16	0.09	0.19
Pepper, sweet (raw)	0.01	0.02	0.01	0.02	0.02	0.01	0.01	0.01	0.01	0.02
Potato										
Baked, then peeled	0.02	0.04	0.05	0.07	0.07	0.02	0.02	0.05	0.05	0.07
Peeled, then boiled	0.02	0.05	0.06	0.08	0.08	0.02	0.02	0.06	0.05	0.08
Pumpkin	0.01	0.03	0.03	0.04	0.05	0.01	T	0.03	0.04	0.03
Radish (raw)	T	0.02	0.02	0.02	0.02	T	T	0.01	0.01	0.02
Spinach	0.04	0.11	0.14	0.21	0.16	0.05	0.03	0.12	0.10	0.15
Squash										
Summer	0.01	0.02	0.03	0.05	0.05	0.01	0.01	0.03	0.02	0.04
Winter, baked	0.01	0.03	0.04	0.05	0.03	0.01	0.01	0.04	0.03	0.04
Sweet potato										
Baked	0.02	0.09	0.09	0.13	0.09	0.04	0.01	0.10	0.07	0.11
Peeled, then boiled	0.03	0.13	0.13	0.20	0.13	0.07	0.02	0.16	0.11	0.18
Tomato (raw)	0.01	0.02	0.02	0.03	0.03	0.01	0.01	0.02	0.01	0.02
Turnip	0.01	0.02	0.02	0.02	0.02	0.01	T	0.01	0.01	0.02
Turnip greens	0.01	0.05	0.04	0.08	0.05	0.02	0.01	0.05	0.03	0.06

Adapted from: Haytowitz and Matthews (1986).

[1] Based upon ½ cup, cooked and drained. Try = tryptophan; Thre = threonine; Isol = Isoleucine; Leu = leucine; Lys = Lysine; Meth = methionine; Cys = Cystine; Phen = phenylalanine; Tyr = tyrosine; Val = Valine; T = Trace

Tryptophan is metabolized and is used in blood platelets, which help prevent bleeding. Cystine and methionine are sulfur-containing amino acids. They help make a S-S bond, which represents about 12% of the amino acids found in hair and insulin. Lysine is deficient in most vegetable proteins except peas and beans.

The essential amino acids should be present in the proper ratio required by humans, and this is known as protein quality (Splittstoesser, 1977). If the amount of one essential amino acid is limited, protein synthesis stops, if that amino acid is required for a certain enzyme or hormone. If a given protein contains only half of the required amount of an essential amino acid, doubling the quantity of protein consumed will not provide optimum growth; but providing the amount of the essential amino acid will. This is suggested as the reason for the pellagra-like condition of humans who consume meals composed mostly of corn, which is deficient in lysine and tryptophan, even though the quantity of protein consumed is adequate to supply the required amount of lysine and tryptophan.

By the proper selection and combination of plant proteins (Tables 6.3 and 6.6), it is possible to obtain the necessary supply of essential amino acids required by humans. Several incomplete vegetable proteins can be combined in a meal to form a complete protein, or a small amount of a complete animal protein can be used as a supplement. Examples would be combinations of legumes (pea, bean, peanut) with grains (rice, corn, wheat) or with nuts. Eggs and dairy products go well with any vegetable protein. Vegetarians who do not consume any animal foods will receive inadequate amounts of Vitamin B_{12}, unless they use a supplement. The diet must still contain enough calories to support ones ideal weight.

VITAMINS

Vitamins are required for normal body functions, and some may be provided in a precursor form. Fat-soluble vitamins are stored in the body, but water-soluble vitamins are not, and must be supplied daily. Several, such as vitamins D and B_{12}, are not found in plants and must be supplied from other sources. Several compounds listed as vitamins, such as vitamin K, pyridoxine, and pantothenic acid, are distributed in so many food sources that its deficiency is highly unlikely. Vegetables (Table 6.4) supply major amounts of the fat-soluble vitamin A and the water-soluble vitamins C, thiamin, riboflavin, and niacin. If vitamin supplements are used, they should be taken with the main meal for maximum absorption.

Carotene, a vitamin A precursor, is orange in color and is found in carrots, sweet potatoes, tomatoes, and leafy vegetables. Carotene is very stable during the handling and storage of most vegetables. Vitamin A is essential for the maintenance of cells. It maintains resistance to infections, increases life, delays senility, and is a necessary part of the reactions that occur in the eye for vision. About 4000 international units (IU–Table 6.4) are recommended daily.

About 60 mg of vitamin C is required daily by humans; which is rather high for a vitamin (Nilsson, 1978). Leafy vegetables, sweet potatoes, and tomatoes are good sources of vitamin C. In many areas, enough potatoes are eaten to supply this vitamin. Vitamin C is essential for good teeth and bone formation and the healing of wounds. Some results suggest that vitamin C also helps reduce stress, prevents colds, and reduces skin disorders.

Thiamin (vitamin B_1) is essential for maintaining a good appetite and normal digestion. It is necessary for growth, fertility, lactation, and normal functioning of the heart and nervous tissue. The thiamin requirement is related to the calories consumed, with large-calorie diets requiring more thiamin. Most vegetables are good sources of thiamin, and about 1 mg is recommended per day.

Riboflavin (vitamin B_2) is involved in numerous functions in cellular metabolism. It is present in eye pigments that are involved in light adaptation. About 0.5 to 3 mg of riboflavin is needed each day. Many vegetables are poor sources, but leafy vegetables, peas, and beans are good sources (Table 6.4).

Niacin is involved in the biochemical machinery of the cells. It is not destroyed in ordinary cooking processes, and about 10 mg of niacin is required each day. Most vegetables are low in niacin, but meat, wheat germ, and yeast are excellent sources.

MINERALS

Many minerals are important in body functions and sodium, chloride and iodine may be supplied by table salt. The most important minerals found in vegetables are calcium, potassium, phosphorus, sulfur, iron, manganese, magnesium and zinc. About 750mg of calcium are recommended per day and this can easily be supplied by milk or cheese; however, many vegetables are excellent sources of calcium (Table 6.3).

Most diets contain adequate amounts of minerals. Animal foods have a low content of potassium and a high content of sodium. The increased use of vegetables in the diet will make the potassium-sodium levels about equal, which may help prevent high blood pressure. Vegetables and fruits supply about 25% of the required magnesium and about 20% of the iron requirement. The iron content of vegetables depends upon the soil and other conditions under which the food was grown. Usually about 10% of the consumed iron is absorbed and results suggest that Vitamin C increases iron absorption. Although spinach is high in iron (Table 6.3), the oxalates in spinach and rhubarb may reduce iron uptake. About 10 mg per day of iron is recommended.

TOXICANTS IN PLANTS

Most of our present-day food plants are genetic selections of wild plants, whose desirable characteristics have been improved. However, many still retain a chemical similarity to their wild ancestors. Natural toxicants occur in most food plants, but these seldom pose a health hazard. Plants with known health hazards are not released or allowed to be imported. The greatest amount and the widest variety of chemical substances consumed are the normal, natural constituents of foods. However, no single food plant has been well-characterized chemically.

Natural Toxicants

Over 150 chemical substances have been identified in potatoes, including solanine, oxalic acid, arsenic, tannins, nitrates, and over a hundred other items with no recognized nutritional significance to humans (Anon. 1975). Solanine is best known. When the tubers are exposed to light, solanine and chlorophyll are synthesized in the epidermis, which gives the tuber a green color and a bitter taste due to solanine (Nilsson 1978). With the annual consumption of potatoes, people consume almost 10,000 mg of solanine. This does not noticeably harm people, as the potatoes are eaten in individual quantities over a year's time. The body's waste disposal system is well able to handle the biological load under normal consumption patterns.

Radishes, carrots, squash, pumpkin, and celery contain naturally occurring cholinesterase inhibitors, which affect nerve impulses. Lima beans contain glycosides which, during cooking or digestion, release about 40 mg of hydrogen cyanide per average consumer per year.

Trypsin inhibitors are common in legumes (peas, beans, etc.), but are also found in many other vegetables. These materials inhibit the hydrolysis of proteins to amino acids, but their effect is destroyed when the vegetables are cooked.

Broccoli, Brussels sprouts, kale, collards, cauliflower, kohlrabi, and cabbage contain goitrogens (thioglucosides) which cause enlargement of the thyroid gland. These thioglucosides are also responsible for their desirable flavor. When the vegetables are chopped, an enzyme hydrolyzes the thioglucoside into allyl thiocyanate, which is goitrogenic. This material only appears to cause problems when excessive amounts are eaten and the iodine content in the diet is low.

Coffee, tea, and cocoa contain caffeine. Spinach contains oxalic acid, which interferes with iron absorption. Antivitamins are found in peas and beans. People consume two teaspoons of nutmeg per year sprinkled on their food as a spice. Thus, they receive 44 mg of myristicin, a strong hallucinogenic drug. Almost every vegetable contains a toxicant of some sort.

Drug Uses

Many of the nonnutritional plant ingredients are also recognized as medicines (Aikman 1975). In fact, some 50% of human prescription drug ingredients still come from natural products, including plants. The jelly from the *Aloe vera* leaf is used as a burn ointment and in suntan lotions.

American pioneers treated fevers with a mixture of willow bark. The active ingredient, salicin, was isolated in the 1820s, and in 1899 a synthetic derivative was produced—which is called aspirin. The true yam (*Dioscorea*) provides the starting material for nearly all cortisone, used for treating arthritis. Many plants are being investigated for possible cancer cures.

The Delaney Clause

The Federal Food, Drug and Cosmetic Act contains three anticancer clauses. These have come to be known generally as the "Delaney Clause." The clause had its origin in the House Select (Delaney) Committee to Investigate the Use of Chemicals in Foods. The area pertaining to vegetables states "that no additive shall be deemed to be safe if it is found to induce cancer when ingested by man or animal or if it is found, after tests which are appropriate for the evaluation of the safety of food additives, to induce cancer in man or animal. . . ."This clause is probably the most controversial food legislation ever produced in the USA.

Only a few substances are known to be human carcinogens. However, a number of methods are used to predict whether substances are potential human carcinogens. Practices that have been adopted in carcinogenesis testing yield a high probability of obtaining positive results. These include selecting animals sensitive to the agent tested, using the maximum tolerated doses and considering tumors as the diagnostic equivalent of cancers. These procedures are usually successful in increasing the incidence of tumors in animals. The argument is that this approach is necessary to provide a conservative prediction of human risk. However, positive responses may be obtained because the dosages used are very large and overwhelm the animal's system. These large dosages have no demonstrable relationship to natural exposures of humans or animals; or to the natural metabolism of the compounds absorbed.

The use of a tumor as indicative of cancer has also been questioned. A tumor is a swelling. Some tumors are new growths or neoplasms, and some are not. Some neoplasms are malignant (cancers), and some are not. All cancers are tumors, but most tumors are not cancers. Some tumors become cancers, but most do not (Anon. 1981). For regulation purposes, no differentiation is made between malignant and nonmalignant tumors.

Carcinogens may be broadly defined to include all substances which might lead to an increase in cancer. This would include cancer promoters. The naturally occurring hormones from the adrenal cortical glands are necessary for life, yet are also cancer promoters (Anon. 1981). The female sex hormone—estradiol 17B is the most potent, naturally produced, human carcinogen (Lemon 1988). It is involved in breast, ovarian, and endometrial cancer in women. These materials and natural toxicants are excluded from the Delaney Clause as follows: ". . . in case the substance is not an added substance, such food shall not be considered adulterated under this clause if the quantity of such substance in such food does not ordinarily render it injurious to health." The Food and Drug Administration can set tolerances for natural substances such as aflatoxin or mercury. Many carcinogens occur naturally in food (such as mushrooms) or are formed during cooking. They are present in such minute amounts that they are not considered a threat to human health.

When the Delaney Clause was passed, substances could be detected at about one part per million. Anything found in a lesser concentration was not detected. Present day analytical methods allow substances to be detected at parts per trillion. One aspirin tablet per adult is about 3.5 parts per billion (baised upon body weight). The concentrations now detectable in many instances are far lower than those that can be demonstrated experimentally to have any effect on carcinogenesis. The question now is not whether carcinogens can be detected, but rather whether the concentration of a carcinogen is significant. The following are some examples of one part per trillion: a six-inch leap on a journey to the sun; one second in 320 centuries; one cent in $10 billion; one pinch of salt on 10,000 tons of potato chips; one bad apple in 2 billion barrels.

The Environmental Protection Agency has begun to apply "negligible risk" in evaluation of pesticides and food additives. Negligible risk is defined as one additional case of cancer per one million, lifetime exposures to a compound. However, almost everything will induce cancer experimentally. The following are the chances (baised upon one million lifetime exposures) of developing a tumor from some readily available substances (Beeler 1979): smoking of cigarettes: 1200 chances per million; eating one quarter-pound of charcoal broiled meat per week: 0.4; drinking water in Miami or New Orleans: 1.2; eating peanut butter containing FDA-approved levels of aflatoxin: 40; exposure to radiation from one transcontinental airplane flight per year: 0.5; sunbathing: 5000 chances in a million.

Decisions are determined by our personal perceptions of values and needs. If the same standards for defining carcinogenecity for saccharin were used, then dietary fats, proteins, and carbohydrates also would be classed as carcinogens (Anon. 1981). Because these nutrients are essential in our food supply, these nutrients are regulated differently. Our aversion to risk is balanced by our perceptions of value, need, and benefit.

Hazards of Natural Toxicants and Food Additives

Toxicity is the inherent capacity of a substance to produce injury or death. Hazard is the capacity of that substance to produce injury under the circumstances of exposure. On the average, each person in the U.S.A. consumes 1500 pounds of food and drink each year. Normal components of natural food products make up all but about 139 pounds of that total (Anon. 1975). Of the 139 pounds, some 129 pounds are additives such as sugar, salt, corn syrup, and dextrose. Another 9 pounds consists of leavening agents such as yeast and substances used to adjust the acidity of food products. This leaves 1 pound per capita divided among 1800 other additives. This includes toxicants and contaminants, both natural and manmade, and accidental amounts of pesticides.

In 1975, the National Academy of Science estimated that each person consumed about 40 mg of pesticide residues yearly, and that this level is decreasing (Rogers 1987), The 40 mg total had a toxicity level about equivalent to the amount of aspirin in one aspirin tablet; or the caffeine in one cup of coffee. In 1986, 5500 crop samples were taken from vegetables packed and ready to sell. Over 84% had no detectable pesticides and 9% contained residues at less than 10% of the legal tolerance (Richardson 1987).

The potential hazard from the 1 pound of these 1800 additives is very low (Anon. 1975). When the additives are eaten, the hazard is still very low due to the way the body handles them.

The amount of any single toxic substance in any single serving of any commonly accepted food is so low that a grossly exaggerated amount would need to be eaten, over an extended period of time, before the toxicity of that substance could be considered a hazard. If one's diet contains a reasonable variety of foods, and no extraordinary amount of any single food, it is highly unlikely that one chemical will be eaten in a toxic amount.

The individual toxicities of each of the thousands of different chemicals present in our diet each day cannot be added up to a total toxicity for that day. If one ate one-hundredth of the toxic dose of each of 100 different food components, the mixture would be harmless. The body readily handles small amounts of many different chemical substances eaten at the same time. The body has many different disposal systems (perspiration, urine, feces, or incorporation into hair; the addition of a sugar molecule, rendering it harmless, and many others).

Antagonistic interactions occur between chemical substances in foods. The toxicity of one material is offset by the presence of an adequate amount of another material. Antioxidants in foods, such as BHA (butylated hydroxyanisole), can reduce the level of stomach cancer. The U.S.A. has the highest consumption of these compounds and the lowest incidence of stomach cancer. Iodine inhibits the action of some goiter-causing agents. The National Academy of Sciences reported that a regular diet of cabbage, Brussels sprouts, broccoli, or cauliflower was associated with a reduction in the incidence of certain cancers (Anon. 1982), even though these plants contain other known toxicants. A lower incidence of lung cancer was noted in both smokers and nonsmokers who ate carrots, green salad, or fruit juice 5–7 times a week (Colditz, 1988).

To consumers and their diet, this means that there is safety in numbers. The wider the variety of foods eaten, the greater is the number of different chemical substances eaten; and since all food is chemical in nature, the less chance there is that any one chemical will reach a hazardous level in the diet.

NUTRIENT COMPOSITION OF ORGANIC VEGETABLES

A sugar or a vitamin has the same identical chemical structure, whether it is natural or manmade. Many long-term studies (Anon. 1974) have shown that organically grown foods are identical in nutrition to those grown by other methods. Fertilizers can increase the size and yield of a crop, but not the composition of the plant, in regard to its major nutritional characters. The amount of Vitamin C in a carrot will always be higher than that found in dry beans, and no fertilizer or chemical stimulant will change this. This is because the nutritional composition of a carrot and a dry bean is determined by the genes of the plant and the maturity of the plant at harvest. Regardless of the type of fertilizer or pesticide used, the nutrient composition of the edible part of the plant, in regard to its content of protein, fat, carbohydrate, or various vitamins, will not be affected.

Fertilizers do influence the mineral composition of plants. The iodine content of the plant will vary with the iodine content of the soil. If greensand is used as an organic fertilizer, the plants will probably also contain fluoride, as greensand contains floride. The plant has no mechanism to selectively remove only certain minerals from the soil.

SELECTED REFERENCES

Adjei-Twum, D.C., Splittstoesser, W.E., and Vandermark, J.S. 1976. Use of soybeans as sprouts. HortScience *11*:235–236.
Aikman, L. 1975. Natures gifts to medicine. Nat. Geog. Sept. 420–440.
Anon. 1971. Nutritive value of foods. U.S. Dept. Agric. Home Gard. Bull. 72.
Anon. 1974. Organic Foods. Food Technology, Jan.
Anon. 1975. Naturally occurring toxicants in foods. Institute of Food Technologists, Chicago.
Anon. 1981. Regulation of potential carcinogens in the food supply: The Delaney Clause. CAST, Ames Iowa.
Anon. 1982. Diet, Nutrition and Cancer. National Academy Press, Washington, DC. 218 pp.
Anon. 1989. USDA forecast of per capita consumption. Washington, DC.
Beeler, D. 1979. Risky Business. Agrichemical Age. June 5C.
Colditz, G.A. 1988. Beta-carotene and cancer. *In* Horticulture and Human Health, p. 150–159. Prentice-Hall, Englewood Cliffs, NJ.
Courter, J.W. 1979. Proceedings Illinois Vegetable Growers Schools. U. ILL Hort Series 12, 124 pp.
Haytowitz, D.B. and Matthews, R.H. 1986. Composition of Foods: Vegetables and Vegetable Products. USDA Agric. Handbook 8–11.
Hoff, M. 1985. Fiber: How much do healthy people need? Minn. Sci. *3:* 9–11.
Lemon, H.M. 1988. Role of estrogens in human caranogenesis and anticarcinogenesis. *In* Sigma Xi NW Region Lectureship program.
Nilsson, T. 1978. Quality of Vegetables. Chronica Hortic. *18*:21–24.
Quebedeaux, B. and Bliss, F.A. 1988. Horticulture and human health. Prentice-Hall, Englewood Cliffs, NJ. 243 pp.
Richardson, L. 1987. Regulating pesticides in food. Agrichemical Age. July, 6–7.
Rogers, R.W. 1987. Foods are chemicals. Sci. Food Agric. *5*:16–19.
Splittstoesser, W.E. 1977. Protein quality and quantity of tropical roots and tubers. HortScience *12*:294–298.

7

Growing Common Vegetables

The use of the term 'common vegetables' is somewhat arbitrary, but includes those vegetables commonly grown by most gardeners with a large number of plants in the garden. This section includes a discussion of each. Suggestions for the control of pests can be found in Chapter 4, on Pest Control.

ARTICHOKE

The globe artichoke *(Cynara scolymus)* is a thistle-like vegetable. It was eaten fresh and preserved for year-round use by the Romans centuries before Christ. The globe artichoke declined in importance as a vegetable with the fall of the Roman Empire. However, during the reign of the Medicis in Italy, it was rediscovered and introduced into France as a gourmet item. When French and Spanish colonists settled in America, they brought the globe artichoke with them.

Plant Characteristics. The artichoke is a perennial that grows to about 5 ft in height with a 5 to 6 ft spread. If the buds are not harvested, 6-in. bluish flower heads appear.

Culture. The globe artichoke grows best in cool, mild climates. It is very sensitive to frost and in these areas may be grown as an annual with a short harvest period in the fall. However, under warm or hot temperatures, yields are reduced and the bud scales become tough. As the bud scales are the edible part, the artichokes are not very palatable. Thus, artichokes are primarily grown along some sections of the California coast and in the South Atlantic and Gulf Coast regions.

Rooted offshoots or divisions from mature plants are planted in early fall. They are planted about 3 ft apart in rows with 5 ft between rows. As they are a perennial, they should be planted at one edge of the garden in a sunny location. They will live and produce for 5 or more years.

After harvest, the plant is cut off at the soil line and removed. It is not watered for several weeks so that it is dormant during part of the summer.

FIG. 7.1. The edible bud of a globe artichoke. *Courtesy of W. Atlee Burpee Co.*

Fertilizer, particularly nitrogen, is added after the dormancy period and the plant is irrigated. Rapid growth occurs, and new stems with new buds develop for fall production.

Green Globe is the cultivar usually grown.

Harvest. In a normal production cycle, new bud production begins in the fall, and reaches a maximum in the spring when warmer temperatures occur. The terminal bud is always the largest one and may be 5 in. in size. Auxiliary buds develop also and, after the initial year, 40 buds per season may develop.

The buds are harvested before the bud scales begin to spread. This spreading is due to the growth of the flower parts within the bud, reducing the edible part of the plant. If the buds are harvested when small, yields are reduced, but the buds are more tender. The bud is harvested by cutting the stem of the plant about 2 in. below the base of the bud. This small length of stem is also edible and tender.

Common Problems. Various types of aphids attack artichokes. These may be controlled with various chemical and botanical insecticides. Slugs and snails may also be a problem.

ASPARAGUS

Asparagus *(Asparagus officinalis)* is a member of the lily family. It originated near the Mediterranean Sea and was considered a delicacy by the ancient Greeks. Methods for growing this vegetable were described in 200 B.C., and cultivated asparagus has changed little since. It was cultivated in England by the time of Christ and brought to America by the early colonists (Garrison and Ellison 1977).

Plant Characteristics. Asparagus is a perennial vegetable that can be grown in the same area for 20–30 years. It has an underground network of fleshy storage roots and underground stems called rhizomes. These roots store food and produce the asparagus spears. The root system is referred to as an asparagus crown. When temperatures are warm, buds develop on the rhizomes, which become the edible spears. If the spears are not harvested, they develop into a green fern-like bush about 4 to 6 ft tall. This foliage produces the food material, which is stored in the storage roots to be used to produce the next year's spears.

Asparagus has separate male and female plants. Bees transfer the pollen from the male plants to the flowers on the female plants, which then produce red berries in the fall. The female plants use some of their energy in the production of these berries, which contain the seeds. Female plants do not produce as well or live as long as the male plants.

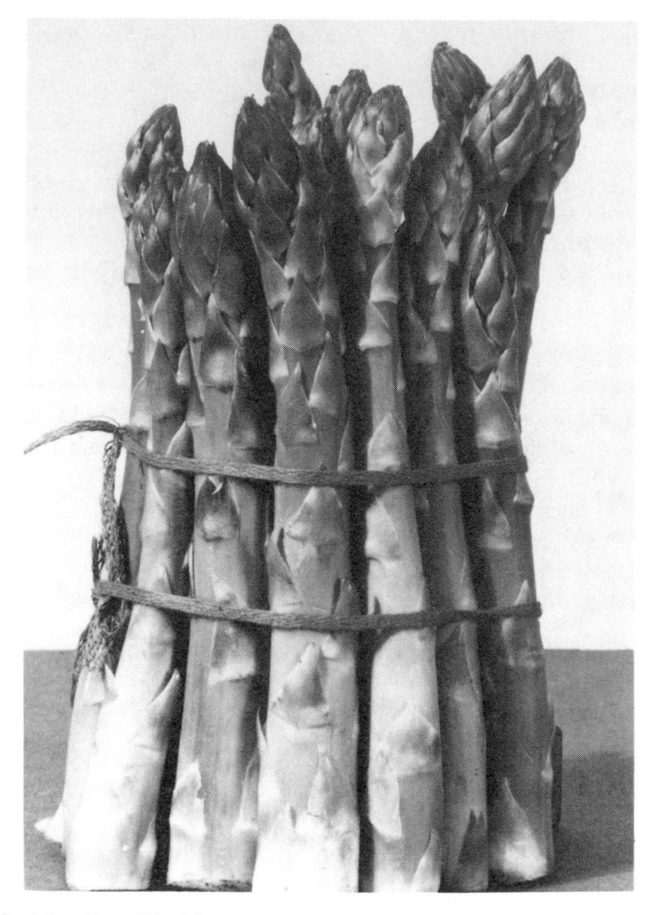

FIG. 7.2. Mary Washington asparagus. *Courtesy of Farmer Seed and Nursery Co.*

Culture. Asparagus is a cool season crop and does best in climates where the soil freezes at least a few inches. In many southern locations and along the Gulf Coast, the temperatures are too warm for asparagus. The plants do not store much material in the roots and few spears are produced. However, asparagus is grown from southern California, where temperatures reach 115°F in the summer, to places where the winter temperature reaches 40°F below zero.

Asparagus should be planted with other perennial crops on the north or east side of the garden so as not to shade other vegetables. It can be planted along a fence in full sunlight. Asparagus can be planted in the spring as soon as the soil can be worked. About 1–2 lb of 5-10-10 fertilizer or organic fertilizer equivalent per 100 sq ft should be worked into the soil before planting. Asparagus grows poorly at an acid pH below 6.0, so the soil should be tested and pH adjusted if needed.

Healthy 1-year-old crowns should be planted in a furrow 8 in. deep. The crowns should be placed a foot apart within the row and 4 ft. between rows. If only one row is planted, at least 3 ft should exist between the asparagus and the rest of the garden. The roots should be spread so the crowns lie flat. The crowns should then be covered with 2 in. of soil, and during the season the furrow should slowly be filled in the remaining 6 in., being careful not to cover the asparagus foliage.

The plants should be watered weekly to wet the soil about 8 in. deep during the first growing season. After a year, the plants have developed an extensive root system, and about 2 in. of water every 2 weeks is sufficient during dry weather.

The second season after planting, the asparagus bed should receive 1.2 lb of 5-10-10 or organic fertilizer equivalent per 100 sq ft both in early spring and late summer. In subsequent seasons, the asparagus bed can be fertilized the same as the garden.

It is important to maintain good foliage growth of the plants after harvest. Allow the asparagus tops to grow until frost. The tops can then be removed or ground and left as a mulch. It is preferable to remove the tops and compost them to decrease problems with asparagus rust.

Most asparagus cultivars came from the Martha Washington and Mary Washington cultivars developed by the U.S. Department of Agriculture (USDA). Mary Washington is the common cultivar grown in the south and the most important cultivar in the United States. In northern areas, Viking may be grown. On the West Coast, Mary Washington and various numbered California strains have been recommended. In the Midwest and East, Rutgers Beacon and various Washington types, such as Waltham Washington, are frequently grown in home gardens. These cultivars are all resistant to asparagus rust.

FIG. 7.3. Asparagus crown should be planted 8 in. deep.

Harvest. Asparagus crowns must grow for two full growing seasons before harvesting the spears. This allows the plants to develop an adequate root storage system to produce the spears. Damage or harvest of the plants before two years may reduce yield for the life of the bed.

When the spears emerge in the spring, all spears should be harvested. The spears should be 7 to 10 in. long. The spears can be cut at the ground surface or snapped off. Except in the cool central valleys of California, asparagus should be harvested for a 2-week period the third year after planting the crowns, 4 weeks the fourth year, and 8 weeks thereafter. In the cool valleys of California, spears may be harvested for 4 weeks the third year, 8 weeks the fourth year, and 12 weeks thereafter.

To harvest white spears, the row should be covered with a ridge of 8–10 in. of soil in the spring when the spears emerge. When the spears break through the top of the ridge of added soil, some soil can be removed from around the spear and a long knife used to cut the spear about 8 in. below the tip.

Asparagus should be eaten immediately. However, it may be washed and the cut ends placed in a small amount of water and immediately stored in the refrigerator for use a day later.

Common Problems. The asparagus bed must be kept free of weeds, and clean cultivation is preferred to reduce rust and beetle problems.

Asparagus rust is a common fungus disease in the East and Midwest and in some areas of California. Water or dew left on the plant for 10 hr helps spread the disease. Resistant cultivars should be planted.

The asparagus beetle may be a serious pest. The adult beetle over-winters in organic material and does not thrive in hot weather. It is not usually a problem in the South. The beetles can be controlled by hand picking, rotenone, malathion, or carbaryl (such as Sevin).

BEANS, SNAP AND OTHERS

The common bean *(Phaseolus vulgaris)* includes a large number of different types, which are called by many names. Those beans grown for the immature pod are known as snap beans, Romano or Italian beans, garden beans, green beans, string beans, French Horticultural beans, French beans, wax beans, and yellow beans. Those grown for the immature seed are known as green shell beans. Those grown for the mature seed, such as kidney beans, are called dry beans. The requirements for growing these beans in the garden are similar.

The common bean originated in Central America and was widely distributed by the Indians of North and South America (Meiners and Kraft 1977). The Indians ate the beans primarily in the green shell or dry bean stage. The original snap beans were stringy, hence the name string bean. Early

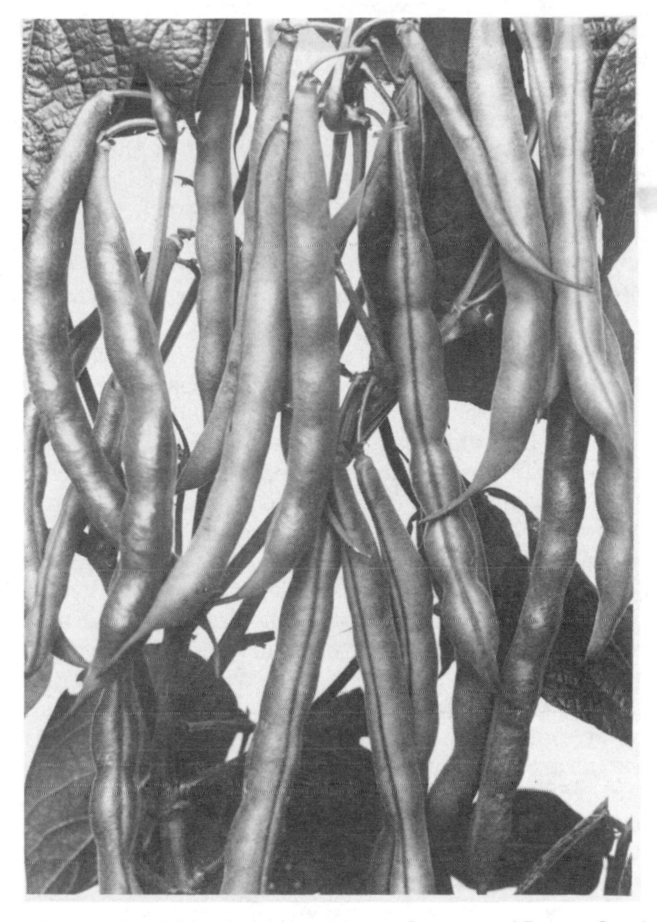

FIG. 7.4. Bush Blue Lake green bean. *Courtesy of Farmer Seed and Nursery Co.*

American settlers introduced the snap bean to Europe. Within the past 25 years, the common bean has been almost entirely redesigned by plant breeders and few cultivars available in 1950 are available today (Anon. 1975).

Plant Characteristics. Garden beans may be bush or pole types. The bush bean is a short erect plant 1 or 2 ft in height. They are used for quick production, and successive plantings every 2 weeks are needed for a continuous supply.

Pole beans develop vines that must be supported by a fence or stakes or be grown on a trellis. Pole beans grow slower than bush beans but produce more beans per plant, and only one planting of pole beans is needed.

Culture. Snap beans are a warm season crop and are grown in all parts of the United States. They should not be planted until the soil is warm. In the South and Southwest, snap beans may be grown during the fall, winter, and spring. In the extreme South, snap beans are grown throughout the winter. During excessively hot, dry, or wet weather, the plants frequently lose the flowers or pods. Snap beans grow best between temperatures of 70° and 80°F and require about 1 in. of water each week.

To reduce diseases, western-grown seed should be used. Snap bean seed should not be soaked before planting as this injures many bean cultivars. The seeds should be planted 1 in. deep in loam soils and 1½ in. deep in sandy soils.

Beans are legumes and as such enjoy an association with bacteria that convert nitrogen from the air into a form the bean plant can use. These bacteria may be purchased in a pea and bean inoculant, available at garden centers and through catalogues. If the area has never grown beans or peas before, some of this inoculant can be dusted onto the *moistened* (not soaked) seed. If the area has grown beans before, it is usually of little value to add additional inoculant. Beans therefore require less nitrogen than some other vegetables, and if they are planted after some other early vegetable is harvested, the residual fertilizer is usually sufficient. Heavy applications of manure or of fertilizers high in nitrogen may induce a large amount of vine or bush growth but frequently delay maturity and yield of the pods.

Bush-type beans are planted 1 or 2 in. apart in 18–30 in. rows. They are thinned to stand 2 to 4 in. apart. Pole-type beans may be planted in rows 2 to 4 ft apart near some support. A trellis made of stakes and string may be used. Pole beans are also planted in hills 3 ft apart with 6 seeds in each hill. The hill is thinned to 4 or 5 plants. Three or four long poles are placed over the hill and tied at the top in wigwam fashion. The vines are then trained to climb around the poles, frequently in a clockwise direction.

There are many cultivars of snap beans. The gardener can choose between bush-type or pole-type snap beans; between green, yellow, or dry bean cultivars. Italian or Romano beans produce large, flat pods that have a distinctive flavor. Horticultural or French Horticultural beans are large-seeded beans that produce colorful pods. The pods are striped and mottled in red. These beans are usually used as green shell beans.

Royalty is a purple-podded bush bean. This cultivar is a selection from seed handed down through the years by a family of New England gardeners (Bubel 1977). It will germinate and grow in cooler soil than regular beans, and it is less bothered by bean beetles. Royal Burgundy is another purple-podded bean.

Harvest. Beans should not be handled or harvested when they are wet as this helps to spread disease. Green and yellow snap beans should be harvested when the pods are still young. The seeds will be small, the pod interior will be firm, but the fiber content of the pod wall will be low. Beans should be harvested every few days as mature pods are of poor quality.

Frequent harvesting induces the plants to continue to produce new pods. Bush types usually produce three or four harvests while pole types produce numerous harvests.

Green shell beans produced by the horticultural cultivars are harvested when the pods change from green to yellow or red. At this time the seeds are about fully grown, but they have not yet dried and become hard.

Dry beans are harvested after the pods are mature. Either the pods are removed or the entire plants pulled up when the leaves have turned yellow. The beans are then dried in a clean area, and when the pods are dry, the beans are removed and stored.

Common Problems. Snap beans have a number of insect and disease problems. Virus diseases and root rots are serious in the West. In the East and South, bacterial diseases infect snap beans. Bacterial diseases and anthracnose can be controlled by using western-grown seed. Virus diseases and rust can be controlled by using disease resistant cultivars. Bacterial disease can be controlled by not handling the plants when they are wet from dew or rain. Root rots can be reduced by rotating the beans to a different spot in the garden each year.

Bean beetles are the major insect pest. Hand-picking or various botanical and chemical insecticides can be used for control.

BEANS, LIMA

Lima beans *(Phaseolus limensis)* are similar to butter beans *(Phaseolus lunatus)* grown in the South. Lima beans are natives to the western hemisphere and probably originated in Central America (Meiners and Kraft 1977). There are two types, a small seeded and a large seeded type. The small seeded type had been used by North American Indians long before the time of Christ. The large seeded types were developed in Peru about the same time. Large seeded lima beans were collected and imported into the United States and called lima beans after the capital of Peru.

Plant Characteristics. Lima beans are large seeded and generally referred to as the Fordhook type, or they are small seeded and referred to as baby limas. Both types are grown as bush or pole plants. The bush bean is a short erect plant 1 or 2 ft in height. They are used for quick production. Pole beans develop vines which must be supported by a fence, stakes, or grown on a trellis. Pole beans grow slower than bush beans but produce more beans per plant.

Culture. Lima beans are a warm season crop that requires warmer temperatures than snap beans. Lima beans need a minimum soil temperature of 65°F for quick germination and then require 3 or 4 months of warm days and nights. Thus, in the states of the United States near the Canadian

FIG. 7.5. Fordhook 242 lima beans. *Courtesy of Harris Seeds.*

border, lima beans are difficult to grow in a garden. Below this area, bush baby limas can be grown as they mature quicker than other lima bean types. Lima beans require such a long growing season that successive plantings are impractical except in the Deep South.

Lima beans grow best in sandy, well-drained soils. The soil should not bake or form a crust. If this occurs, sand, vermiculite, or leaf mold should be placed over the beans instead of soil when planting.

To reduce diseases, western-grown seed should be used. Bean seed should not be soaked before planting as this injures many bean cultivars. The seeds should be planted 1 in. deep in loam soils and 1½ in. deep in sandy soil.

Beans are legumes and as such enjoy an association with bacteria that convert nitrogen from the air into a form the bean plant can use. These bacteria are available in a pea and bean inoculant, available at garden centers and through catalogues. If the area has never grown beans or peas before, some of this inoculant can be dusted onto the *moistened* (not soaked) seed. If the area has grown beans before, it is usually of little value to add additional inoculant. Beans therefore require less nitrogen than some other vegetables, but the area should receive an application of a general garden fertilizer. Large amounts of manure or of fertilizers high in nitrogen may induce a large amount of vine or bush growth but frequently delay maturity and yield of the pods.

Bush-type beans are planted 1 or 2 in. apart in 18–30 in. rows. They are thinned to stand 2 to 4 in. apart. Pole-type beans may be planted in rows 2 to 4 ft apart near some support. A trellis made of stakes and string may be used. Pole beans are also planted in hills 3 ft apart with 6 seeds in each hill. The hill is thinned to 4 or 5 plants. Three or four long poles are placed over the hill and tied at the top in wigwam fashion. The vines are then trained to climb around the poles, frequently in a clockwise direction.

During excessively hot, dry, or wet weather, the plants frequently lose the flowers or pods. Lima beans grow best between temperatures of 70° and 80°F and require about 1 in. of water each week.

Pole-type and bush-type cultivars are available; and large or small seeded lima bean cultivars or butter bean cultivars may be selected.

Harvest. Beans should not be handled or harvested when they are wet as this helps spread disease. Lima beans should be harvested when the pods are filled and are light green in color (creamy white for butter bean types). The end of the pod should feel spongy when squeezed. By frequent harvesting more pods are produced. Bush types are harvested for about 3 weeks and pole types for about 4 weeks or until frost.

Common Problems. Lima beans are susceptible to mildew in the mid-Atlantic states, and anthracnose and nematodes in the South. These can be controlled by using disease-free seed, rotating the beans in the garden area, using resistant cultivars, and not handling the plants when they are wet with rain or dew.

Bean beetles are the major insect pest. Hand-picking or various botanical and chemical insecticides can be used for control.

BEANS, FAVA

The fava bean *(Vicia faba)* is often called broad bean, Windsor, or horsebean. It is not a bean but is a type of vetch. The broad bean is native to northwest India and was used by ancient Greeks, Egyptians, and Romans. They introduced it into China about 2800 B.C. It was brought to America by early settlers from southern Europe.

Plant Characteristics. The plants are erect and medium tall and produce pods about 7 in. long. The pods contain 5 or 6 large beans.

Culture. Fava beans are grown under even, warm temperatures. In most areas. the summers are too hot and the winters too cold for it to be planted in the autumn and overwintered. In some parts of the South and Pacific Coast areas, it may be planted in the fall and grown through the winter for a spring crop. Other culture requirements are similar to snap beans.

Harvest. Fava beans are harvested as green shell or dry beans. The pods contain 5 or 6 large flat, oblong bans and are harvested similar to green shell or dry snap bean varieties.

Common Problems. Many of the problems are similar to snap beans. As fava beans can withstand cool weather, they should be planted in early spring (about the same time as peas) to avoid aphid damage during summer.

An allergy known as favism sometimes occurs in people who have previously ingested the bean or inhaled the flower pollen of fava beans.

BEANS, GARBANZO

The garbanzo bean *(Cicer arietinum)* is neither a pea nor a bean. It is also called chickpea, Egyptian pea, gram, chestnut bean, and ceci or cece bean. It is a native of the Mediterranean area and was widely used by Greeks, Romans, and Egyptians. Today the gram is one of the important food sources in India. Ceci or cece beans are important in Italy, and in Spain they are called garbanzos. Settlers from these countries brought this plant to America.

Plant Characteristics. The garbanzo bean is an erect bush-type plant which grows about 2 ft high. It produces one, and at most three, beans in a puffy little pod. The bean has a chestnut-like flavor.

Culture. Garbanzo beans require 100 days to mature and are grown in warm climates. The seeds are planted 1 in. deep in loam soils and 1½ in. deep in sandy soils. The plants are thinned to stand 3 in. apart, in rows 2 ft apart. Other culture methods are similar to snap beans.

Seed catalogues usually do not list a cultivar. They are usually listed as Chickpea or Garbanzo.

Harvest. They are harvested similarly to dry beans. The mature pods are picked and allowed to dry. These pods are hairy and often exude a substance that irritates the hands. Often gloves are used when picking. After the pods are dry, the beans are shelled and stored in airtight containers.

BEETS

Beets (*Beta vulgaris,* Crassa Group) are also called garden beets and table beets. Beets originated in the Mediterranean area of North Africa, Europe, and West Asia. Swiss chard belongs to the same species as the garden beet and was probably the original wild beet without a fleshy root. Beets were first cultivated about the third century A.D. and were first used for their root in the 1500s. However, it was not until the late 1800s that much interest was shown in beet cultivars.

Plant Characteristics. Beets are a biennial that may be grown either for the roots or for the tops as greens. Various cultivars have been developed that are classified according to the shape of the root and the time of maturity. Beet roots are most often round but may be flat or elongated. Their fleshy roots are usually red, but golden cultivars are available. The "seed" of beets is actually a dried fruit containing several tiny true seeds (Reynolds 1977).

Culture. Beets are adapted to all parts of the country, being tolerant of both hot and cold temperatures. They will not withstand severe freezing and are grown in the spring, summer, and fall in the North. In warm climates they are planted in the fall and winter for harvest in the spring. Temperatures below 50°F for 2 or 3 weeks may induce seed stalks to develop instead of the fleshy root.

Beets are planted one-half inch (loam soils) to 1 in. (sandy soils) deep in rows 12 in. apart. Heat, drought, or crusting of the soil surface may interfere with seedling germination and emergence. The seeds can be covered with sand, leaf

FIG. 7.6. Both leaves and roots of beets may be used as a vegetable. The cut beet shows the alternating rings of food—and water—conducting (white rings) tissue. *Courtesy of Harris Seed Co.*

mold, or vermiculite instead of soil to prevent this problem. The beets should be thinned to stand 2 or 3 in. apart in the rows. Successive plantings, 3 weeks apart, are needed to ensure a continuous supply of young beets.

Flat, globular, and elongated cultivars are available. There are early-, medium-, and late-maturing cultivars, and red or yellow-colored roots available. A number of cultivars with large leaves have been developed for producing beet greens.

Harvest. Beets are harvested when desired but have the highest quality when they are less than 2 in. in diameter. Beets left in the garden will withstand mild freezing. For storage, fall-maturing beets should be used. The tops are removed 1 in. from the roots and stored at 95% humidity at 32°F.

Common Problems. Beets have few insect and disease problems in the garden. In western areas a leafhopper may transmit curly top virus disease. Beets should be planted early to allow them to mature before the virus develops in the summer.

In some areas, a lack of boron in the soil results in beets containing bitter black spots in the roots. Only small amounts of boron are needed, and one-half teaspoon of household borax in 12 gal. of water per 100 ft of row is sufficient (Bubel 1977). Composted leaves of plants that accumulate boron may also be used, such as muskmelon (cantaloupe) and sweet clover.

BROCCOLI

Broccoli (*Brassica oleracea,* Italica group) is a member of the mustard family and its culture and problems are similar to cabbage. Broccoli is also known as calabrese, Italian broccoli, and green sprouting broccoli. Broccoli developed from various leafy cabbage forms in southern Europe. They were brought to America by early immigrants from that area.

Plant Characteristics. Sprouting broccoli is the type grown in home gardens. The plant forms a loose flower head on a tall, green, branching stalk. The center flower head is from 5 to 10 in. across, and the plant may reach 3 to 5 ft in height.

Culture. Broccoli is grown similar to cabbage. It can be planted as a transplant or from seed. Transplanting saves 3 or more weeks growing time, but fall crops can be direct seeded.

Broccoli is adapted to all parts of the United States. In northern areas with short growing seasons, it can be planted in the spring for summer harvest or for early fall harvest. In areas with a longer growing season, it can be planted in early spring for a summer harvest and again during the summer for a fall harvest. Where mild winters occur, late summer and early fall plantings are made for harvest during the winter. Broccoli is hardy and will withstand some frost.

In western areas, broccoli is frequently grown on a raised bed. This is an advantage if drainage of water is slow during fall and winter seasons and furrow irrigation is used. The beds are formed several days in advance to allow the soil to settle. These beds are raised 6 in. and 2 rows 1 ft apart are planted on the beds.

Broccoli transplants are set in holes that are deep enough that the stem of the plant is slightly below ground level. Transplants should be 4 or 5 weeks old, and a starter fertilizer should be used after planting. Older plants have more food reserve; if they receive temperatures of 50°F for 2 weeks, they will produce a small flower stalk immediately. Small plants should be planted during cool weather.

For direct seeding, the seeds should be planted one-fourth to one-half inch deep about 1 in. apart. They are later thinned to stand 1 to 2 ft apart in the row. Broccoli must develop rapidly and requires adequate water. Once established, it requires 1 to 1½ in. of water weekly.

Broccoli cultivars are generally classified by the length of time they take to mature. They mature in 50–85 days after transplanting, depending on the time of year planted and the cultivar. There are early-, medium-, and late-maturing

FIG. 7.7. Premium crop broccoli can be grown in all parts of the country. *Courtesy of Harris Seed Co.*

cultivars. Royal Purple Head is a type of broccoli which produces a compact purple head, somewhat like cauliflower. It turns green when cooked. This cultivar is often found listed as a purple cauliflower.

Harvest. The central head together with 6 in. of stem is harvested. These flower buds and the attached stem are the edible part. It should be harvested before the flower buds begin to develop into bright yellow flowers. After the center head is removed, the smaller side shoots develop into new heads. This can extend the harvest a month or more.

Common Problems. The insect and disease problems of broccoli are similar to cabbage. Wilt, blackleg, and black rot can cause severe damage. These can be controlled by using disease-free seed or transplants and by using resistant cultivars.

Insect pests are rather local, but aphids, cabbageworms, and flea beetles are common through the United States. They can be controlled with various botanical, chemical, and biological control agents.

BRUSSELS SPROUTS

Brussels sprouts (*Brassica oleracea*, Gemmifera group) belong to the mustard family. They originated in Europe and have been grown near Brussels, Belgium (whence the name) for hundreds of years.

Plant Characteristics. Brussels sprouts are a type of nonheading cabbage that develops sprouts or small heads in the axils of the leaves (the upper area near where the leaf stem joins the main stalk). The plants grow slowly and require 80–100 days from transplanting for the first sprouts to mature. Brussels sprouts are biennial, requiring a cold treatment to flower.

Culture. Brussels sprouts are a cool season crop and can withstand some frost. They are best transplanted during the summer or fall for a fall or winter harvest. In some areas they may be grown as an early spring crop.

Brussels sprouts for transplanting are seldom available when needed, and gardeners will need to grown their own. The seeds can be planted in a protected area of the garden 4 to 6 weeks before they should be transplanted. If only a few plants are needed, the seeds can be planted in peat pellets. Seeds are planted about 1 in. apart in rows 3 in. apart. At transplanting, they are placed in the garden, 2 ft apart in rows 2 to 3 ft apart. A transplant fertilizer should be used and additional nitrogen applied when the plants are a foot high to stimulate their growth. Brussels sprouts require good fertility and moisture. They should receive 1 to 1½ in. of water a week during dry weather.

Harvest. The sprouts are harvested when they are 1 or 2 in. in diameter and plants may be harvested for a month or more. The leaf below the sprout is removed, and the sprout is then cut or broken off. Sprouts are harvested upward on the stem until the sprouts are too small. These sprouts are left for future harvests.

In warm weather, the sprouts become loose and are not firm. When cool weather occurs, the sprouts firm up and become milder in flavor. In areas where an early frost occurs, plants 15–20 in. tall may be debudded. The top growing point is cut out in August or September to force all the plants' energy into the developing sprouts. All the sprouts will then be ready about the same time.

FIG. 7.8. Brussels sprouts are a cool-season vegetable and can withstand some frost. *Courtesy of Harris Seed Co.*

Some gardeners remove the lower leaves of Brussels sprouts as the sprouts develop. The top leaves should not be removed. This practice makes harvest of the sprouts easier but it is not required.

Common Problems. Brussels sprouts have insect and disease problems that are similar to cabbage. They should not be grown where other members of the cabbage group (cabbage, cauliflower, and others) were grown the previous year. Most diseases can be reduced by using disease-free transplants and crop rotation.

Aphids, cabbageworms, and flea beetles are common pests in all areas of the United States. They can be controlled by various botanical, chemical, and biological control agents.

CABBAGE

Cabbage (*Brassica oleracea,* Capitata group) developed from wild, leafy, nonheading types still found growing in Europe. Wild cabbage is found along the chalk coasts of England, Denmark, northwestern France, and various other areas from Greece to Great Britain. Cabbage was in general use by 2000 B.C. It is said that the ancient Egyptians worshipped cabbage. Cabbage was used throughout Europe by 900 A.D., but the head type was first described in 1536. Shortly thereafter, these types were brought to America by the settlers.

Plant Characteristics. Cabbage types vary widely. They range in color from green to purple; in leaf character from smooth to savoy leaves; in head shape from flat to pointed; and in maturity from early (55 days after transplanting) to late (130 days or more after transplanting). The green, round-headed types are most common. Cabbage is a biennial that requires a cold treatment to flower.

Culture. Cabbage is a cool season crop that tolerates frost but not heat. In the far South, it can be grown in all seasons except summer. On the Pacific coast, it can be grown all year long. In areas with mild frosts, it can be planted in the fall and become one of the first crops harvested in the spring from the garden. Cabbage will adapt to cool temperatures and can withstand short periods of temperatures below 0°F. In the North, it can be grown as an early summer crop or a late fall crop.

Quality in cabbage is dependent upon rapid growth. Cabbage responds to liberal applications of nitrogen fertilizer. It should be fertilized before seeding and a starter fertilizer used when transplanting. Cabbage can be fertilized once or twice during the growing season at least 3 weeks apart.

For early spring planting, transplants should be used. If these transplants are hardened, they can be transplanted into the garden when the earliest vegetables are planted. As cabbage does not tolerate hot weather, cultivars that mature before this time should be transplanted. Cabbage transplants are spaced 12–18 in. apart in rows about 2 ft apart. Fast-growing vegetables such as green onions, early lettuce, and radishes can be planted between the cabbage plants to utilize the space. These other plants should be harvested before the cabbage starts to spread out.

The stem of cabbage transplants should not be larger than the size of a standard pencil. Larger plants have more food reserve; and if they undergo low temperatures of 40°–50°F for 3 or 4 weeks, these plants will produce a seedstalk. Large transplants that experience a low temperature during the winter will flower in the spring. Heads from overwintered plants should be harvested early.

For late planting, seeds are planted one-fourth to one-half inch deep during midsummer. The plants should later be thinned. Summer plantings will produce the head during the cool fall weather. Late plantings may follow early potatoes, beets, peas, or spinach. Frequently, cabbage is planted before the potatoes are harvested to conserve space.

Cabbage requires adequate moisture. From 1 to 1½ in. of water a week are required.

To provide for a harvest over a long interval, different cultivars with different maturity dates can be used. Of the numerous cultivars available, many are resistant to one or more diseases. There are early and late cultivars of standard red, green, and savoy cabbage cultivars available.

FIG. 7.9. A standard green cabbage cultivar. *Courtesy of Harris Seed Co.*

Harvest. Cabbageheads are harvested any time after the head has become fairly firm. Heads from fall-maturing cultivars can be harvested and stored for several months at 40°F and high humidity.

Cabbage planted early for summer harvest will produce a second crop of small heads. The center head is cut, leaving as many leaves on the remaining plant as possible. Small heads will develop on the stem near the base of the leaves. These heads are quite edible and should also be picked when fairly firm (Minges 1977).

Common Problems. The heads of early cultivars frequently split in warm weather. This is caused by too much water moving into the mature head. To reduce this problem, either cabbage should not receive water once the head is mature or the roots of the plant should be pruned (Doty 1973). To prune the roots, the plant can be twisted to break some of the roots or the roots can be cut with a shovel or knife.

There are several major diseases of cabbage. Fusarium wilt, commonly referred to as yellows, is a disease found in the upper South and the North. The lower leaves become yellow and turn brown. In areas where the disease is a problem, yellows-resistant cultivars should be grown.

Black rot and blackleg are diseases spread by diseased seeds or transplants or by insects. Black rot causes yellowing of the leaves and blackening of the veins. Blackleg causes dark sunken areas on the stem of young plants. The control is to use disease-free transplants and to purchase hot water-treated seed. Soaking the seed in water at 112°F for 27 min will destroy the organisms. Black rot-resistant cultivars should be grown if this disease is a problem. Once it infects the garden, no members of the cabbage family (cauliflower, broccoli, cabbage, and others) should be grown on the area for three years. Crop rotation, destroying (not composting) diseased plants, and insect control are recommended to help in disease control.

Clubroot attacks the roots of plants of the cabbage group. This causes the roots to become club-like in appearance. Once the fungus is established, it lives in the soil for 15 years or more. The disease is transmitted by diseased plants and infected soil.

The major insects affecting the cabbage family are cabbageworms of various types. The adults are grey, brown, or white moths. The adult commonly seen is a white moth hovering over the plants in the garden. The worms are light green or dark green and can be controlled with chemical or biological control agents. If given a choice, cabbageworms prefer savoy cabbage, and attack red cabbage the least.

The cabbage aphid is a problem in some areas. The aphid has a waxy covering, and a small amount of detergent is usually needed with the chemical or botanical insecticide for control. These aphids need to be controlled before the heads begin to form and serous damage has occurred.

CANTALOUPE (*See* MUSKMELON)

CARROTS

Carrots (*Daucus carota* var. sativus) are native to the northwest India area. The wild carrot had a thin, tough root, and roots resembling present-day cultivars were developed in France. Carrots were well established as a food in Europe by the thirteenth century. They were brought to America by the early settlers where they became a popular food among the Indians.

Plant Characteristics. The carrot is a biennial that is grown for its root. A carrot root cut crosswise shows two distinct areas, an outer area and an inner area or core. High quality carrots are those with a large outer area compared with the inner one. The outer area contains more sugar and vitamins than the inner one.

Carrot produces a large root its first year of growth. During the second year, a large seedstalk 2 to 3 ft high is produced.

Culture. Carrots are moderately frost tolerant. They are grown in the fall, winter, and early spring in the South and Pacific Southwest. In the North, they are grown from early spring to fall.

FIG. 7.10. Carrots with short roots should be grown in shallow soils. *Courtesy of W. Atlee Burpee Co.*

Carrots require deep soil to produce a long root. If the garden contains a hardpan or if only the top 3 or 4 in. are worked, short-rooted cultivars should be grown. Long-rooted cultivars will be misshapen or forked if grown in poorly prepared soil. Carrots are easily grown in containers or in raised beds. A simple raised bed can be made from 2 × 8 in. lumber 4 ft or less wide. This bed is then filled with a container-type soil mixture containing fertilizer. Soil in this raised bed warms up quicker than the surrounding soil and carrots may be planted earlier in the spring. By being above the surrounding soil, the bed drains rapidly, preventing the accumulation of water from fall rains. This reduces root decay problems.

Unfinished compost or manure used as a fertilizer for carrots induces rough and branched roots. These materials should be well composted before being added to soil where carrots are grown.

Carrots are seeded about one-half inch deep or less in the early spring. They should be covered with sand, vermiculite, sawdust, or fine peat moss instead of soil. Two to 4 seeds per in. are planted. About half of the seeds will germinate. Carrots can be planted in a bed, a wide row, or in single rows 1 ft apart. Successive plantings about every 3 weeks will ensure a continuous supply.

Carrots cannot tolerate deep planting or a dry seedbed. The seedbed must be kept moist during the germination period, which frequently means that the area is sprinkled lightly with water each day. Seeds may take 2 weeks to germinate. Gardeners often water the area after seeding and then cover the row with a clear plastic sheet. This will warm the soil and conserve moisture. The sheet should be removed when the seedlings emerge.

The top of the carrot root becomes green when exposed to sunlight. To prevent this, the plants should grow rapidly so that the leaves shade the roots. Gardeners frequently place additional soil around the roots 40 days after planting, when the roots begin to enlarge, to prevent the green areas from occurring.

Cultivars need to be selected for the soil condition. All cultivars grow well on deep sandy or loose soils. On heavy or impermeable soils, the short-rooted cultivars are best. Short-rooted cultivars are easily grown in containers.

Harvest. Harvest of carrots can begin when they are pencil size to thin the plants in the row. The more space the remaining roots have, the larger they will become. Continued thinning and harvesting can continue for 3 to 6 months.

Carrots can be harvested, the tops removed, and stored in a cool, moist place. They should not be washed until needed. However, carrots are best left in the garden until a severe frost is expected. A heavy mulch can be placed over them and the roots harvested until the ground freezes. In areas where mild winters occur, they can be left in the ground until needed.

Common Problems. The most frequent problem encountered is a poor stand. This can be solved by shallow planting, frequent light watering, and preventing the soil from crusting during germination.

Carrots have few insect and disease pests in the garden. In the Southeast, carrot yellows may be a problem. This is a virus disease carried by leafhoppers.

In the Northeast and coastal regions of the Pacific Northwest, the carrot rust fly may cause damage. The larvae burrow into the roots. For control, carrots can be grown during cool weather; the plants should be grown to maturity as rapidly as possible. Carrots should be rotated within the garden area to reduce damage.

CAULIFLOWER

Cauliflower (*Brassica oleracea*, Botrytis group) is often called heading broccoli. It is a type of cabbage that originated with cabbage in southern Europe. Cauliflower is often called the aristocrat of the cabbage family because of its delicate growing requirements.

FIG. 7.11. The leaves of cauliflower are sometimes tied loosely around the head (background) to produce a whiter head. *Courtesy of Harris Seeds.*

Plant Characteristics. Cauliflower is grown for its white head, called the curd. The head is formed from shortened flower parts at the top of the plant.

Culture. Cauliflower is a cool season crop that requires more attention than cabbage or broccoli, its close relatives. Too much heat prevents the cauliflower head from forming, and the plant is more sensitive to cold than cabbage. In the South and Pacific Southwest, it is grown in the early spring, fall, and winter. In the North, it can be grown as a spring or fall crop. At high elevations and a few other areas where the climate is favorable, it is grown during the summer as well.

Cauliflower must be grown rapidly through its entire life, from seedling to harvest. Anything which slows or delays its growth, such as insects, lack of water, or excessive heat or cold, will prevent development of the head.

To ensure vigorous growth in the seedling stage, cauliflower is usually grown as a transplant for all planting dates. For spring crops, the plants are started indoors and are ready for use after 6–8 weeks. The plants are then transplanted to stand 15 in. apart in rows, which are 15–24 in. apart. They should be planted 2 weeks before the average frost-free date. A starter fertilizer should be used when transplanting.

For fall plantings, transplants or direct seedings are used. For transplants, the seeds are planted one-fourth to one-half inch deep in a protected area of the garden. Seeds are seeded thickly 4 weeks before they are to be used. In hot, dry weather the area is watered immediately and may require additional water to ensure germination. Fall plantings are transplanted 2 ft apart in rows. For direct seeding, seeds are planted 2 weeks before the time cauliflower transplants are set in their permanent garden location. The fall crop must be planted so that it will form the head during cool weather. Direct-seeded plants are thinned to 2 ft apart in rows 30 in. apart.

Cauliflower plants can be induced to produce the head prematurely. If the plants are large enough, and exposed to temperatures below 50°F for several weeks, a small head will form. This frequently occurs in areas with mild winters where the cauliflower is transplanted too early in the spring or too late in the fall. In the north, transplants are frequently set out in the garden too late, and excessive heat causes the small head to form.

Adequate fertilizer and moisture are essential. The area should be fertilized before planting and the plants should receive additional nitrogen fertilizer when they are half grown. In dry weather, the plants should receive 1 to 1½ in. of water each week.

When the head is 2 to 3 in. in size, it is often protected from sunlight. This keeps the head white, protecting it from sunscald and frost injury and preventing the resulting off-flavor. The top leaves are tied loosely over the head, with care taken not to cramp the head. Many gardeners use spring-type clothespins to tie the cauliflower plants for blanching. The leaves are clipped together with one or two clothespins. This makes it easier to adjust the outer leaves as the plant grows. The heads are ready to harvest 1 to 2 weeks after covering. If the heads are produced under conditions where sunscald and frost injury are not a problem, slightly yellowish heads are produced. These heads are generally of a quality equal to the covered heads (Minges 1977).

Various types of Snowball cultivars mature early and are most practical in the garden. Late maturing cultivars take 4 to 6 months to mature and are difficult to grow in the garden.

Self-Blanche is a cultivar reported to take 70 days to mature. However, at hot temperatures it stops growing, unlike other cultivars which produce a small head under these conditions. Thus Self-Blanche may mature fairly late. When grown under cool conditions, Self-Blanche produces leaves that curl naturally over the head. When grown at warm temperatures, tying may be necessary to produce a white head.

Royal Purple Head is often listed as a purple cauliflower. However it is a special type of broccoli and may be grown in place of cauliflower. See Broccoli discussion for its culture.

Harvest. The heads are ready to use when they reach suitable size, usually 6 in. or more in diameter. Heads should be harvested before they are overmature and the flower parts separate. The heads are harvested by cutting the main stem and should be used soon after harvest.

Common Problems. The most frequent problem is the production of small heads. This is induced by cold, heat, drought, plant damage, lack of fertility, hardening the transplants, or other factors that interrupt the continuous growth of cauliflower.

Disease and insect problems are similar to those of cabbage. Cabbageworms and black rot are most common. Cabbageworms can be controlled with a chemical or biological control agent. To help control disease, cauliflower should not be planted where members of the cabbage family (broccoli, cauliflower, and others) grew the year before.

CELERY

Celery (*Apium graveolens* var. dulce) is native to marshy areas of Europe. It can be found growing wild from Scandinavia to Egypt, in parts of Asia, and in the mountains of India. It was not grown for food until relatively recently, about 1600.

Plant Characteristics. The first cultivated types were similar to the wild types. These plants had hollow, fibrous, strong-flavored stalks. Modern cultivars are of two types, the green types and the golden or self-blanch types. Celery is grown for its long fleshy edible petiole.

Culture. Celery is a cool season crop that can be grown in most of the United States. It is grown during the winter in the Pacific Southwest and

FIG. 7.12. Celery is usually grown by experienced gardeners. *Courtesy of Harris Seeds.*

the lower South. It can be grown as an early spring or late fall crop in the upper South and the North. Celery requires rich, moist soil and mild, even temperatures. Any condition that interrupts its growth causes problems. Celery requires 120 to 140 days to produce a crop. Because celery is slow growing and a delicate crop to produce, it is usually grown only by experienced gardeners who desire a challenge.

Celery is usually produced from transplants. At high temperature, the seed becomes dormant and will not germinate. Celery must be started 4 months before spring planting time. The seed requires 15–21 days to germinate, but soaking the seed overnight before planting reduces the time somewhat. The seeds are sown one-sixteenth of an inch deep in flats or peat pellets. They must be kept moist and at a temperature of 60°–70°F until the seedlings appear. The flats or peat pellets are covered with moist burlap.

Once the sprouts have appeared, the plants are moved into the sunlight at a slightly cooler temperature. Seedlings are transplanted or thinned to stand 2 in. apart and kept in full sunlight until danger of frost is past. The plants are then placed outdoors to harden them. If the plants receive several cold nights, they will produce seedstalks instead of the edible petioles.

Celery requires rich, moist, well-drained, deeply prepared, mellow soil. Large amounts of well-composted material or peat can be added to the garden soil to simulate this condition. The transplants are spaced 6 to 10 in. apart in rows 2 ft apart. If possible, they should be transplanted on a cool cloudy day to reduce water consumption. The plants should be watered after transplanting if the soil is at all dry.

Celery requires a continuous supply of fertilizer and water. It should be fertilized with a complete fertilizer every 2 or 3 weeks and plants should be watered thoroughly and often. When the plants become large, soil should be placed around the outside of the plants to keep them upright. If soil gets into the center of the plant, it will rot. About 120 days after transplanting, the plants will be ready for harvest.

Blanching to produce white petioles is usually not necessary with the self-blanch cultivars. However it can be done by excluding light from the petioles by wrapping them with paper or placing a drain tile around the plant. Blanching induces a change in petiole color but little change in flavor.

Harvest. Celery is harvested by cutting the taproot below the ground. It is usually ready to harvest 90–120 days after transplanting. Overmature plants contain cracked and pithy petioles.

Celery may be left to overwinter and receive a cold treatment in mild areas. The plant then produces large amounts of celery seed, which can be harvested and used in flavoring.

Common Problems. A number of leaf-eating worms and aphids attack celery. These can be controlled with a chemical or botanical insecticide. Blight and mildew can be controlled with a fungicide.

CHARD OR SWISS CHARD

Chard (*Beta vulgaris*, Cicla group) or Swiss chard is a type of beet that was developed for its large crisp leaves. It was first reported in the Mediterranean region and Canary Islands. Chard was popular as long ago as 350 B.C. It is a favorite crop of the people of Switzerland. Swiss settlers introduced it into the United States in 1806.

Plant Characteristics. Chard is grown for its large green crinkly leaves and fleshy leafstalks. The leafstalks and midribs are usually white but red cultivars are available. Chard does not produce seedstalks during hot weather as do lettuce and spinach. Chard is frequently used in place of these vegetables during the summer.

FIG. 7.13. Swiss chard is a tasty summer-to-fall green.
Courtesy of W. Atlee Burpee Co.

Culture. Chard is grown similarly to beets. Seeds are planted in the fall to spring in the South and Pacific Southwest. In other areas it is planted from early spring to midsummer.

Chard grows best in rich mellow soil. Seeds are planted one-half inch deep, 1 to 2 in. apart in rows about 18 in. apart. Plants emerge in 8–10 days. Each seedball of chard contains 1 to 6 seeds and the plants will need to be thinned until they are 12 in. apart. For rapid and continuous growth, additional nitrogen fertilizer can be applied about a month after planting. Both white- and red-ribbed cultivars are available.

Harvest. The small chard plants can be harvested when the row is thinned and used as greens. Large leaves and stems are harvestable 50–60 days after planting. The outer leaves are cut off an inch above the ground with a sharp knife, while they are still young and tender. Care should be taken not to injure young leaves or the center bud. Leaves should be harvested continuously to stimulate the production of young new leaves from the center bud. Just before the first hard freeze, the entire plant may be dug up, placed in a pail, and stored in a cool place. Plants are placed upright with roots still containing some soil. They should be watered lightly to prevent wilting. The leaves can then be harvested into the winter months.

Common Problems. In the garden, chard is generally free of insect and disease problems. Occasionally cabbageworms, aphids, beet leaf miner, and flea beetle are a problem. Worms may be hand picked; and on smooth-leaved cultivars, the aphids can be washed off with a spray of water from the garden hose. Various chemical, botanical, or biological control agents may also be used. In some warm climates, curly top virus disease causes severe damage. This can be avoided by planting in late winter or early spring.

COLLARDS

Collard (*Brassica oleracea*, Acephala group) is sometimes called nonheading cabbage. It originated in the British Isles and Western Europe, where it has been used for food for more than 4000 years. The present cultivated form has changed little in the past 2000 years. Collard was brought to America by early settlers and was a common garden vegetable by 1670.

Plant Characteristics. Collard is a nonheading type of cabbage. It forms a large rosette of blue-green leaves. Plants grow 2 to 4 ft in height.

Culture. Collard is a hardy, cool season vegetable. It withstands summer heat and short periods of cold as low as 10°F. It is grown throughout the entire year in the South, and as a spring and fall crop in the north.

Collard may be transplanted or seeded. It can be seeded 4 to 6 weeks before the last spring frost for summer harvest; or in midsummer for fall harvest; or, in mild climates, in the fall for winter harvest. Seeds are planted one-fourth inch deep, 1 in. apart in rows 30 in. apart. They are later thinned to 6–8 in. apart depending upon the type of harvest.

The plants require 1 to 1½ in. of water each week and respond to nitrogen fertilizer. They should be sidedressed with nitrogen 1 month after planting.

Harvest. Collards may be harvested by three methods. (1) The entire young plant may be cut off at ground level and used; (2) the entire mature plant may be cut off at ground level when the plant is 6 to 10 in. high and used; or (3) small plants can be harvested to thin the row and used. The remaining plants are allowed to mature and the bottom leaves harvested, beginning when the plants are 10–12 in. high. The bud will produce more leaves, and the bottom leaves can

FIG. 7.14. Collards are a cool season vegetable grown for both spring and fall crops. Harvested leaves are in the foreground.

be harvested continuously throughout the growing season. Successive plantings are not required for a continuous supply. The last harvest procedure produces a plant with a bare stalk containing a group of young succulent leaves at the top. These plants may require staking to keep them from falling over.

Frost improves the mild cabbage-like flavor.

Common Problems. The common insect and disease problems are similar to cabbage. Cabbageworms, aphids, and harlequin bugs are the common insects. Downy mildew and blackleg diseases may also be a problem.

CORN, POP

Popcorn (*Zea mays* var. praecox), like sweet corn, originated in North America and is a mutation of field corn. Archaeological evidence has shown (Anderson 1954) that popcorn was among the most primitive types of corn used for food and was used in 5000 B.C.

Plant Characteristics. Popcorn has very hard starch in the kernels. This starch explodes when heated, producing fluffy white popcorn.

The type used by the Indians of North America was called "rice corn" (Anderson 1954). These kernels are sharply pointed and are produced on a relatively small plant. The rounded or "pearl" kernels are produced on larger plants of more recent origin.

Culture. Popcorn can be grown for its popping qualities or as an ornamental. It can be grown wherever sweet corn is grown, but early maturing cultivars must be selected in northern areas.

Popcorn is planted about 1 in. deep, 4 seeds per ft, in rows 30 in. apart. Seeds are planted when the soil temperatures have warmed to 50°F or

FIG. 7.15. Strawberry popcorn may be used for popping or as an ornamental. *Courtesy of Harris Seed Co.*

above. When the plants are established, they are thinned to stand 6 in. apart. Popcorn should be planted in 3 or 4 short rows to ensure good pollination, and not in one long row. If popcorn is planted near field corn or sweet corn, it will be pollinated by these corns. This does not affect popping quality. Seeds from these cross-pollinated plants should not be saved and planted the next year as the popcorn produced seldom pops.

Popcorn, like sweet corn, requires adequate fertility and moisture (see sweet corn discussion for culture requirements).

There are many cultivars available in several colors. Red, purple, and black-colored cultivars produce small ears and are sometimes used for ornamental purposes. Standard cultivars are yellow or white, hulled or hulless, and early- or late-season producers.

Harvest. Many cultivars of popcorn produce 2 or 3 ears per plant, particularly the cultivars that produce small ears.

The kernels should mature on the plant before frost to about 20% moisture. Unless the kernels contain 35% moisture, they will not be damaged by a light frost. Freezing does not damage kernels with less than 20% moisture.

The ears should be harvested, the husks removed, and the ears stored an additional 3 weeks in a sheltered dry place outdoors. The kernels will dry naturally to about 13% moisture, which is ideal for popping. The kernels can be removed from the cobs with a moderately aggressive twisting motion. The kernels are then stored in airtight sealed containers for later use. When popped, the corn should expand 20–40 times its original volume.

FIG. 7.16. Indian ornamental corn is frequently grown for a fall decoration. *Courtesy of Harris Seeds.*

Common Problems. Popcorn has many of the same pest problems as sweet corn.

Popcorn frequently fails to pop. If the popcorn has been stored in a heated area in an open container it will dry to about 7% moisture. This is too low for good popping. The moisture content can be increased by about two percentage points by adding a small amount of water to the dry popcorn. One tablespoon of water is added to 1 qt of popcorn; and the container is sealed, shaken, and allowed to absorb the water for a week.

CORN, SWEET

The word "corn" is applied to oats, wheat, barley, or rye in Europe. In the Americas, it refers to Indian corn or maize. Corn was grown in North America before 2000 B.C. The type of corn known as sweet corn (*Zea mays* var. rugosa) is believed to be a mutation of field or dent corn. It was grown by the Indians and first collected and described by settlers in about 1780. By 1900, over 63 cultivars had been described. Today, over 2000 cultivars and hybrids are available.

Plant Characteristics. Sweet corn is an annual grass plant. It differs from all other types of corn. Sweet corn produces and retains large amounts of sugar in the kernels, hence its name "sweet." The skin of the kernels is thinner, making them tender. The seed is wrinkled when dried, and the plant has a tendency to produce "suckers" or additional shoots at the base of the plants.

The male part of the corn plant is the tassel and the female part is the ear. The ear is pollinated by pollen blown by the wind, and sweet corn ears can be pollinated by other types of corn, making them less sweet.

Some newer sweet corn hybrids (extra-sweet cultivars) produce kernels that contain more sugar, and convert sugar to starch less rapidly, than standard cultivars. These new hybrids are sweeter initially and remain sweeter for a longer period of time. They also contain fewer water-soluble polysaccharides than standard corn, and this is what gives sweet corn a creamy texture. This lack of creaminess is not noticeable if sweet corn is eaten fresh, but it reduces the quality of canned or frozen sweet corn.

Culture. Sweet corn is a warm season crop. It is easily killed by frost and requires a soil temperature above 50°F for the seed to germinate. It can be grown successfully in all states. In the South, sweet corn is grown from early spring until fall. In the North and Northeast, it is grown from late spring until fall. In south Florida, Texas, and some parts of the Southwest, it is grown in the fall, winter, and spring. In the Pacific Northwest, an early maturing cultivar will need to be grown during the summer.

Sweet corn requires a lot of space and sunlight and is frequently grown only in larger gardens. It is wind-pollinated and should be planted in 3 or 4 short rows rather than a single row. If the prevailing wind is blowing across,

FIG. 7.17. Sweet corn requires a lot of garden space. *Courtesy of W. Atlee Burpee Co.*

rather than down this single row, the pollen will not land on the silks and few kernels will develop on the ear. Different types of corn should not be planted together. Pollen from field corn or popcorn will land on the silks of sweet corn, causing the kernels to be less sweet. The extra sweet and standard cultivars should not be planted at the same time. The pollen will intermingle, and the quality of sweet corn produced by both cultivars will be reduced. Cross-pollination between standard yellow and white cultivars will change the colors of the kernels produced but not affect the flavor or quality. Different types of corn should be separated by planting 400 yards apart in distance of 1 month apart in time.

Seeds are planted about 1 in. deep in moist soils and 1 to 2 in. deep in light sandy soils after the soil has warmed to 50°F or above. Seeds of many of the extra sweet cultivars do not germinate in cool, wet soil and should be planted when the soil is 65°F or above. Seeds should be planted about 5 or 6 in. apart in rows 30–36 in. apart. Once the plants are well established, they should be thinned to stand 10–16 in. apart. The greatest reduction in quality and ear size in gardens is caused by lack of thinning or by planting the seeds too close together.

To conserve space, sweet corn is often planted next to vine crops, such as cucumber, pumpkins, and muskmelons. The vines can be trained to grow between the corn plants and utilize this space.

Plant nutrients are especially important for growing sweet corn. Once the plant is stunted, it never fully recovers. Fertilizer is generally broadcast and worked into the soil before planting. Additional amounts are applied when the plant is half grown. The fertilizer is then applied 3 in. to the side of the row and worked into the soil about 4 in.

On light, sandy soils, such as found in the Great Plains and near the Atlantic and Gulf coasts, 20 lb of 10-10-10 fertilizer or equivalent is applied per 1000 sq ft over the entire area before planting. An additional one-half pound of actual nitrogen per 100 ft of row is added when the plants have produced 6 to 8 leaves.

On soils of average fertility, such as found in the Northeast, 15 lb of 5-10-5 fertilizer or equivalent is added before planting. An additional 3 lb per 100 ft of row of the same fertilizer is added beside the row at planting.

On fertile soils, such as the Midwest, some western valleys, and the Pacific Northwest, the fertilizer is added at planting time. About 3 lb per 100 ft of row of 5-10-5 is applied beside the row.

Sweet corn needs a continuous supply of moisture. After the tassels are produced, sweet corn requires 1 to 1½ in. of water each week. This ensures pollination and production of the ear.

Successive plantings can be made to provide a continuous supply of sweet corn. The number of days to maturity is not a good index. Sweet corn responds to warm temperature; and the warmer it is, the faster sweet corn grows and matures (up to 100°F). Successive plantings are made when plants in the previous planting contain 3 or 4 leaves. An early cultivar and a full season cultivar may be planted at the same planting date to provide for a longer harvest.

The number of suckers a sweet corn plant produces depends upon the cultivar. These should not be removed. Their removal does not increase yield, and it may actually reduce yield.

Sweet corn cultivars are affected by the length of day. Early maturing cultivars grown in the north are adapted to long, cool summer days. These cultivars do not grow well in the South. Cultivars adapted for the South respond to short summer days. If these are planted in the North, these southern cultivars will not flower (produce the tassel and ear) until short days occur in the fall. These plants may become 10 ft tall or more and are sometimes grown as a novelty. However, they seldom produce a satisfactory ear before they are killed by frost.

There are many sweet corn cultivars. These include early, full season, and late cultivars; yellow, white, and bi-color; and standard sugar and extra sweet types. There is a wide assortment for the gardener to choose from.

Harvest. Each sweet corn plant should produce at least one large ear. Many cultivars produce a smaller second ear which develops later. The ears should be harvested at prime maturity. This occurs 17–24 days after the silk has emerged from the ear. When grown under warm days and nights, prime maturity occurs 17–18 days after silking. If grown during cool weather, prime maturity may occur 22–24 days after silking.

Sweet corn of prime maturity is in the milk stage. The silks are brown, and the ear has enlarged to fill the husk tightly to the tip. The kernels are plump, soft, tender, and filled with a milky juice. Prime quality lasts only 4 or 5 days. It is not a good practice to remove part of the husk to determine if the ear is at prime maturity. Birds and insects then frequently attack the kernels. Placing a paper bag tightly over disturbed ears usually prevents damage, however.

Sweet corn is removed from the plant by snapping the ear from the stalk. The ear is bent downward, twisted to one side, and pulled off, all in one quick movement.

Harvested ears should be eaten as soon as possible after harvest. The sugar content rapidly declines. The ears may be kept in a refrigerator for 2 or 3 days with only a small loss of quality.

Common Problems. Diseases are usually not a serious problem in gardens. Corn smut is sometimes prevalent, particularly on white cultivars. The smut is not poisonous and is characterized by large, fleshy galls on the stalks, tassel, or ears. Smut can be controlled by avoiding injury to the plants, avoiding areas where smut occurred the year before, and removing and destroying the smut. The smut should be removed before the galls break open and release new spores to infect other plants. They should be destroyed and not left in the garden area.

Root rot frequently occurs on seedlings planted on cool moist soil. Most sweet corn seeds sold commercially have been treated with a mild fungicide to prevent this problem.

Stewart's bacterial wilt is sometimes a problem in the mid-South. Early infected plants wilt and die. Later-infected plants are stunted and contain yellow streaks on the leaves. The disease is spread by the corn flea beetle and is more of a problem after mild winters have occurred. Most newly developed full season corn cultivars are resistant to Stewart's wilt.

Insects which attack sweet corn during its early growth include southern corn rootworm, cutworm, white grub, wireworm, and flea beetle. See Chapter 4 on Pest Control for control measures if these are a problem in the garden.

Once the sweet corn plant is established, it can withstand considerable insect damage. Corn borer damage is usually ignored in the garden as the crop is harvested before severe damage occurs.

Corn earworms can be a problem. Early plantings in the North are often not infested, but later plantings are usually affected. Corn earworm control must be timely. The corn earworm moth lays eggs on the silk. These eggs hatch in 24 to 48 hr, and the small caterpillars move down the silk into the ear, where they feed on the tip. Once inside, they cannot be controlled. Insecticides must be applied while the caterpillar is still on the silk, and that means every 2 or 3 days. Thus, most gardeners ignore the problem and just remove the damaged ear tip when the sweet corn is husked. Tightening the end of the husk around the ear to restrict the caterpillars' entry is sometimes effective. Placing a rubber band around the tip after the silk has appeared may reduce damage.

CUCUMBER AND GHERKIN

Cucumber *(Cucumis sativus)* and gherkin *(Cucumis anguria)* are natives of Asia and Africa where they have been grown for food for at least 3000 years. They were introduced into China in 100 B.C. and into France in the ninth century. They were grown by the Indians before Columbus, and other additional cultivars were introduced by the early settlers.

The gherkin is well established in tropical Central and South America and grown widely in the West Indies. The West Indian gherkin is also called burr cucumber. In the produce trade, the term "gherkin" refers to any immature cucumber fruit, usually pickled.

Plant Characteristics. The cucumber and gherkin plants are similar. They are vine plants with large leaves and long petioles. Some cultivars produce short vines for growing in a small garden or containers. The vine contains separate male and female flowers, with the first flowers produced being males. A tiny cucumber develops at the base of the female flowers, and these produce the edible fruit.

Gynoecious cucumbers are special hybrids that have all female flowers. These plants usually produce fruit earlier than other cultivars and are usually sold with a few seeds of a standard cultivar for use as a pollinator.

Special types are sold for production of fruits in greenhouses. These types are the long, narrow, seedless, English cucumbers frequently seen in the produce market. These types are usually parthenocarpic, that is, they produce fruit without pollination. However, in the garden, these flowers will cross with other cucumbers and produce fruits with seeds.

Cucumbers do not cross-pollinate with muskmelons or watermelons. The flavor of the fruit is unaffected by pollen from these plants.

Cucumber fruit is usually green and warted. However, white and yellow cultivars and smooth-skinned green cultivars are available. Gherkin fruit is generally round, more warted than cucumbers, and only 1 to 3 in. long.

FIG. 7.18. Cucumbers ready to pick for use as pickles. *Courtesy of Harris Seeds.*

Culture. Cucumbers are a warm weather vegetable and are very sensitive to cold. They can be grown during the winter only in a few extreme southern locations.

Cucumbers grow best in a fertile mellow soil with large amounts of compost. The compost should not be made from cucumbers, pumpkins, melons, or squash as these plants have similar disease problems.

Fertilizer should be added before planting and again when the plants begin to produce vines.

Cucumbers love warm soil and the use of plastic mulches is common. They may be started as transplants, particularly in the Pacific Northwest and the Rocky Mountain states. In warm areas they are planted directly in the garden. Temperatures should be 60°–75°F when the plants or seeds are planted.

Seeds are planted 1 in. deep in rows 3 or 4 ft apart. They may be planted and thinned to stand 12 in. apart as single plants, or 3 plants together in a hill every 2 or 3 ft.

Cucumber vines and fruit are lightweight and easily trained onto a trellis, fence, or cage similar to that used to control tomato fruit rots. When vines of the long-fruited cultivars are supported, the fruits hang free and

FIG. 7.19. The West Indian gherkin produces small spiny fruit.

are long and straight. When space is limited, or the plants are grown in containers, the various small vining cultivars may be used.

Cucumbers are relatively shallow-rooted and require irrigation in most parts of the country. If alternate methods of irrigation are available, sprinkler irrigation is not recommended. Moisture on the leaves from rain, dew, or irrigation encourages diseases. If the plant does not receive adequate water, it stops growing but will resume growth when water is applied.

Cucumbers are generally used for slicing and pickling. The pickling cultivars produce smaller fruits but all cultivars may be used for both purposes. There are numerous cultivars available.

Common cultivars include: standard cultivars for slicing, standard cultivars for pickles, gynoecious cultivars for slicing, gynoecious cultivars for pickles, burpless cultivars that produce long fruits with skin that is tender and free of bitterness, small vined cultivars used in small gardens or in containers, and greenhouse cultivars. The gherkin listed is commonly called the West Indian Gherkin.

Harvest. Cucumbers are usually harvested immature before the seeds become hard. They are of highest quality when dark green, firm, and crisp. The large burpless and greenhouse types should be about 1½ in. in diameter and less than 10 in. long. To force the plant to produce continuously, fruits should be picked every other day, even if the fruit is not needed.

Common Problems. Cucumbers are susceptible to a number of diseases. Resistant or tolerant cultivars should be used wherever these diseases are likely to be a problem. Do not handle or harvest the plants when they are damp to help prevent spread of disease.

Anthracnose. This disease is particularly a problem in the Southeast. It attacks cucumbers, muskmelons, and watermelons. Small, round, water-soaked spots appear on the fruits. Infected leaves have a scorched appearance. To control it, plant resistant cultivars and treated seed. Remove all vines from the garden at the end of the growing season and rotate the crops in the garden.

Mildew. Downy mildew is a fungus that grows rapidly under warm, moist conditions. It is particularly a problem in the Atlantic and Gulf states. Low humidity, high temperatures, and lack of water on the leaves prevent its growth.

Symptoms appear on leaves as small, yellow spots with irregular edges. The tissue in the middle of each spot dies and spots appear in large numbers. The leaves die and curl upward. The problem develops about the time the plants begin to set fruit. Use resistant cultivars if possible.

Powdery mildew is a problem with cucumbers, muskmelons, and pumpkins but not watermelon. Small white spots develop on the underside of older leaves. The leaves then become covered with a white mold growth. The leaves die and leave fruit exposed to sunlight to become sunburned. In muskmelon, the fruit ripens prematurely and is of poor quality. This fungus requires high temperature and a lot of sun. Rain and low sunlight reduce its growth.

Mosaic. This disease is caused by a virus. Watermelon mosaic is not seedborne and is carried by aphids from various ornamental plants around the home. Squash mosaic is seedborne and carried by cucumber beetles. Symptoms appear as light-green mottling of the leaves. The young leaves and flowers are malformed and small. The leaves are dwarfed and the vines fail to grow. Poor quality fruit is produced. Late infections are usually mild and do little damage. For prevention, control aphids and cucumber beetles and grow resistant cultivars.

Scab. Scab fungus attacks the fruit and is particularly a problem with cucumber. Sunken, dark brown spots, irregular in shape, appear. A gummy material exudes from these spots. It is most serious in the northern states. Using resistant cultivars and destroying vines after harvest usually controls this problem.

Aphids. Aphids are a sucking insect on the underside of leaves. See Chapter 4 for identification and control.

Cucumber Beetle. Cucumber beetle spreads bacterial wilt. The eastern type has 3 black stripes down its wing covers. The western type has 12 black spots on yellow wing covers. Bacterial wilt causes the plants to wilt and die. The disease plugs the water transporting system, and when symptoms are noted it is usually too late for control of the wilt. Remove infected plants and see discussion in Chapter 4 for control of the beetle.

EGGPLANT

Eggplant *(Solanum melongena)* is probably a native of India, where it grows wild. It was recorded as growing in China in the fifth century. Eggplant was introduced into Spain during the Moorish invasion, where it spread throughout Europe. The Spanish explorers introduced purple and white cultivars into America. The fruit of eggplants grown in the sixteenth century were small and were similar to the types now grown as an ornamental.

Culture. Eggplant is grown in the late spring, summer, and fall in most of the North and upper South. In the extreme South, it can be grown in the winter. It grows poorly in many parts of the Pacific Northwest.

FIG. 7.20. Eggplant is a warm weather plant. *Courtesy of Harris Seeds.*

Eggplant is a warm weather plant requiring 100–140 days from seeding to produce fruit. In most areas it is therefore grown as a transplant. Anything that stops its growth is detrimental to eggplant. See Transplants in Chapter 2.

Eggplants should be fertilized before planting and grow best in well-drained soil with ample organic matter. The plants should receive a starter fertilizer and be fertilized every 6 weeks over the season. If eggplant stops growing, inferior fruit is produced.

Plants are planted in the garden in direct sunlight 2 weeks after all danger of frost has passed. Eggplants grow best at 78°F day and 68°F night temperatures. A common mistake is to set the plants out too early. If there is any danger of frost, the plants need to be covered with paper or plastic coverings or boxes. The covers should be removed during the day.

Plants are spaced 12 in. apart in rows 30 in. apart. Eggplants are easily grown in containers or minigardens. Plants grow best with 1 in. of water each week. They will tolerate dry weather once the plants are established but they need water during extended dry weather. Eggplant grows well under all types of irrigation systems.

Many cultivars are available different shapes and colors. Most fruits are purple, but white- and green-fruited cultivars are available. Fruit size ranges from large to small oval or long slim fruits. Japanese eggplant cultivars bear long, slim fruits.

Harvest. Fruits are harvested about 70 days after transplanting when the fruits are one-third to two-thirds mature. Large cultivars will be 6 to 8 in. long. The fruit skin should have a glossy shine and fruit can be removed with pruning shears or a knife. Part of the stem should be left on the fruit. When the fruit skins become dull or brown, they are mature and too tough for good quality. Fruits should be eaten soon after they are harvested.

Common Problems. Phomopsis rot may be a problem. Symptoms are large sunken tan or black areas on the fruit. Canker-like lesions may appear on the stem. Leaf spots appear and enlarge, the centers die, and the leaves turn brown. The disease is often carried over-winter by previously infected plants of eggplant. To control, remove and discard diseased plants, use resistant cultivars, and use clean seed.

Verticillium wilt is common in cooler areas. The fungus lives in the soil almost indefinitely and is also carried by tomatoes and peppers. Resistant cultivars and crop rotation are the best control methods.

Insects that attack eggplant include flea beetles, aphids, and sometimes Colorado potato beetle and red spiders. See Tables 4.3 and 4.5 for control procedures.

ENDIVE AND ESCAROLE

Endive and escarole *(Cichorium endivia)* probably originated in East India. They were used for food by the Egyptians and Greeks in 200 B.C. and later by the Romans.

FIG. 7.21. Endive is a common fall crop. *Courtesy of Harris Seeds.*

Plant Characteristics. Endive has finely cut, loose, narrow, medium green, fringed, and curly leaves. The leaves have a slightly bitter taste.

Escarole, sometimes called Batavian endive, is a selection of endive. These plants have broad, thick, smooth leaves with a white midrib. The plant forms a loose head with partly blanced (white) inner leaves. Escarole has a milder flavor than endive.

Culture. The culture of endive and escarole are similar and both are grown like lettuce. They are planted in the winter in the south and in the early spring, summer, and fall in the north. The plants produce seedstalks with blue flowers in hot weather and a fall crop is best. They grow best at a temperature of 60°–70°F.

Endive and escarole may be planted indoors for later transplanting. The seeds are planted one-fourth inch deep in flats or peat pellets 6 to 8 weeks before planting in the garden. Plants are thinned to stand 2 in. apart and are transplanted when they are 3 in. high. A starter fertilizer helps establish the plants.

Seeds may be planted directly in the garden in early spring, 2 or 3 weeks before the frost-free date. They are planted one-fourth inch deep. Both seeds and transplants are placed in rows 18–30 in. apart and thinned to stand 8 to 10 in. apart. Summer plantings produce a fall crop that matures in cool weather, making the leaves milder in flavor. Summer plantings are frequently interplanted as transplants into other rows of vegetables.

Endive and escarole respond to fertile soil and a uniform supply of water.

Summer plantings will require frequent irrigation during dry periods. The plants are hardy and withstand temperatures of 28°F.

When the leaves are 10 in. long, they can be gathered together and tied so the inner leaves or heart will become white (called blanching). For winter use, plants may be dug with a ball of soil and placed in a cold frame or cool place where they do not freeze. The leaves may be blanched and plants used as needed. Blanching requires 2 to 3 weeks.

Common cultivars of endive include Deep Heart, Green Curled, Ruffic, and Salad King. Common escarole cultivars include Broad Leaved Batavian, Florida Deep Heart, and Full Heart Batavian.

Harvest. The entire head is cut at the base when leaves are partly or wholly blanched. The outer leaves are usually tough and bitter and discarded.

Common Problems. Endive and escarole have few insect and disease problems. Slugs and snails sometimes are a problem.

Rotted centers sometimes occur. This is frequently due to over-crowding, water splashing in the center, or tying the leaves for blanching when the leaves are wet.

GOURDS

The word gourd is commonly used to describe warm season vining crops which produce fruits for decoration, novelty items, or household utensils. They are members of the same family and various types are found in *Cucurbita pepo, Cucurbita maxima, Lagenaria siceraria,* and *Luffa cylindrica.* The *Cucurbita* species originated in America and were widely distributed in North and South America long before Columbus. They were used for receptacles, household utensils, and food. The *Luffa* species originated in tropical Asia and had spread into China by 600 A.D.

Plant Characteristics. The gourds have sprawling vines with large leaves. The flowers are large and usually yellow. The bottle gourds have white flowers. The plant contains separate male and female flowers on the same plant and may cross-pollinate with squash and pumpkin, as shown in Table 2.8. Any pumpkins, squash, or gourds within the same genus and species can cross-pollinate with each other. The *Lagenaria* and *Luffa* genera do not contain any pumpkins or squashes commonly grown in America and do not cross-pollinate with the *Cucurbita* species shown.

Culture. Gourds are tropical or semitropical plants that are extremely frost tender. They are not able to grow below 60°F. They can be grown in all areas if planted during the late spring or summer. Plants may be transplanted as are pumpkins or seeded directly. Plants are transplanted and seeds are planted after all danger of frost has passed. Seeds are placed 2 in. deep, 6 to 12 in. apart, in rows 6 to 8 ft apart. The plants are thinned to stand 3 ft apart in the row. With ample water the vines grow rapidly and cover the entire area in 2 months. When garden space is limited, vines may be trained onto a fence or trellis.

Gourds grow on almost any good, well-drained garden soil containing a generous amount of organic matter.

The *Lagenaria* species produce the best-shaped gourds for dippers, birdhouses, and rattles. They should be grown on a fence or trellis so the fruits hang free and the fruit is long and straight.

Some *Cucurbita* cultivars of gourds are shown in Fig. 7.22. These gourds usually have thick shells and are difficult to cure. Their color usually begins to fade in 3 or 4 months. Some, like Turk's Turbin (Fig. 7.23), appear in the catalogues as winter squash.

Lagenaria cultivars of gourds include Calabash, Cave Man's Club, Large Bottle, Dolphin, Dipper, Drum, Hercules Club, and Swan Gourd. One edible cultivar called New Guinea Bean, Guinea Bean, or Italian Edible Gourd grows to 5 ft in length and weighs 15 lb.

The *Luffa* cultivar is called Vegetable Sponge, Sponge Gourd, or Dishrag Gourd.

Harvest. The *Lagenaria* cultivars can be eaten like squash or eggplant when immature. The fruits are harvested about 1 week after they blossom. The *Luffa* cultivar is eaten like cucumber or cooked like a vegetable when the fruits are 4 in. long or less. Turk's Turbin can be harvested mature and used as an ornamental or eaten like squash. The large blossom end is cut where it joins the rest of the fruit. The seeds are removed, the cavity is filled

FIG. 7.22. Various ornamental gourds of *Cucurbita pepo*. Left to right—White pear, warted, pumpkin, devils claw. Foreground—Bicolor. *Courtesy of J.S. Vandemark II.*

FIG. 7.23. Turk's Turbin is often called a gourd, but it is edible and tasty. *Courtesy of Harris Seeds.*

with rice, favorite seasonings, and ground beef, the blossom end is replaced, and the entire fruit is baked similar to winter squash.

Fruits that are not eaten are usually allowed to remain in the garden until frost kills the vines. The *Cucurbita* cultivars will have brightly colored, hard, glossy shells. When harvesting, 1 or 2 in. of stem is left on these gourds. The gourds should be handled carefully as bruises discolor them and cause them to soften and decay. The *Cucurbita* gourds are cured for several days in the shade under warm dry conditions.

The *Lagenaria* cultivars are thin-shelled and are harvested after a frost or when the shells begin to harden, the fruits become lighter in weight, and the tendrils on the vine nearest the fruit begin to dry and shrivel. These gourds may require 6 months in a warm, dry place to cure. Large dipper and birdhouse gourds may be made into wren houses. After harvest, the very thin skin is scraped from the shell of the gourd with a knife and the gourd is then dried. A hole the size of a quarter is drilled into the dipper end for the wren and a small one at the stem end (for a hanging wire). The surface is painted with shellac or polyurethane.

Luffa or sponge gourd is harvested when the *Lagenaria* cultivars are harvested. The fruit is placed in a tub of running water, and the outer skin is peeled off, as with an orange. This exposes the inner core or sponge. The sponge is squeezed to remove the inner contents.

Common Problems. Weeds can be a problem late in the season. If weeds are controlled until the vines cover the ground, many weeds will be shaded. Late in the season the crop is mature and little damage is caused by weeds.

Gourds have insect and disease problems similar to those of cucumber. Cucumber beetles should be controlled from the time the plants emerge.

The squash bug occasionally attacks gourds. See control measures in Chapter 4.

KALE

Kale (*Brassica oleracea,* Acephala group), or borecole, is a member of the mustard family. It is native to Europe and was recorded as being used for food as early as 200 B.C.

Culture. Kale is a hardy, low-growing plant well adapted to growing in the fall in both the North and the South. It will live overwinter as far north as northern Maryland, southern Pennsylvania, and areas having similar winter climates. It withstands heat but is usually grown as a fall crop. Plants grow best on well-drained fertile soils with ample moisture. They are grown similarly to collards.

Seeds are planted in early spring for a summer crop; in early summer for a fall crop; or, in areas with mild winters, in late summer for a winter crop. Kale is frequently planted to replace vegetables harvested in midsummer. Seeds are planted one-fourth to one-half inch deep in rows 18–24 in. apart. Seedlings are

FIG. 7.24. Kale is a hardy plant grown both spring and fall.

thinned to stand 8–14 in. apart. Kale is easy to grow and has few insect and disease problems when spring or fall planted. Cabbageworms can be a problem during the summer.

Tall and dwarf cultivars are available.

Harvest. The flavor of kale is usually improved by frost. Kale is harvested two ways: (1) the entire plant is cut off at ground level 40 days after seeding or (2) the lower leaves are periodically stripped off. The inner leaves and bud are left to produce new leaves for a continuous supply. Leaves are harvested before they become old, tough, and stringy.

KOHLRABI

Kohlrabi (*Brassica oleracea*, Gongylodes group) is a cool season vegetable. It is a relative of cabbage that developed in northern Europe about 1500. It was imported into the United States about 1800.

The plant develops an enlarged stem just above the ground line, and the leaves grow out from this stem. It can be eaten raw or cooked like turnips.

Culture. It is grown as a cool season vegetable in all parts of the United States. It can be grown in the early spring but is usually grown as a fall crop in the North and a winter crop in the South. Kohlrabi is normally planted for fall use after all the frost-tender vegetables are gone. It grows easily in containers.

FIG. 7.25. Kohlrabi is delicious raw or cooked. *Courtesy of W. Atlee Burpee Co.*

Seeds are planted one-fourth to one-half inch deep in rows 12–18 in. apart. Plants are later thinned to stand 4 in. apart.

Kohlrabi grows best in fertile soil with 1 in. of moisture each week, similarly to cabbage. Seeds are frequently planted in midsummer between vegetables that are maturing or already harvested. The seeds will need frequent watering to establish the plants. Slow or checked growth of the plants results in tough, woody stems. As kohlrabi must grow rapidly in the spring before hot weather to be of good quality, most gardeners plant it for a fall or winter crop.

Kohlrabi may have insect and disease problems similar to cabbage. Early spring, fall, and winter crops are usually not attacked severely by cabbageworms.

Common cultivars are the white-skinned Early White Vienna and the purple-skinned (but white flesh) Early Purple Vienna.

Harvest. Kohlrabi is ready to harvest in 50–60 days after planting. It has the best flavor when it is 2 or 4 in. in size and the flesh is still tender. Large kohlrabi have woody, stringy, tough stems. The leaves of young plants may be used like spinach.

LEEK

Leek (*Allium ampeloprasum*, Porrum group) is a biennial, native to the Mediterranean region. It has been cultivated for food since prehistoric times. Leeks are related to onions and garlic, but they are milder in flavor and do not form a bulb. The thick, fleshy stalk is the same diameter as the base and resembles a large green onion.

Culture. Leeks can be grown in any garden that produces good onions. Leeks are grown as an annual from seed in all parts of the United States. It is either transplanted or seeded directly into the garden. Leek requires about 120 days from seeding to harvest or 80 days from transplanting to harvest. In areas with short growing seasons, seeds are planted in a hotbed 2 or 3 months before transplanting. The plants are set out in the fall in the South and Southwest and in early spring in the North. Leek transplants are handled similar to onion transplants.

Seeds are planted one-half to 1 in. deep, 1 in. apart, in rows 12–18 in. apart. The plants are thinned to stand 4 in. apart. When the plants are at least the size of a pencil, soil is placed around the edible portion. This makes the edible part whiter and longer at harvest. Placing soil around the edible part (leaf bases) when the leaves are small may cause the plants to rot and die.

There are several cultivars of leeks listed in seed catalogues, but certain cultivar names are synonyms. Thus, the popular American Flag cultivar is also known as London Flag. Musselburg is similar to American Flag but has longer stems. Carentan has short but thick stems that may become 2–3 in.

in diameter. Odin and Tivi are early Danish cultivars for fall production, and Conqueror is a hardy strain that holds up well in the garden during the winter in the northern states.

Leeks have few insect and disease problems. In hot, dry conditions, onion thrips may stunt the plants' growth. The problem can be reduced by harvesting all affected leeks and onions and destroying weeds that provide winter protection for the thrips.

Harvest. Leeks can be harvested whenever they are three-fourths of an inch in diameter, but they can be left to grow to almost 2 in. in diameter. Plants can be removed with a shovel or fork until the ground freezes. They may be mulched heavily and harvested during the winter whenever the ground is not frozen. Plants may also be left mulched or unmulched for spring harvest. Plants are removed in early spring, before the leeks begin spring growth and send up a seedstalk.

The roots and all but 2 in. of the green leaves are cut from the harvested leeks. Soil may be caught in the layers of the growing leek. The plant may be cut lengthwise and thoroughly washed to remove this soil.

FIG. 7.26. Leeks may be harvested throughout the winter whenever the ground is not frozen.

LETTUCE

Lettuce *(Lactuca sativa)* is native to Europe and Asia and has been grown for over 2500 years. It was grown by Persian kings in 500 B.C. The various forms of head lettuce appeared in the 1500s. Columbus brought lettuce to America, and it was one of the first plants grown by the early settlers.

Plant Characteristics. Lettuce is extremely sensitive to high temperatures. When seeded at temperatures of 80°F, lettuce seed does not germinate and the seed becomes dormant. High temperatures induce a seedstalk to form in all types and results in internal tipburn of the leaves of crisphead types.

Four general types of lettuce are recognized as subspecies. Gardeners, however, generally recognize five types.

Butterhead. Butterhead (*L. sativa* var. capitata) or bibb lettuce is a head type in which the leaves are loosely folded. The inner leaves are cream or yellow, and the outer leaves green. Butterhead types bruise and tear easily. Cultivars include Bibb, Buttercrunch (resistant to heat), Butter King, Dark

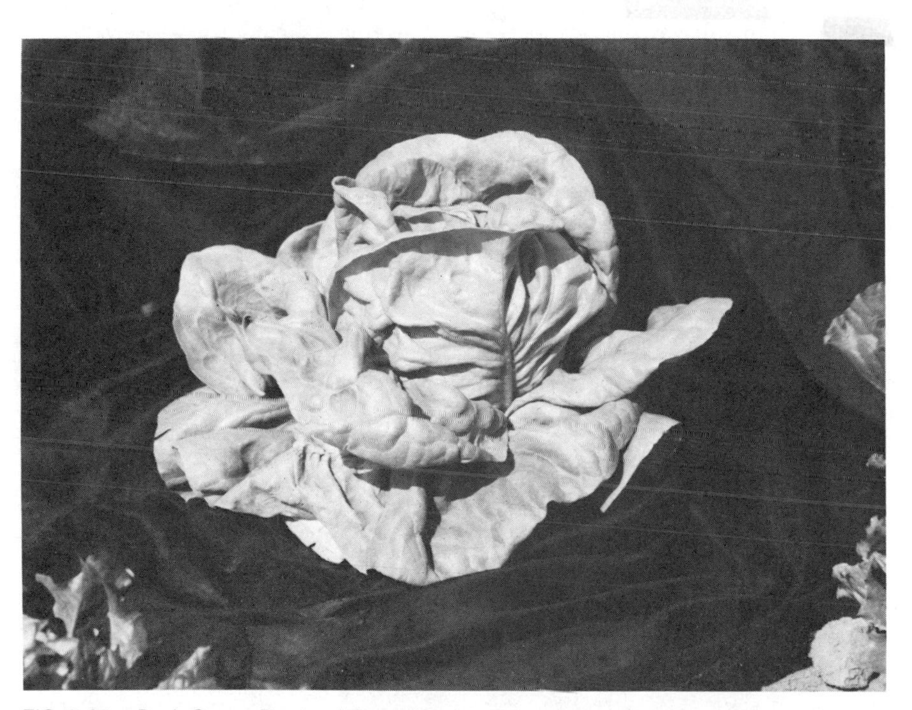

FIG. 7.27. Dark Green Boston lettuce is a butterhead type. *Courtesy of Harris Seeds.*

Green Boston, Deer Tongue (also called Matchless), Summer Bibb (resistant to heat), Summerlong (resistant to heat), and White Boston. Tom Thumb is a small cultivar easily grown in containers.

Cos. Cos or romaine (*L. sativa* var. longifolia) lettuce is an upright plant that grows about 10 in. high. The outer leaves are smooth and green, the inner leaves whitish green. Leaves are more crisp than other heading types. Cultivars include Paris Island, Paris White, Valmaine, and White Paris. Sweet Midget is a small cultivar easily grown in containers.

Crisphead. Crisphead (*L. sativa* var. capitata) or Iceberg is the common lettuce found in the produce market. Leaves are thin and crisp and frequently have curled and serrated edges. Heads are hard and durable. Cultivars include Calmar, Fairton, Great Lakes types, Iceberg, Ithaca, Mesa, and Pennlake.

Leaf. Leaf lettuce (*L. sativa* var. crispa) is also called looseleaf or loosehead lettuce. Plants do not form a head and leaves may be serrated, deeply lobed, or crinkled. Leaf color varies from light green to red. Cultivars include Black-Seeded Simpson, Grand Rapids types, Green Ice (resistant to heat), Oak Leaf, Prizehead, Salad Bowl, Slobolt (resistant to heat), and Walsmann's Green. Ruby is a red cultivar that grows well in containers during the winter.

FIG. 7.28. Leaf lettuce grows rapidly and is a popular salad vegetable. *Courtesy of W. Atlee Burpee Co.*

FIG. 7.29. Celtuce is grown for its edible stem. *Courtesy of W. Atlee Burpee Co.*

Stem. Stem (*L. sativa* var. asparagina) or asparagus lettuce can be used like celery and lettuce. The young leaves can be used like lettuce. The plant produces an edible seedstalk that is eaten raw like celery or cooked in Chinese dishes. The cultivar commonly listed is Celtuce and often indexed as such.

Culture. Lettuce grows best at cool temperatures. It is grown in late fall, winter, and spring in the south and Pacific Southwest. In the North, lettuce is limited to spring and fall.

Lettuce is planted one-quarter inch deep, 1 seed per in. in rows 12–18 in. apart. It can be planted as soon as the soil can be worked in the spring. The plants are thinned to stand 6 to 8 in. apart for butterhead and cos types, and 4 to 6 in. apart for leaf types. Several spring plantings may be made.

All types of lettuce may be grown as transplants, and crisphead types should be grown as such, particularly in the northern, central, and midwestern areas. A fall crop of crisphead types cannot be grown in these areas. Seeds are started indoors, in a cold frame or hotbed (see Chapter 2 on Plant Growth). Plants are hardened and planted in early spring. They may be planted up to a month before the frost-free date as hardened lettuce is usually not harmed by temperatures as low as 28°F. The crisphead types are set out 10–12 in. apart, in rows 18 in. apart. For fall crops, seeds are planted directly in the garden, as transplanting does not save time.

Lettuce has a shallow root system, and frequent watering causes the leaves to develop rapidly. On heavy soils, too much water may lead to burning at the edges of the leaves and disease. Summer and fall plantings will require frequent irrigation to establish the plants. All cultivation around lettuce should be shallow, due to its meager root system.

Harvest. Leaf lettuce can be harvested whenever the leaves are large enough to use. Plants that have been thinned from the row are also eaten. The inner leaves of the leaf cultivars are of high quality, and the entire plant can be harvested in 50–60 days. If the plants are crowded, many plants are formed, but the leaves are small and of low quality.

Butterhead and cos types are usually ready to harvest in 60–70 days. Crisphead types are harvested when the heads are firm and full.

Lettuce may be washed, dripped dry, and stored in an airtight container in the refrigerator.

Common Problems. Lettuce produced in the garden is often free of disease and insect problems. In many areas, cabbageworms attack the plants. If the plants are grown closely together or mulched with organic material, garden slugs and snails will need to be controlled.

MUSKMELON OR CANTALOUPE

Muskmelon (*Cucumis melo*, Reticulatus group) is called cantaloupe in the produce trade. It is native to Iran and India, although it has never been found growing in the wild. There are many types of these melons in addition to the netted skin, musky type of muskmelon. The honeydew (green flesh) and casaba (white flesh) melons (*Cucumis melo*, Inodorus group) are larger, smooth-skinned, late-ripening winter melons. Cultivars of these melons were grown by the early settlers in the 1600s.

Plant Characteristics. The fruit of all, except some later-maturing cultivars, have a musk aroma, and this character gave the plants their name. The plants have separate male and female flowers on the same vine and are cross-pollinated by bees. Only garden plants within the same genus and species will cross-pollinate and muskmelons will cross-pollinate only with other muskmelons. They do not cross-pollinate with cucumbers, gourds, watermelons, squash, or pumpkins. The bitter flavor sometimes noticed is due to cloudy weather during ripening, too high temperatures, or too much or too little water.

Culture. Some types of muskmelons may be grown in nearly all parts of the United States. The casaba, honeydew, and other large-maturing cultivars may require 130 days to mature and grow best in the South and parts of California. Plants may be grown in the garden from transplants or by seeding directly. Transplants must be grown in individual containers to prevent root damage.

The cultural requirements for muskmelon are about the same as for

FIG. 7.30. Burpee Hybrid muskmelons. *Courtesy of W. Atlee Burpee Co.*

cucumber, except that muskmelons are less tolerant of high humidity and rainy weather. In northern and midwestern states, the rain and humidity hasten disease problems, and resistant cultivars should be used.

Garden soils should be well drained and contain large amounts of compost or well rotted manure. Sandy soils warm up quickly and are preferred for an early crop. In northern locations and on heavier soils, black plastic mulches (see Chapter 3 on Soils and Plant Nutrition) can be used to warm the soil. The mulch is installed before planting and holes made every 2 or 3 ft for the transplants or seed.

Seeds or transplants (a starter fertilizer should be used) are placed in the garden when all danger of frost has passed and temperatures have warmed to 60°F. Plants grow best at temperatures of 60°–85°F. Muskmelons are of the highest quality when grown at these temperatures, when the vines remain healthy, and when the weather is relatively dry when the fruits are ripening.

Seeds are planted 1 to 1½ in. deep, 1 ft apart in rows. Seeds may also be planted in hills, 2 plants every 3 ft or 3 plants every 4 ft. Rows are spaced at least 5 ft apart. The vines will fill in the space between rows. Crowding the plants results in a large leaf cover that reduces the opportunity for bees to pollinate the flowers, and yields are reduced. In gardens where space is limited, the vines can be trained on a fence or trellis. The fruit will need to be supported with a mesh bag or cheesecloth sling.

Muskmelons should receive an application of fertilizer about the time the plants begin to vine. Late-maturing cultivars also respond to a second application when the plants begin to set fruit. Vines require ample water when they are growing rapidly and the fruits are developing and should not be irrigated when the fruits are ripening as this may cause them to split open. Furrow or trickle irrigation is best, as moisture on the leaves from any source encourages disease.

A large number of cultivars are available. Very early maturing ones may be grown in the far North and the very late cultivars grown in the South and Far West. In areas of high humidity and rainfall, mildew-resistant cultivars should be grown.

Harvest. At maturity, the fruit changes from green to yellow or tan, and an abscission layer forms between the stem and the fruit. The stem breaks away cleanly from the fruit with slight pressure, which is called the "full slip" stage. Muskmelons should not be harvested before this stage. They do not increase in sugar when picked green.

Casaba, crenshaw, honeydew, and Persian cultivars do not develop an abscission layer and "slip." They are harvested when the fruits turn yellow (casaba) or white (honeydew) and the blossom end of the fruit begins to soften. These melons can be stored for a short time.

Muskmelons should be harvested every other morning once they begin to ripen. At the peak season, they should be harvested daily. Harvesting when the plants are wet with dew or water helps spread disease, and the plants should be dry.

Common Problems. Insect and disease pests are similar to cucumber. In humid, wet regions, resistant cultivars should be selected.

OKRA

Okra *(Abelmoschus esculentus)* or gumbo is a perennial grown for its seedpods. It probably originated in Africa and was in use by the Egyptians in the twelfth century. It was being grown in America by the early 1700s.

FIG. 7.31. Okra is grown for its pods. *Courtesy of W. Atlee Burpee Co.*

Plant Characteristics. Okra is grown as an annual, and cultivars differ in their size from dwarf plants of 3 ft to tall plants of 5 to 10 ft. Pod shapes range from round to ridged and short to long. The plant and pods may have small spines on them that create allergies in some people. Spineless cultivars are available. The cultivars vary in pod color and in flower color and shape and may be grown as an ornamental.

Culture. Okra is a warm season plant similar to cucumber and tomato. It may be grown in all areas of the United States. In the North, the growing season is short and yields will be reduced. In the South and Pacific Southwest, it withstands midsummer heat and produces when many other vegetables do not.

Before planting, the seeds are soaked in water overnight. The seeds that have swollen are planted after all danger of frost has passed. Seeds are planted one-half to 1 in. deep in rows 3 ft (dwarf types) to 5 ft (tall types) apart. Plants are thinned to stand 1 to 2 ft apart. Okra is usually planted in full sunlight at one edge of the garden where it does not shade other vegetables.

Okra grows well on any well-drained garden soil. It needs an application of fertilizer high in phosphorus before planting. As okra has a long growing season, an additional application is beneficial at the time the pods begin to form. A fertilizer high in nitrogen should not be used as this stimulates the vegetative growth and reduces the number of pods produced.

When plants become too large in the south, they are sometimes cut off 6 to

8 in. above the ground and allowed to regrow. These plants should receive some fertilizer high in nitrogen to stimulate vegetative growth.

Okra requires irrigation during dry weather, about 1 to 1½ in. of water per week.

The size of the garden often determines the cultivar to choose. Dwarf plants are preferred in small gardens, as they will still grow 5 feet tall. Cultivars with ribbed or smooth pods and spineless plants are available.

Harvest. Pods are harvested by cutting them from the plant with a knife or shears. They are harvested 4 to 7 days after the flower has opened, and the pods are not fibrous (pods 2 to 4 in. long). Pods should be harvested every other day, and mature pods should be removed and discarded as they reduce the plant's growth and decrease yield. The pods contain a mucilaginous material that makes some okra dishes seem slimy to some people. This is the material that is valuable in thickening soups and gumbos.

Pods may be stored for several days in an airtight plastic bag in the refrigerator.

Common Problems. Nematodes may be a problem in gardens that have been in the same location for several years. The corn earworm may eat into the pod and stinkbugs cause damage late in the season.

ONION

Onion (*Allium cepa,* Cepa group) is probably a native of Asia. It has been grown for food since recorded history and is mentioned in the Bible. There are over 300 species of onion, some of them native to North America. The domesticated types were brought to America by Spanish explorers.

Plant Characteristics. Onions can be grown over a wide range of soil and climate and are grown in nearly all parts of the United States. The production of the onion bulb is dependent upon day length. Early or southern cultivars require 12 hr of daylight to bulb, while late or northern cultivars require 15 hr. The bulbs begin to form regardless of the size of the plant. Onion cultivars adapted to the South grow little and form small bulbs in the North. Some bunching cultivars, such as Evergreen, are used for green onions produced from seed. These cultivars are a different species and do not form bulbs at all. See also under Welsh onion in Chapter 8.

General Care. Onions continuously produce new roots and develop best in cool, damp conditions. They produce the bulbs in warm weather. Onions require about twice as much fertilizer as other vegetables. Onions respond to additional fertilizer 40–60 days after transplanting or seeding. About 1 lb of 10-10-10 or equivalent per 25 ft of row should be worked into the soil about 2 in. deep and 3 in. to the side of the row.

FIG. 7.32. Onions are grown over a wide range of climates. *Courtesy of Harris Seeds.*

Onions develop best in loose, crumbly soil. Hard soils induce the bulbs to be small and irregularly-shaped. One pound of compost per square foot of soil helps loosen the soil.

Onions grow in the South during the fall, winter, and spring. In the North, they are grown from early spring to fall. Onions may be grown from seed, sets, or transplants. Seeds are least expensive but require a longer time to grow.

Onions may be bothered by onion thrips. In the North, root maggots may burrow into the bulbs. Maggots can be controlled with a soil insecticide applied before planting. Onions are poor competitors and weeds must also be controlled.

Onions from Sets. Sets are small, dry onions that have been grown the previous year specifically for starting plants. They may be used to produce green onions or mature, dry bulbs.

Culture. Sets are usually yellow or white, but occasionally red sets are available. A cultivar is seldom listed. Sets should be purchased early while they are still firm and dormant. The round sets mature into flat-shaped bulbs, and elongated or tapered sets mature into round bulbs.

The sets should be separated into two sizes, sets smaller than three-fourths of an inch (the size of a dime) and larger sets. The large sets are used for green onions as they frequently form a seedstalk instead of a bulb. The small sets are used for mature, dry bulbs.

For green onions, sets are planted 1½ in. deep, close enough to touch each other, in rows 12 in. apart. When the plants are 4 in. high, soil can be placed around the stems to produce long, white stems.

For dry onions, the sets are planted 1 in. deep, 2 in. apart in rows 12–16 in. apart. Soil is not placed around dry onions as this may induce the bulb to rot in storage.

Harvest. Green onions may be harvested whenever the plant is 6 in. or more high. If a plant begins to produce a seedstalk, it should be harvested. It will not produce a bulb for storage. Green onions become stronger in flavor as they become older.

Dry onions are harvested when half or more of the onion tops have fallen over naturally. Tops should not be broken over early as this reduces bulb size and may introduce diseases, causing rots in storage. Mature bulbs are pulled, the tops placed over the bulb to prevent sunburn, and are air-dried in the garden for a day or two. Bulbs are air-dried in a well-vented place for an additional 2 to 3 weeks. Tops may then be cut off about 1 in. from the bulb and the bulbs stored under dry cool conditions in a mesh bag.

Onions from Transplants. Transplants are an easy way to produce an early crop of large, mature bulbs. Practically none of the transplants form seedstalks.

Culture. Transplants are planted in the fall in the South and Southwest and as soon as the soil can be worked in the North. Transplants are set 4 in. apart in rows 12–18 in. apart. Plants are set 1 to 1½ in. deep and a starter fertilizer applied. Plants also receive a side dressing of fertilizer similar to sets.

Transplants are not always available to the gardener at the correct planting time in the South. Transplants may then be grown from seed and planted (see Chapter on Plant Growth). Cultivars can be the same as those grown from seed in the South.

In the North, transplants are more readily available. Both red- and white-bulbed cultivars are often available.

Harvest. Mature dry bulbs are ready to harvest when the tops have fallen over, similar to those grown from sets. Bulbs from transplants usually do not store as well as bulbs from sets and should be used by early winter.

Onions from Seeds. Onions may be grown from seed to produce both green (or bunching) onions and mature, dry bulbs. The bunching onions are a different species from the onions produced from sets or those used to produce bulbs. Most of the bunching onions do not form bulbs and are used only for green onions.

Culture. For best results, seed for mature, dry bulbs should be sown indoors and transplanted (see Chapter 2 on Plant Growth). Seed may also be planted directly in the garden. In the North, seed is planted as soon as the garden can be worked. In the extreme South, seed can be planted in late September to produce bulbs in June or planted in January for an August harvest. The seed of bunching onions is planted directly in the garden. There is no value in growing it as a transplant.

Seed is planted 1 in. deep, 1 seed per in., in rows 12–18 in. apart. When the plants are 4 in. high, they are thinned to stand 1 in. apart for green onions, 2 in. apart for medium-sized onions, and 4 in. apart for large onions. The thinned plants are used as green onions.

Many cultivars are available for green and dry onions.

Harvest. Green onions can be harvested when desired. Mature, dry bulbs are harvested similar to those produced from sets. Bulbs produced from seed can frequently be stored until spring.

PARSNIP

Parsnip *(Pastinaca sativa)* is native to the Mediterranean area and was known to the ancient Greeks and Romans. It was introduced to America by the early colonists. Parsnip can be grown in all parts of the country. It does not grow well in midsummer in the South and is usually grown and used by early summer. In mild areas, it can be planted in the fall and grown during the winter.

Culture. For spring planting, seed is planted 1 or 2 weeks before the frost-free date. The soil and weather should be warm at planting time, but seeds germinate poorly during the summer. Fresh seed should be used as old seed germinates poorly.

Seeds are planted one-half inch deep, 2 seeds per in. in rows 18 in. apart. The seed may be covered with sand, peat, or vermiculite instead of soil to prevent crusting and poor seedling emergence. Plants are thinned to stand 3 in. apart in the row. Crowding the plants at this spacing forces the plants to produce smaller, more tender roots.

Culture is similar to carrots. Plants grow best in lighter soils and stony soils cause the plants to produce rough, forked roots. Parsnips have few insect and disease problems.

Harvest. Parsnips are harvested before they mature and the root has become woody. They are harvested in midsummer in the South and after a frost in colder climates. Frost induces some of the starch in the roots to be converted to sugar, and plants have a sweet nutlike flavor. Roots left in the

FIG. 7.33. Harris Model Parsnip. *Courtesy of Harris Seed Co.*

ground are remarkably resistant to freezing injury and rots and may be harvested and used at any time. Plants may be mulched with straw so roots can be dug throughout the winter. In areas with mild winters, the spring-seeded plants continue to grow more slowly throughout the winter and the roots become woody. In these areas, the roots should be harvested in early winter and stored. Neither chilled nor nonchilled roots are poisonous as sometimes supposed.

PEA: ENGLISH, EDIBLE-PODDED, AND SNAP

Peas (*Pisum sativum* var. sativum) originated in central Europe, Asia, and northwest India, and were grown in the Stone Age. They were an important crop in England by the eleventh century. English peas are types that originated in England and are called English peas to distinguish them from Southern peas, which are really beans. The edible-podded pea (*Pisum sativum* var. macrocarpon) is a special type grown for its pod rather than its seeds. Snap peas are a special type in which both the seed and pod are eaten, similar to a green bean.

Plant Characteristics. The pea is a hardy, cool season, tendril climbing annual. It is best grown in the garden on a trellis, fence, or some other

FIG. 7.34. Both the pod and the seeds of snap peas may be eaten. Above—Sugar Snap. *Courtesy of Harris Seeds.*

support to conserve space. Peas are related to beans, and their roots will support nitrogen-fixing bacteria. If peas or beans were never grown in the garden before, the seeds should be inoculated with these bacteria. The inoculum is available in garden centers and from catalogues. Once the bacteria are established in the garden, it is of little value to inoculate each year.

There are several types of peas the gardener may grow, and their culture is similar. English peas (garden peas or sweet peas) are grown for the seed. Edible-podded peas (sugar peas, snow peas, or Chinese peas) are grown for the long pod. The walls of the pod are brittle, succulent, tender, and fiber-free. Snap peas look like English peas, but the entire pod (including seeds) is used, similar to the edible-podded types. However, the pods are harvested after the peas have developed. Snap pea pods are smaller than the edible-podded types.

Culture. Peas grow and mature best in cool weather and can be grown in all parts of the country. In the South and lower California, they are grown during fall, winter, and early spring. Further north they are grown as both a spring and a fall crop. In the far North and at high elevations, they may be grown from spring to fall. Pea plants do not withstand summer heat and the seeds mature and become hard rapidly.

Peas grow in any well-drained, moderately fertile soil. They grow poorly in wet or water-soaked gardens. Excessive nitrogen causes the plants to produce large vines, but fewer peas are produced. Inoculated peas on loam soils seldom respond to nitrogen fertilizer in a garden. They usually produce more on light, sandy soils when a light amount of a complete fertilizer is applied before planting.

Seeds are planted in the spring whenever the soil temperature is above 45°F. Fall crops are planted 60–70 days before the date of the first freeze. Seeds are planted 1 to 1½ in. deep in heavy soils and 2 in. deep in sandy soils. Peas are planted 8 to 10 seeds per ft in a single row 18 in. apart or in a double row 8 to 10 in. apart. A support can be placed between the double row.

There are now over 2000 distinct cultivars of peas (Anon. 1975). Smooth seeded peas have a starchy flavor, even when young. Wrinkled seeded cultivars are sweeter and the sugar does not convert to starch rapidly; these types remain sweet past their prime quality. There are large vined, small vined, or bush types and include Mighty Midget, a cultivar for containers that grows vines only 6 in. long. Cultivars vary in disease resistance and heat tolerance. Cultivars should be chosen to fit the gardener's location and needs.

Edible-podded cultivars include Snowbird, Dwarf Grey Sugar, Giant Melting, Melting Sugar, and Oregon Sugar Pod. Sweetness and starch are unimportant in these cultivars as they are eaten for the pod before the seed attains any size.

The original snap pea cultivar was Sugar Snap, a wilt-resistant All-America winner. Newer cultivars are available.

Harvest. English peas, edible-podded peas, and snap peas are all harvested differently.

English Peas. The pods are harvested when they are well filled but before they harden and fade in color. The peas should not be hard and starchy. Peas are best picked and shelled just before cooking as the sugar content decreases rapidly after harvest. Two or three pickings are made, as all the pea pods do not mature at the same time. The pods should be carefully pulled from the vine to prevent the plants from being uprooted. At the last harvest, the plant may be pulled up and all the pods picked.

Edible-Podded Peas. These peas are picked when the pods are long and the peas just developing. Pods, 3 to 5 in. long, are produced 5 to 7 days after flowering, and the pea seeds are slim and small. Pods need to be picked every other day to prevent them from developing large seeds and fibrous pods. If the seeds develop, they may be used similar to English peas.

Pea pods can be stored 10 days in plastic bags in the refrigerator without a loss in quality. Pods are stir-fried or briefly steamed to prevent overcooking.

Snap Peas. The pods are picked when the peas are large and the peas and pod are full size. The entire pod is used raw in salads or with dips. The pods may be snapped like green beans or used whole and steamed similar to English peas. They may be stir-fried similar to edible-podded peas; but because of the large pea seed, some gardeners would not consider them a replacement for the edible-podded type.

Common Problems. Pea roots are easily damaged by hoeing, and shallow cultivation should be practiced. Peas have few pests in a garden. On cold, wet soils, fusarium wilt may be a problem and resistant cultivars should be grown. Pea weevil is a pest in the West that feeds on the blossoms; the larvae enter the developing peas. Pea aphids may cause the plant to wilt (see Tables 4.3 and 4.5 for control measures).

PEANUT

Peanut *(Arachis hypogaea)*, also known as earth nut, groundnut, and goober pea, originated in South America and was taken to Africa by the Portuguese. Peanuts were shipped to North America as on-board food for slaves. It was well established here by the year 1800.

Plant Characteristics. Peanut produces bright yellow flowers aboveground, which are self-pollinated. The ovary, called a "peg," emerges and grows downward until it enters the soil. The peanut then develops in the soil. Peanuts are divided into several types. The Virginia type has 2 large seeds per pod and is better adapted to short growing seasons. Spanish types have 2 or 3 small seeds per pod.

Culture. Peanuts are popular in the South, Southeast, and Southwest but can be grown in any area having 110–120 days of warm weather. They grow poorly under a long, cool growing season. As the "nut" develops underground, coarse-textured sandy soils are best, particularly for Virginia types. Spanish types also grow well on fine-textured or heavy soils. Soils should have a good supply of calcium to prevent the production of empty pods. If the soil is low in calcium, 2 lb of gypsum per 100 ft of row should be applied when the plants begin to flower.

Individual seeds are planted 1 in. deep in heavy soils and 2 in. deep in sandy soils, 2 weeks after the frost-free date in the spring. They can be grown as transplants in peat pellets if the growing season is short. Spanish types are placed 6 in. apart in rows 24 in. apart. Virginia types are spaced 8 in. apart in rows 36 in. apart. Plants may also be grown in large containers.

Plants should be irrigated during dry weather when the plants are flowering and the "pegs" are entering the soil. Near harvest time they should not be watered as this may cause the peanuts developing underground to sprout.

FIG. 7.35. Peanuts produce a yellow flower above ground, but the ovary elongates and enters the soil, causing the seed to develop underground.

The soil should be cultivated to control weeds and keep the soil loose to help the "pegs" penetrate the surface. Once pods develop in the soil, it is difficult to cultivate without harming the plants. Do not place soil around the plants or cover up the spreading branches.

Harvest. When the leaves have begun to turn yellow, the plants may be harvested and should be dug before a hard freeze. The plants containing the peanuts are air-dried in a warm, dry place for 2 weeks. The moisture content should then be about 15% and the peanuts can be removed from the plants. They can be shelled and roasted in a shallow pan at 350°F for 20 min. Unshelled peanuts may also be roasted. The oven is preheated to 500°F, the peanuts placed in the oven in a colander or wire basket, and the oven turned off. They are roasted when the peanuts have cooled.

Common Problems. Corn earworm, cutworms, armyworms, and caterpillars may feed on the leaves. The white-fringed beetle eats the belowground parts.

Stem rot may attack the stems, roots, pods, and pod stems. Leafspots cause infected leaves to become yellow as though the plant were mature. To help control these diseases, peanuts should not be grown in the same garden location each year.

PEPPER

Pepper is native to tropical America and was grown in North and South America over 2000 years ago. The small hot peppers were taken back to Europe by Columbus, where they became popular. Pepper cultivars used by gardeners are grouped in *Capsicum annuum* var. annuum and are different from the red hot tabasco pepper and the household pepper.

Cultivars. There are many cultivars that may be classfied as sweet, mild, or hot depending on the amount of capsaicin present. This material gives pepper its hot or pungent taste.

Bell, Perfection,* or *Pimiento types. These plants (*Capsicum annuum* var. annuum, Grossum group) produce two types of fruits. Bell types have large, blocky fruits, 3 in. wide and 4 in. long. The fruit may be yellow, deep purple to black, or dark green. The dark green fruit turns red when mature. Fruits are stuffed, eaten raw, or in salads, and many cultivars are available.

The Perfection or Pimiento type fruit is sweet, round, and slightly pointed;

FIG. 7.36. Bell types of pepper are a common garden vegetable. *Courtesy of Harris Seeds.*

FIG. 7.37. Long red cayenne peppers are used in chili. *Courtesy of W. Atlee Burpee Co.*

often 2 in. wide and 6 in. long. Fruits are canned and eaten in salads. Green- and yellow-fruited cultivars are available.

Cayenne and Chili Types. These plants (*Capsicum annuum* var. annuum, Longum Group) produce fruits that are slim, pointed, and slightly curved. Fruits are 2 to 12 in. long and usually red when mature, although yellow types are available. Fruits are dried and ground in a blender to make chili powder (hot types) or paprika (sweet types). Many cultivars of cayenne and chili types are available, ranging from sweet to hot and extra hot (pungent).

Celestial Types. Plants are usually one foot tall and are grown as ornamentals in the home. One plant is usually sufficient. They include plants (*Capsicum annuum* var. annuum, Fasciculatum group) which produce small, slender, erect fruits up to 3 in. long. Fruit color ranges from yellow to red and purple to orange. More than one colored fruit may appear on the same plant. Fruits are usually very pungent and may be dried and ground for chili powder. Cultivars are often found under "ornamentals" or "flowers" in a catalogue.

Cherry Types. Plants (*Capsicum annuum* var. annuum, Cerasiforme group) are usually grown as an ornamental houseplant. Fruits are erect or declined, cherry-shaped, 1 in. in size, yellow, orange, red, or purple in color and range from mild to very pungent. They may be used as chili powder, similar to celestial types.

Culture. Peppers are a warm weather crop. Flowers fall off when night temperatures are below 60°F or above 75°F. Plants do not grow when night temperatures are below 55°F. Peppers are grown in most sections of the country as transplants (see Plant Growth, Chapter 2). They are easily grown in containers.

Transplants should be set out after all danger of frost is past and planted 18 in. apart in rows 24 in. apart. A transplant fertilizer should be used at the same time.

In areas with a long growing season, seeds may be thickly planted one-fourth inch deep to ensure a good stand. To prevent crusting, seeds may be covered with sand, peat, or vermiculite instead of soil. Plant and row spacings are similar to transplants.

Peppers respond to a small amount of additional fertilizer when the plants have set several fruits. One to 1½ in. of water are needed during the growing season.

Harvest. Fruits may be harvested 70–130 days after planting. Fruits are normally broken off from the plant with part of the stem attached to the fruit. Bell types are usually harvested when they are 3 or 4 in. long. Hot peppers are harvested when they are red and mature. The jalapeno is usually harvested when it is green in color.

Common Problems. Blossom-end rot can be a problem. It is caused by drought stress and a deficiency of calcium (limestone or gypsum). Tobacco mosaic is a problem on peppers handled by gardeners who use tobacco. Mosaic-resistant cultivars should be grown.

Leaf miners, aphids, flea beetles, weevils, cutworms, and root maggots can be a problem. They can be controlled with various insecticides.

POTATO

Potato (*Solanum tuberosum*) originated in Peru and was used by the Incas. It was taken by Spanish explorers to Europe about 1540 and was a major source of food in Ireland from 1600 to 1845. As Irish settlers brought it back to America in 1719, it is commonly called the Irish potato. The potato is the only vegetable among the five principal world food crops (Splittstoesser 1977).

FIG. 7.38. Potatoes can be produced in a straw mulch for easier harvesting.

Plant Characteristics. Potato develops the edible part, the tuber, underground. Long days, warm temperatures, and high moisture and fertility promote vegetative growth. After the plants have reached a certain size, tubers are formed. Short days, cool temperatures between 60° and 70°F, lower moisture, and less fertility promote tuber development. Tubers do not form when the temperature is above 80°F.

Potato plants often produce flowers and fruits or seedballs, which resemble small tomatoes. These fruits are not edible and are not the result of cross-pollination with tomatoes. Flowering has nothing to do with tuber formation. Nonflowering potato plants will form tubers.

Culture. Potatoes are a cool season crop and do not grow well in the South during midsummer. They can be grown as an early crop in small gardens and vine crops (pumpkins, squash) later planted between the rows to utilize the space once the potatoes are harvested.

Potatoes are grown from seed pieces or small whole potatoes. For seed pieces, the potato is cut into 1½ to 2 oz blocks, about the size of a hen's egg. Potato eyes, sold in some garden centers, frequently weigh less than an ounce and are too small to produce strong plants. The seed piece is planted immediately after cutting. Small whole potatoes are best for planting. They have less disease and decay and produce better plants.

Planting dates vary, but potatoes are generally planted in early spring in the North and November to February in the South. They should be planted early for highest yields, and light frost will not damage the tubers. Prolonged wet and cold weather may cause the seed pieces to rot, however. Early maturing potatoes do not keep as well as those that mature later. Midseason and late cultivars are frequently planted in late spring or early summer in the North for winter storage. Southern-grown potatoes are usually harvested during hot weather when not fully mature, and they do not store as well as northern-grown potatoes.

Potatoes are planted 2 or 3 in. deep, 10 in. apart in rows 24 in. apart. Plants should shade the area and prevent high temperatures which inhibit tuber development. When the plants are 5 in. high, soil should be drawn around the plants in a ridge to cover the stems. This prevents green areas from developing on the tubers.

Potato tubers develop in darkened areas and will easily produce tubers aboveground in a mulch. This method reduces scab disease, keeps the soil 10°F cooler, thereby increasing yields, and makes harvesting easier. The potatoes are planted at or 1 in. below ground level. When the plants are 1 in. high, the entire area is covered with a 6-in. layer of mulch, such as straw or pine needles. The tubers will be produced in the straw, which can be carefully moved aside to harvest early potatoes. The straw can be replaced and the plants left to produce more potatoes until the vines die.

Potatoes grow best in fertile, well-drained soil. Misshapen potatoes develop in hard, compact soil, and organic matter should be added to such soils. Fresh manure increases the incidence of scab disease, and well-rotted manure should be used.

Plants should be irrigated weekly during dry weather. If the soil becomes dry and is then irrigated after a period, the tubers begin a second growth period and knobby potatoes are produced. Alternate wet and dry periods also cause potatoes to develop a cavity in the center of the tuber.

There are over 100 potato cultivars available. Those commonly grown in North America have white flesh and light brown (called white) or red skins. To reduce disease problems only certified seed potatoes should be used. Home-grown potatoes may become infected in a single season. Potatoes purchased for eating often have virus diseases present and may have been treated with a material to prevent them from sprouting. Potato cultivars are early, midseason, and late maturing.

Harvest. Potatoes may be harvested before the vines die, but as long as the vines are alive, the potato size will increase. The tubers are usually 1 or 2 in. in size when the plants are flowering. Potatoes are usually harvested when the vines die and should not be bruised or injured when dug. Potatoes are stored for a week at 65°–70°F in the dark to help heal bruises. Tubers are then stored at 35°–40°F at high humidity.

Common Problems. Numerous insect and disease pests attack potatoes. Certified seed and resistant cultivars should be used where the disease in question is prevalent. Crop rotation should be practiced.

Blights can be controlled with a fungicide. Colorado potato beetle, flea beetle, and leafhoppers may also be a problem.

PUMPKIN

Pumpkins originated in tropical America and were used by the Incas. Pumpkins belong in three species and are defined as the edible fruit of any

Cucurbita species (see Chapter 2) that is harvested mature and usually not used as a baked vegetable.

Plant Characteristics. Pumpkins have large leaves on sprawling vines. The plant produces male and female flowers on the same plant and must be cross-pollinated. The first flowers are males and do not produce fruit. Pumpkins do not cross with cucumbers, watermelons, or muskmelons. They readily cross with any plant of the same species, however. Table 2.8 gives the cultivars of squash, pumpkins, and gourds that freely cross-pollinate within a given species.

Culture. Pumpkins are sensitive to cold and also do not thrive in the South during midsummer. Pumpkins are transplanted (see Plant Growth, Chapter 2) or direct seeded 2 weeks after the frost-free date. They grow best between 65° and 85°F. Plants require a lot of space but can be grown near an early crop of corn or potatoes and utilize this space when the plants are harvested. The pumpkin vines may be trained on a fence or trellis, but the fruits will need to be supported with a mesh bag or cheesecloth. Large pumpkins produce fruits that are usually too heavy to grow on supports.

The vining pumpkins are planted in hills, 2 in. deep. Large vining types are thinned to 3 plants every 5 ft with 10 ft between rows. Semi-vining types are thinned to 2 plants every 4 ft with 8 ft between rows. Bush types are planted 2 in. deep and thinned to 1 plant every 3 ft in rows 6 ft apart.

Pumpkins are grown similar to cucumbers and muskmelons. They grow best on well drained, fertile soil, with a large amount of organic matter.

There are many cultivars available (see Table 2.8). All pumpkin seeds are edible, and pumpkins were originally grown for their seeds, not the flesh. Seeds are still sold like peanuts in Mexico. The naked seed cultivars have a large seed without a seed coat. The flesh of these pumpkins is coarse in texture and flavor. It can, however, be baked, pureed, and used to thicken soups.

FIG. 7.39. Three *Cucurbita* species used as a pumpkin. Left to right—*C. maxima* cv Golden Delicious. *C. moschata* cv Cheese. *C. moschata* cv Dickinson Field. *C. pepo* cv Connecticut Field. *Courtesy of J.S. Vandemark II.*

Calabaza *(Cucurbita moschata)* is the common name of several strains of pumpkin, sometimes called Cuban pumpkin or Cuban squash. It is grown similar to pumpkin. The fruit weighs 5–7 pounds, and is round and somewhat flattened on both ends. The fruit is mottled green or yellow and buff-cream. The light-yellow colored interior is used similar to wax gourd or winter melon. Seeds are seldom found in catalogues, but may be saved from fruits purchased in southern Florida.

Harvest. Pumpkins can be harvested whenever the rind is hard and they have developed a deep solid color. Some types have the best quality if they are harvested after the vines are senescent or have been killed by frost. When the pumpkin is cut from the vine, 3 or 4 in. of stem should be left attached to the fruit. Pumpkins without stems do not store well. All pumpkins should be harvested before they are injured by a heavy frost. They are stored in a dry, well-ventilated area between 50° and 60°F.

Common Problems. Disease and insects are described in the section on Cucumber. Mildews can be controlled with a fungicide. Cucumber beetles should be controlled from planting onward. Any insecticide should be applied in the early evening when bees are not active. Bees are needed for pollination.

Pumpkins may also be attacked by the squash bug and squash vine borer, particularly east of the Rocky Mountains. If only a few plants are infected with the squash vine borer, the wound can be located on the stem just above the soil line. The stem can be slit with a sharp razor blade, puncturing the borer. Moist soil is placed around the stem to a height just above the wound. New roots will develop, compensating for the injury.

RADISH

Radish *(Raphanus sativus)* is a native of Asia and was being used by the Egyptians at the time of the Pharaohs. They are hardy to cold but cannot withstand heat. In the South, they are grown in the fall, winter, and spring. In the North, they are grown in the spring and fall. In regions with cool summers, they can be grown throughout the growing season.

Soils should be prepared and fertilized before planting. Spring radishes grow rapidly and conditions must be favorable for rapid growth. Slow-growing radishes have a hot or pungent flavor. Overfertilization can result in excessive top growth with no root enlargement.

Spring Radishes. Spring radishes mature in 20–30 days and can be grown throughout the season in cool climates and all but the hottest months in warmer areas. Successive plantings are made very 10–14 days beginning in spring as soon as the soil can be worked and again a month before frost. In the South and Pacific Southwest they are planted in the fall. Seed is

FIG. 7.40. Red King is a red globe-type of radish. *Courtesy of Harris Seed Co.*

planted one-fourth to one-half inch deep in rows 12 in. apart. They are frequently grown between slow-maturing vegetables such as cabbage, pepper, and tomato. Plants are thinned to stand 1 in. apart. Overcrowding results in poor root development. During hot weather, the roots become pungent and plants produce seedstalks. Radishes are pulled when they are 1 to 1½ in. in size. Mature radishes become pithy.

Radishes can be mixed with carrots, beets, and parsnips to mark the row of those vegetables that grow slower.

Many cultivars are available, including red or white round types, and red or white icicle (long root) types. Cultivars with a red top and white bottom on round or icicle roots are also available. Champion is a red, round root type developed to be grown under low light intensity, such as occurs in a greenhouse during the winter months.

Winter Radishes. Winter radishes (*R. sativus*, Longipinnatus group) grow slower, are larger, and keep longer than spring types. They require a 45–70 day growing period and are usually grown as a fall crop. The lower temperatures and shorter daylengths discourage the plants from forming seedstalks. The cultivar All-Season can be grown either early spring to summer or fall to winter.

Seeds are planted one-half inch deep, 4 to 6 in. apart, depending on the cultivar. They are usually planted in the space in which early corn, onions, or potatoes were grown. Cultivars range in pungency from mild to very hot and include All-Season (white, Japanese "daikon" type), Chinese Rose Winter (white, medium hot), Round Black Spanish (black skin, white flesh), and White Chinese (also called Celestial, white skin). Sakurajima is a very large type that may grow to 50 lb. Winter radishes may be harvested and stored similar to other root crops (beets and carrots).

Common Problems. The cabbage maggot is the only problem most gardeners will encounter. For control, a soil insecticide is worked into the area before planting.

RHUBARB

Rhubarb *(Rheum rhabarbarum)* is sometimes called pie plant or wine plant and is a native to Central Asia, probably Siberia. It was grown extensively in Italy in the sixteenth century and was brought to America from there in the eighteenth century.

Culture. Rhubarb grows best in areas that have cool, moist summers and winters that freeze the ground several inches. It does not grow well in most parts of the south.

FIG. 7.41. Rhubarb is a perennial grown for its long petioles.

Rhubarb is a perennial and should be planted at one edge of the garden. The area should contain a heavy application of manure or compost, which is worked into the soil 12–18 in. As rhubarb is planted 3 or 4 ft apart, each area can be prepared separately. Good drainage is essential for rhubarb.

Crowns from established plants are planted with the crown bud 2 in. below ground level in early spring as soon as the soil can be worked. Plants are set 3 or 4 ft apart in rows 3 ft apart. Seeds are not recommended, as the seedlings vary greatly from the parent plant.

Crowns can be purchased or obtained from established plants. Established plants are dug in early spring and split into pieces with one large bud to each section of crown and roots. Vigorous, healthy plants should be dug and the crowns planted immediately. To maintain vigorous plants that are not overcrowded and producing small petioles, they should be dug and re-established every 8–10 years after planting.

Plants should be fertilized in the spring and mulched with compost or well-rotted manure in late fall. Plants require abundant water during the growing season. Under conditions of low fertility or overcrowding or with old plants, seedstalks may form. These should be cut at the base of the plant and discarded so the energy will go into the roots for the next year's crop.

Rhubarb cultivars with mainly green petioles include German Wine, Sutton's Seedless, and Victoria. Cultivars that produce red petioles include Canada Red, Crimson Wine, McDonald, Ruby and Valentine.

Harvest. Rhubarb produces long leafstalks or petioles from the food stored in its fleshy roots from the previous year. The first year after planting, the stalks or petioles should not be picked to allow the roots to enlarge. The second year, the petioles may be harvested for 2 weeks only. The third year after planting, the petioles may be harvested for 6 weeks or until they become small, indicating that the food reserve is gone. One-third of the petioles or leafstalks should remain on the plant after each harvest.

The stalks are pulled from the plant and the leaf blade is removed. Rhubarb leaves contain soluble oxalic acid, which is toxic, and the leaves should not be eaten. Leafstalks or petioles from frosted plants should not be eaten, as the leaf oxalic acid may have moved into the petiole.

Common Problems. Rhubarb is relatively free of insect and disease problems. Rhubarb curculio may be a pest in the eastern half of the country. It is a large, usually reddish, snout beetle about three-fourths of an inch long. The beetle punctures the leafstalks or petioles, leaving black spots. The beetle also feeds on curly dock, and this weed should be destroyed.

Crown rot also affects rhubarb in the eastern half of the United States. The petiole develops lesions at the base, and the entire leafstalk or petiole collapses. There is no effective control, and rhubarb should not be grown in the infected area for 5 years.

SALSIFY

Salsify *(Tragopogon porrifolius)* is also called oyster plant or vegetable oyster because of its faint oyster-like flavor. It originated in the Mediterranean area and is distinct from black salsify and Spanish salsify, which are common in Europe.

Culture. Salsify is a hardy biennial which can be grown in most regions of the United States. It is grown similarly to parsnip but needs 150 days to mature from seed. Salsify is more cold tolerant and is planted in the spring as soon as the soil can be worked. The soil should be loose and crumbly, at least 18 in. deep, so that long straight roots will be produced.

FIG. 7.42. Salsify is often called oyster plant because of the faint oyster-like flavor found in the roots. *Courtesy of W. Atlee Burpee Co.*

Fresh seeds are planted one-half inch deep in rows 18 in. apart. Old seed germinates poorly. Seedlings are thinned to stand 3 in. apart. Plants should be irrigated until they are established, at which time they will withstand some dry conditions.

The cultivar usually listed is Mammoth Sandwich Island. It has few insect or disease problems.

Harvest. Roots are harvested in the fall after a freeze, which improves their flavor. Plants may be left in the ground and harvested as needed. Salsify withstands hard freezing and can be mulched so roots can be dug during the winter. These plants should be harvested before they begin to grow the following spring or else a long-stemmed, purple flower will be produced.

SOUTHERN PEA

The unqualified word "pea" in the South usually refers to the Southern pea *(Vigna unguiculata)* or cowpea. There are three types of Southern pea, each with its own distinct flavor. These types are blackeye pea, cream pea, and crowder pea; Southern peas are sometimes referred to by these names. Yard-long bean (*Vigna unguiculata* subspecies *sesquipedalis*) or asparagus bean is a pole-type subspecies of Southern pea.

The Southern pea is not a pea but a bean and originated in southern Asia and India. It was then taken to Africa. It was carried to the Spanish settlements in

FIG. 7.43. Pods of Pink Eye Purple Hull Southern peas. *Courtesy of Otis Twilley Seed Co.*

the West Indies by slave traders, where it has become an important food. The Southern pea was then brought to the United States in the 1700s (Meiners and Kraft 1977).

Plant Characteristics. Southern peas are pole and bush types of short or long season maturity. The bush types mature quickly and are best for gardens outside the Deep South. Crowder types have the seeds crowded tightly together in the pod. Cream types produce cream seeds and blackeyed types produce cream seeds with a black embryo area.

The yardlong bean or asparagus bean is a pole type. It produces a bean pod that may grow to 3 ft or more if left on the vine to mature. It is grown like Southern peas but is less productive than Southern pea cultivars.

Culture. Southern peas are a warm season crop. In the South, they can be grown during the summer between the harvest of early spring crops and before planting fall crops. Southern peas are very susceptible to cold and are planted later than snap beans. Southern pea, including yardlong bean, can be grown in northern climates wherever lima beans can be grown.

Southern peas grow best without a heavy application of nitrogen fertilizer, and they are particularly heat and drought tolerant. Excess moisture reduces yield.

Seed should be planted 1 in. deep, 2 to 4 in. apart in the rows. The rows should be 2 or 3 ft apart. Pole types, including yardlong bean, will need some type of support, as do pole snap beans. In the South, successive plantings can be made 3 weeks apart until midsummer to provide a continuous supply.

Cultivars of Southern pea may be found in garden catalogues under pea, bean, blackeye peas, or Southern pea. Many cultivars are available.

The Yardlong or Asparagus Bean is usually found in garden catalogues listed as such.

Harvest. Southern peas are harvested as green shell or dry peas. They are harvested when the deep green pod color changes to light yellow, red, purple, or silver depending on the cultivar. This is 55–80 days after planting. The pods should be picked twice each week, with the lower pods maturing first. For dry beans, the pods are allowed to remain on the plant and mature. The pods are then picked and the peas removed, dried, and stored for later use.

Yardlong or asparagus beans can be used like snap beans, green shell, or dry beans. Often they are harvested when the pods are 10–12 in. long, cut into sections, and used in Chinese cookery. The pods may be left on the plant; when it begins to change color, the seeds are harvested as green shell peas. The pod can also be left to mature on the plant and the seeds then used like Southern peas.

Common Problems. Various insects and disease attack Southern peas. The major pest is the cowpea curculio, which feeds on the pods and seeds.

FIG. 7.44. Asparagus or yardlong beans have very long pods. *Courtesy National Garden Bureau, Inc.*

Fall-planted Southern peas are often attacked by stinkbugs and cornworms. To reduce diseases and nematodes, resistant seed should be used. Purchased seed is more likely to be true to type and carry this resistance than seed saved from the garden.

SPINACH

Spinach *(Spinacia oleracea)* is native to Asia and was used over 2000 years ago in Iran. It was brought to Spain in 1100 A.D. and taken to America by the early colonists.

Plant Characteristics. The edible part of spinach is a compact rosette of leaves. Spinach produces a seedstalk easily in response to long days, and hot temperatures further hasten the flowering response. Thus, planting times are critical, and other leafy vegetables such as New Zealand spinach or Swiss chard are frequently grown instead of spinach.

Culture. Spinach is grown in the North as soon as the soil can be worked (5 weeks before the last frost) and again in late fall (7 weeks before the first frost). In the South, it can be grown from fall to spring as the plant is a hardy, cool season crop.

Seeds are planted a half inch deep and an inch apart in rows 12 in. apart. Plants are thinned to stand 4 in. apart before they become crowded in the row. Two or three successive plantings can be made.

Spinach grows well on any fertile, well-drained soil. It grows poorly on soils that crust easily. Spinach is shallow rooted, and 1 to 1½ in. of water are needed each week to ensure rapid and continuous growth.

Cultivars have thin or thick, smooth or savoyed, dark green leaves. Spring cultivars do not produce seed stalks as rapidly as fall cultivars.

Harvest. The entire spinach plant can be cut off just at ground level about 35–45 days after seeding, when plants have 5 or 6 leaves. An alternate method is to harvest the outer leaves when they are 3 in. long, leaving

FIG. 7.45. Winter Bloomsdale spinach. *Courtesy of Harris Seed Co.*

the remaining leaves for later harvest. The entire plant should be harvested when a seedstalk begins to form.

Common Problems. Mosaic and mildew may be a problem, and resistant cultivars should be used for control. Leaf miners, aphids, or cabbageworms may be a problem. (see Tables 4.3 and 4.5 for control measures.)

SPINACH, NEW ZEALAND

New Zealand spinach *(Tetragonia tetragonioides)* is not a true spinach but is used in the same manner. It is native to New Zealand, Japan, and Australia and was introduced to England in the 1700s.

Culture. New Zealand spinach grows 2 or more feet tall with a spreading, branching type of growth reaching 4 to 6 ft. It is a warm season crop and is planted after all danger of frost is past.

Seeds germinate slowly and seed may be soaked 2 hr in warm water before planting. Seeds are planted 1 in. deep, 4 to 6 in. apart in rows 3 or 4 ft apart. Plants are thinned to stand 12 in. apart in the row. Transplants may also be used.

New Zealand spinach is simply listed as such with no named cultivars. It has no insect or disease problems of consequence.

Harvest. Harvest begins about 70 days after seeding by cutting off 3 in. from the tips of the branches. Leaves and stems are both used. About one-half to two-thirds of the tips may be harvested at any one time. Plants can be harvested until frost.

SQUASH

Squash is native to tropical America and was widely used by the Indians. The male flowers are still eaten as a batter-dipped, fried fritter in Latin America. Squash is defined as the edible fruit of any *Cucurbita* species that is harvested either immature or mature and eaten as a cooked vegetable. Some squash fruit is eaten both as a vegetable for the main part of the meal and for dessert as "pumpkin" pie.

Plant Characteristics. Squash plants may be vining, semi-vining, or bush types of plants. They have separate male and female flowers on the same plant and belong to three separate species. Thus, squash will cross-pollinate with another squash, pumpkin, or gourd within the same species (see Table 2.8). It does not cross-pollinate with cucumber or watermelon.

FIG. 7.46. Zucchini summer squash. *Courtesy of W. Atlee Burpee Co.*

FIG. 7.47. Three species of *Cucurbita* used as a winter squash. Left to right—*C. moschata* cv butternut. *C. pepo* cv acorn. *C. maxima* cv Golden Turban. *Courtesy of J.S. Vandemark II.*

Culture. Squash grows well in practically all regions of the United States. It does not stand frost but in the far South may be grown in winter. Plants may be transplanted (see Plant Growth, Chapter 2) or seeded directly in the garden after all danger of frost is past and soil temperatures are 60°F or more.

Seeds of vining types of squash are planted 2 in. deep, 4 or 5 seeds per hill. Hills are placed 5 ft apart in rows 8 ft apart. Hills are thinned to 2 or 3 plants.

Seeds of semi-vining types are planted 2 in. deep, 4 or 5 seeds per hill. Hills are placed 4 ft apart in rows 6 ft apart. Hills are thinned to 2 plants per hill.

Seeds of bush types are planted 2 in. deep, 10 in. apart in rows 3 ft apart. Plants are thinned to stand 30 in. apart in the row.

Squash plants grow best on fertile, well-drained soil well supplied with organic matter. Plants should be irrigated during dry weather. Trickle or furrow irrigation is better than sprinkler irrigation as any moisture on the leaves increases the incidence of leaf diseases.

There are many cultivars of squash available, and Table 2.8 lists some of them.

Summer Squash. Summer squash is a bush-type plant that produces many fruit. The fruit are harvested immature before the rind hardens. Many shapes and colors of fruit are produced by the various cultivars. Fruits are usually white, green, or yellow. Fruits may be round and thin (pattypan type), club-shaped (zucchini type), or with a constricted, curved (crookneck type) or straight (straightneck) neck.

Squash fruits should be harvested when they are small and tender. Pattypan types are harvested when they are 3 or 4 in. in diameter and other types when they are 6 to 8 in. long and 2 or 3 in. in diameter. Larger fruit is not of the best quality. All large fruit should be picked and discarded to encourage new production. The immature fruits bruise easily, and they should be handled with care. Summer squash fruits are best used immediately and should be harvested 2 or 3 times a week.

Winter Squash. Winter squash plants are vining, semi-vining, or bush types. The fruits vary widely in shape and color. The fruits are harvested when they have become a solid color and the rind is hard. The acorn types are harvested when a yellow-orange color has developed on the fruit in the area where it was in contact with the ground.

Winter squash is usually baked but may be used to make pumpkin pie. Many winter squashes make better pie than some pumpkins.

Winter squash is frequently harvested after a frost has killed the vines but before a hard frost. The fruits are cut from the vines with 2 in. of stem left on the fruits. Fruits which have been bruised or frost injured or are immature do not store well. They may be stored in a cool, dry place at a temperature of 50°–55°F.

Squash Flowers

Squash blossoms may be dipped in a light mixture of flour and water, fried, and eaten as a vegetable. "Butterblossom" is a squash cultivar specifically designed for producing squash blossoms to be eaten. It is a bush-type plant and is grown similar to summer squash. The plant produces a large number of male flowers, which should be harvested when they are 8-10 in. long. Female flowers may also be used, but the center pistil should be removed. However, flowers from any pumpkin or squash plant may be harvested and used as squash flowers.

Common Problems. Squash plants have problems similar to those of cucumber, and the cucumber beetle should be controlled from planting onward. Squash may also be attacked by squash bugs and squash vine borer similar to pumpkin. If an insecticide is used, it should be applied in the late afternoon to prevent damage to bees. Bees are needed for pollination.

SUNFLOWER

Sunflower *(Helianthus annuus)* is native to North America, where it grows wild. The giant types can be grown at the edge of the garden to form a screen or windbreak. The seeds of these giant types are eaten for food or used as birdseed.

Culture. Seeds are planted in early spring, 1 in. deep and 18–24 in. apart for the giant types. Plants grow best in well-drained, fertile soils in full sunlight. During dry periods they should be irrigated.

Harvest. Plants take 80–120 days to mature. As the seeds form they attract birds, and a paper bag can be placed over them for protection. The head will be brown and the seeds plump at maturity. The head is cut off and the seeds rubbed loose and dried. They can be eaten raw or roasted, similar to pumpkin seeds.

SWEET POTATO

Sweet potato *(Ipomoea batatas)* is native to Central and South America. It was taken to Europe by Columbus and its first recorded use in the United States was in the early 1600s. There is no evidence to suggest it was used by the North American Indians.

In many African countries, the true yam, which is a different plant (*Dioscorea* spp.), was grown instead of sweet potatoes. When the African

FIG. 7.48. Sunflower plants grow very tall.

blacks were brought to America, they called the sweet potato a yam *(nyami)*, since the two species grow and are eaten in a similar manner. In the Deep South, moist-fleshed cultivars were grown, while other regions grew dry-fleshed cultivars, and many people thought of these plants as two different vegetables. The sweet potato and southern yam are in the same species. The orange- and yellow-fleshed cultivars are rich in vitamin A, but many tropical cultivars are white-fleshed.

Culture. Sweet potatoes grow best with 4 or 5 frost-free months but will produce smaller roots under shorter growing seasons. They thrive under the hot conditions of the South but can be grown as far north as southern

FIG. 7.49. Sweet potato *(Ipomoea batatas)* storage roots of a single plant. *From Yamaguchi (1983).*

Michigan and in the mild climates of the Pacific Northwest.

The edible part of sweet potato is a root, and plants should be grown in fertile, loose, sandy, crumbly soil. In heavy or wet soils, long and stringy roots are produced. Thus, in many areas, sweet potatoes are grown on a ridge of soil 8–15 in. high and 3 or 4 ft apart. This ridge is made before planting. In many parts of the world, the tips of the sweet potato vine are eaten as "greens."

Sweet potatoes are a vine-like trailing perennial, but plants are typically grown as an annual from stem cuttings called "slips." These may be purchased or produced by the gardener. Only disease-free plants or roots should be used. To grow slips, disease-free roots are placed in a hotbed about 1 in. apart and covered with 3 in. of sand or potting soil 5 or 6 weeks before transplanting. Soil temperatures should not exceed 85°F, and the area should be kept damp but not waterlogged. The slips are removed from the roots by grasping them firmly and rapidly pulling them from the soil. The slips should be planted immediately.

Slips are transplanted 1 or 2 weeks after the last frost. They are planted 2 or 3 in. deep, 12–16 in. apart on the rigid rows. A starter fertilizer should be used.

Plants are hoed or cultivated to control weeds and prevent the vines from rooting and forming small roots. These roots would prevent the main storage roots from developing rapidly.

Once established, the plants need about three-fourths of an inch of water per week when small and more when growing vigorously. Excessive water after a dry period causes the roots to split. These usually heal with no loss in eating or storage quality. The plants should not be irrigated the last 2 weeks before harvest.

Harvest. Sweet potatoes do not ripen or mature, and time of harvest is determined by root size. They should be harvested before the first fall frost, but a frost which damages only the vines will not harm the roots if they are dug immediately. Roots bruise easily and should be dug carefully. Roots, removed from the vines, are left to dry on the ground for 2 or 3 hr and then cured at 85°F and 85% humidity for 10 days. Roots can then be stored at 55°F and high humidity.

Common Problems. Sweet potatoes have few insect and disease pests in the garden. Various diseases can be controlled by using disease resistant cultivars and crop rotation in the garden.

Soil stain or "scurf" causes black spots to appear on the skin of the roots. The roots are still edible once the skin is removed.

SWISS CHARD (*See* CHARD)

TOMATO

Tomato *(Lycopersicon lycopersicum)* is native to the Andes mountains of South America, where it was used long before Columbus. It was taken to Europe where it was a popular vegetable by the 1500s. The tomato did not gain wide acceptance in the United States until the mid 1800s. Today it is the most popular vegetable grown in gardens.

Culture. Tomatoes are a warm season crop and require only a small growing space. They are grown in the extreme South in the winter and the upper South and the North from spring to fall. Tomato does not thrive in cold weather and will not set fruit at temperatures below 58°F. Tomatoes are grown practically everywhere during some part of the growing season.

Tomatoes should be grown where they will receive at least 6 hr of direct sunlight. They are best started as transplants (see Plant Growth, Chapter 2) and planted after all danger of frost has passed. By using plant protectors during cold periods, the plants may be set out somewhat earlier. The hot, dry, midsummer weather in the south is not favorable for planting tomatoes. Tomatoes do not set fruit above 85°F. Tomato transplants should not have fruit on them when planted, as this stunts their growth and the plants fail to

FIG. 7.50. Jet Star is a popular main crop type of tomato. *Courtesy of Harris Seeds.*

develop sufficiently for a good yield. The transplants should be planted to about the same depth as they were growing indoors. Soil should be pressed firmly around the plant after transplanting and a starter fertilizer added.

Dwarf tomato plants are spaced 12 in. apart in rows 3 ft apart; wirecage or uncaged plants are spaced at 24–36 in. in rows 4 ft apart; and staked plants are spaced 18 in. apart in rows 3 ft apart.

Plants may be allowed to grow on the ground. A mulch should be used under the plants to prevent diseases and conserve water.

Plants may be grown in wire cages to keep the fruits off the ground. This will help control diseases and make cultivation and harvesting easier. A 5 ft length of 6 in. mesh, concrete reinforcing wire can be made into a cage 18 in. in diameter (see Fig. 4.22).

Plants may also be pruned and either staked or trained onto a trellis. Some cultivars are not suited for this procedure. The stakes should be pushed 2 ft into the ground near the plants soon after transplanting. Twine or cloth is tightly tied to the stake, 2 in. above a leaf stem, and then loosely tied around the main stem just below the base of the leaf stem. The plants may also be trained onto a trellis or supported by the side of a building, instead of stakes. Training the plants to a stake or trellis prevents fruit rots and conserves space but increases fruit cracking and blossom-end rot.

Tomatoes may be produced 2 weeks earlier by pruning the staked plants, but total yield is reduced. The tomato plants can be pruned to 1 or 2 stems. Once a week, the small shoots that appear where the leaf stem joins the main stem are removed. The shoot is bent sharply to one side until it snaps and is then pulled off in the opposite direction.

Tomatoes grow best on fertile, well-drained soil. Plants can be side-dressed when the plants have several fruits. About 3 lb of 10-10-10 fertilizer or equivalent per 100 ft of row can be used. During the growing season, tomatoes need 1 in. of water each week and container-grown plants will need daily irrigation. However, large amounts of water and fertilizer stimulate vegetative growth and the plant may not set fruit.

Cultivars. There are hundreds of tomato cultivars available. Fruits come in a number of shapes, sizes, and colors. Plants produce an early, main

FIG. 7.51. Roma VF is a paste type of tomato. *Courtesy of Harris Seeds.*

season, or late crop; and the cultivar should be adapted to the gardener's growing season. In short growing seasons, early or midseason types may have to be grown. Cultivars should also carry resistance to diseases, and in the South and West they should carry nematode resistance. Resistance to verticillium wilt disease, fusarium wilt disease, and nematodes is frequently indicated by the letters V, F, N following the named cultivar. In order to grow the cultivar desired, the plants may have to be grown as transplants by the gardener.

Main Crop Cultivars. The main crop types produce medium to large fruit of various maturities. Most cultivars listed are the main crop types. New and more disease-resistant cultivars are continuously being produced. See your state agricultural experiment station for cultivars adapted in your specific location (see Table 1.3 for addresses).

Orange or Yellow Fruit Cultivars. These cultivars are frequently considered to be lower in acid, but the acid content is similar to main crop cultivars. The orange and yellow fruits do have a higher sugar content and taste sweeter. Caro Rich, Golden Boy, Jubilee, and Sunray are cultivars.

Container Cultivars. Any tomato that can be grown in the garden can also be grown in containers and supported with a trellis. However, there are a number of cultivars that produce small red fruit under 2 in. in diameter on plants less than 2 ft high. Such cultivars include Patio, Pixie, Presto, Early Salad, Small Fry VFN (vines 3 ft high), Tiny Tim, Toy Boy, and Tumbling Tom.

Paste Cultivars. These cultivars contain less water in the fruits and are eaten fresh or used for paste, catsup, or canning whole. Roma VF and San Marzano are cultivars.

Salad Tomatoes. These are often called cherry tomatoes and vary in size up to 1½ in. in diameter. Fruit is yellow or red and produced on standard type plants. Cultivars include Basket Pak (red), Cherry (red), Gardeners Delight (red), Pear (yellow or red), Plum (yellow or red), Red Cherry (red), and Sugar Lump (red).

Harvest. To obtain the best flavor, tomato fruits should be harvested when they are fully ripe and firm. They ripen best at 68°F. When temperatures are near 90°F, the tomatoes become soft before the color has fully developed. At this time, tomatoes should be picked every other day when the fruits have turned pink and ripened indoors at 60°–75°F. Light will increase the color somewhat, but it is not required for ripening.

Before frost, green tomatoes can be harvested, wrapped in newspaper, and stored at 50°–60°F for several months. They will ripen when placed at room temperature. Tomato vines containing the fruit can be hung in a warm shelter to finish ripening for immediate use.

Common Problems. Insect that attack tomatoes include flea beetle, tomato fruitworms, hornworms, aphids, leaf miners, Colorado potato beetles, whitefly, and spider mites. (See Tables 4.3 and 4.5 for control measures.)

The two most common diseases are fusarium and verticillium wilts. Resistant cultivars should be grown and crop rotation practiced. Blossom-end rot is caused by a calcium deficiency and water stress. This is more common on staked or pruned plants. More uniform irrigation helps.

Sunscald occurs on fruits exposed to high temperatures in direct sunlight. A large, whitish area appears on the exposed fruit and occurs when little foliage is covering the fruit.

Tobacco mosaic virus (TMV) can be transmitted to tomato by gardeners who use tobacco. Plants should not be handled unless hands and tools are washed, and gardeners should not smoke around the plants. The virus is transmitted by direct contact.

TURNIP AND RUTABAGA

Turnip (*Brassica rapa,* Rapifera group) originated in western Asia and grows wild in Siberia. It is related to Chinese cabbage, and turnip was originally a selection of Chinese cabbage grown for its oil seeds, which have been used since prehistoric times. The turnip root has been cultivated in China since at least the fifth century A.D. Rutabaga (*Brassica napus,* Napobrassica group) or Swede turnip was developed in the Middle Ages from a cross between cabbage and turnip.

Plant Characteristics. The two plants are generally considered close relatives, but they are different plants. Turnips are a biennial grown as an annual for its roots. Roots vary in shape from flat, to long, to globe shaped. The flesh is usually white or yellow, while the skin is white, yellow, red, black, or purple. Turnip plants grow rapidly and have rough, hairy leaves. Rutabaga roots are generally yellow-fleshed and more solid and have a longer storage life. Rutabaga plants have smooth, waxy leaves and need a month longer to develop roots of similar size to those of turnips.

Culture. Both plants are cool season vegetables and are the most widely adapted root crops in North America. They are grown in the South during fall, winter, and spring and in the North as a spring or fall crop. Rutabagas grow best in northern regions while turnips grow best in latitudes south of Indianapolis.

Plants are hardy to cold and susceptible to heat and should be planted as late as practical for use in the fall before a hard freeze. Turnips are easily grown in the south during winter and spring. In the North, turnips and

FIG. 7.52. Above—Tokyo Cross turnip. Right—Purple-Top Yellow rutabaga. American purple top rutabaga has light yellow flesh. Rutabaga has a thick leafy neck; turnips have little or no neck. *Photos Courtesy of W. Atlee Burpee Co.*

rutabagas are best planted for a fall crop after cabbage, peas, early potatoes, or sweet corn. If turnips are spring planted, they should be planted as soon as the garden can be worked. However, if they are exposed to temperatures of 40°F for long periods, they may flower. Rutabagas are often difficult to grow as a spring crop due to their longer growing season.

Seeds are planted one-half inch deep in rows 12–18 in. apart. The area should be irrigated during summer to germinate the seeds and establish the seedlings. Turnip is thinned to stand 3 in. apart and rutabaga thinned to 6 in. apart.

Cultivars of turnip include Early Purple Top Strap Leaf, Just Right (also for greens), Purple Top White Globe, Shogoin (also called Foliage and also for greens), and Tokyo Cross (also for greens).

Most rutabaga cultivars are types of American Purple Top. Laurentian, Purple Top Yellow, and Red Chief are also frequently available.

Harvest. Turnip greens may be harvested a month after seeding. In hot weather, turnip roots become bitter and pithy and must be used before warm weather occurs. Fall-harvested roots may be 2 to 4 in. in diameter for turnips and 3 or more inches thick for rutabagas. Both turnips and rutabagas may be left in the ground like parsnips and harvested when needed in mild climates. They are hardy to frosts and can be mulched to allow harvest through early winter in northern regions.

Common Problems. Similar to radish, the cabbage root maggot can be a pest. If a problem exists, an insecticide is needed before the crop is planted.

WATERMELON

Watermelon *(Citrullus lanatus)* originated in Africa and was taken to Europe in the sixteenth century. Early settlers brought the watermelon with them to America.

The citron (*C. lanatus* var. citroides) is a type of watermelon that is used differently. It is not edible when raw and is eaten cooked, pickled, or candied. This type of melon was used by the North American Indians before Columbus and is possibly of American origin. Catalogues list it under Watermelon.

FIG. 7.53. Sugar Bush watermelon. *Courtesy of W. Atlee Burpee Co.*

Plant Characteristics. Watermelon requires a lot of space to grow its large vines. Separate male and female flowers occur on the same plant, but they do not cross-pollinate with cucumbers, squash, or pumpkins. The citron melon is grown similar to watermelon.

Culture. Watermelon can be grown in nearly all regions of the country. It requires more summer heat than muskmelon, but if muskmelon growns well in the area, the small icebox watermelon cultivars can be grown. Under short growing seasons, the plants do not produce enough sugar and the melons produced are not sweet. In these areas, early cultivars, transplants, and mulches should be used.

Watermelons are grown similar to cucumber but require much more space. Transplants (see Plant Growth, Chapter 2) or seeds are planted after all danger of frost is past and are set apart in hills. Seeds are planted 1 in. deep and thinned to 3 plants per hill. They may also be planted as single plants, 1 plant every 2 or 3 ft. Rows are spaced 6 or 7 ft apart. For transplants, a starter fertilizer should be used.

Plants grow best on light sandy soils. Watermelon plants are deep-rooted and once established will not require irrigation if the soil stores 9 in. of water. At this minimum amount of water, yields are increased by irrigation.

There are a number of cultivars available that produce red- or yellow-colored flesh in round, oval, or long fruits. Seedless cultivars require a standard cultivar for pollination as they do not produce viable pollen. Seeds of such a cultivar are generally included when seedless cultivars are purchased. Seeds of seedless watermelons are expensive, so they are usually started as transplants.

Harvest. A number of indicators can be used to determine if a watermelon is ripe. (1) Fruits have a ground spot where they lie on the soil. This spot becomes light green or yellow as the fruit matures. (2) The curly tendril on the leaf near the stem dries and becomes brown when the melon is ripe. However, in some cultivars, the tendril dries up 10 days before the fruit is ripe. (3) The skin becomes rough as they mature. (4) The melon can be rapped with the knuckles. A sharp metallic sound indicates the melon is immature. A dull sound indicates it is mature. This test is best done in early morning, as melons thumped in the heat of the day and those that have been picked and stored all sound ripe.

Watermelons are cut off the vine. The rinds may be pickled or candied.

Common Problems. Insect and disease problems are similar to those of cucumbers and pumpkins. Cucumber beetles need to be controlled from planting onward. Anthracnose disease is a problem in warm moist weather, particularly in the Southeast. If this disease is a problem, resistant cultivars should be grown; treated seed should be used; the crop should be rotated in the garden; and all old vines should be removed since the disease overwinters in these vines.

Birds may make holes in the tops of the watermelons. The dark-skinned cultivars, like Sugar Baby, are not as attractive to them.

SELECTED REFERENCES

Anderson, E. 1954. Plants, Man and Life. Andrew Melrose, London.

Anon. 1975. Seed for today. Asgrow Seed Co. Descriptive Catalog of Vegetable Varieties *21*.

Banadyga, A.A. 1977. Greens or "potherbs"—Chard, collards, kale, mustard, spinach. New Zealand spinach. *In* Growing Your Own Vegetables. U.S. Dep. Agric. Bull. *409*.

Bubel, N.W. 1977. Vegetables Money Can't Buy but You Can Grow. David R. Godine Publisher, Boston.

Doty, W.L. 1973. All About Vegetables. Chevron Chemical Co., San Francisco.

Garrison, S.A. and Ellison, J.H. 1977. Asparagus starts up slow but goes on and on; rhubarb also takes its own sweet time. *In* Growing Your Own Vegetables. U.S. Dep. Agric. Bull. *409*.

Gibbons, E. 1973. Feast on a Diabetic Diet. David McKay Co., New York.

Meiners, J.P. and Kraft, J.M. 1977. Beans and peas are easy to grow and produce a wealth of food. *In* Growing Your Own Vegetables. U.S. Dep. Agric. Bull. *409*.

Minges, P.A. 1977. Play it cool with cole crops (cabbage, etc.); they attain best quality if matured in fall. *In* Growing Your Own Vegetables. U.S. Dep. Agric. Bull. *409*.

Reynolds, C.W. 1977. The complex art of planting. *In* Growing Your Own Vegetables. U.S. Dep. Agric. Bull. *409*.

Sims, W.L. *et al.* 1977. Home vegetable gardening. Univ. Calif. Agric. Sci. Leafl. *2989*.

Splittstoesser, W.E. 1977. Protein quality and quantity of tropical roots and tubers. HortScience *12,* 294–298.

Yamaguchi, M. 1983. World Vegetables: Principles, Production and Nutritive Values. AVI Publishing Co., Westport, CT.

8

Growing Special-Use Vegetables

The selection of special-use vegetables is somewhat arbitrary. Vegetables in this category include those in which a limited number of plants are grown in the garden or those that are usually not grown by the majority of gardeners. However, these vegetables are not novelties grown for fun when extra space is available in the garden. These vegetables possess many valuable characteristics that make them worthy additions to the garden. Most special-use vegetables are easy to grow and produce high yields. If they are included in the diet, most are substantially cheaper to grow in the garden than to purchase commercially. Suggestions for control of pests can be found in Chapter 4, on Pest Control.

AMARANTH

Amaranth (*Amaranthus,* spp.) is also known as pigweed, edible amaranth, Chinese spinach, and tampala. It was an important grain crop of the Aztecs of Mexico in pre-Columbian times. Red-leafed *Amaranthus tricolor* is used for greens in Southeast Asia, while the green form of *A. gangeticus* is grown for boiled greens in the USA. *A. candatus* is used as a grain crop in Ethiopia. It is grown in all parts of the world.

Plant Characteristics. Amaranth is an erect plant growing 1–3 feet tall. Some plants are large and broadleaved (6 inches) while others are small and narrow leafed. There are many shapes and colors of the flowers, and plants flower readily under short-day conditions.

Culture. Amaranth is grown from seed in light, well-drained soils. It may be sown indoors and transplanted in the garden 3 inches apart. Plants are thinned to 12 inches after 4 weeks. Frequent cutting for greens increases productivity and delays flowering. The grain can be harvested in the fall, but some shattering will occur from the early maturing part, as seeds ripen progressively. Plants respond to a balanced fertilizer (such as 10-10-10) and irrigation during dry periods to produce maximum yields. Amaranth is grown in all climates, from tropical to artic, although yields vary.

Harvest. About 4 weeks after transplanting, the entire plant may be harvested for greens. The young stem tips and leaves may be harvested every 2–3 weeks. The more mature stems should be cooked separately. Seed heads should be harvested when seeds begin to ripen in the fall. Seed heads may be placed in a paper bag and dried. Seed may be removed by rubbing the seed head between one's hands.

Common Problems. Damping-off fungus is a problem during early spring. Many chewing insects attack the plant. When grain is desired, the garden should not contain common pigweed, as amaranth and this weed may hybridize, creating weeds next year. Usually the garden amaranth does not become a serious pest the next year.

ASPARAGUS BEAN (*See* SOUTHERN PEA, Chapter 7)

ASPARAGUS LETTUCE OR CELTUCE (*See* LETTUCE, Chapter 7)

BEANS, MUNG

The mung bean (*Vigna radiata*) is widely known in the Orient. It originated in northwest India. It is used primarily for bean sprouts, raw in salads, or as a cooked vegetable in Oriental dishes.

Plant Characteristics. The mung bean is a bush type plant 1 or 2 ft tall. The bean pods are produced at or above the foliage level which makes harvesting easy.

Culture. The mung bean has about the same climatic and soil requirements as the Southern pea and is grown in a similar manner. It requires 90–100 days to mature and is very susceptible to cold. It can be grown in the middle to deep South.

The seeds are planted 1 in. deep, about 2 in. apart in rows 2 ft apart. They are planted after all danger of frost has passed.

Harvest. The plants are pulled when the pods have matured and are allowed to dry in a clean area. The pods shatter easily, and careful handling is needed to prevent loss of beans. When the pods are dry, the beans can be removed and stored in an airtight container.

Common Problems. Mung beans are resistant to most of the insect and disease problems that are found on Southern peas, snap beans, and soybeans.

Use as Sprouts. The beans are soaked overnight and then placed in a container that has holes in the bottom for drainage. The container needs to be large enough to allow the beans to expand and swell as they sprout. The container is placed in the dark in a warm place. The beans are moistened three times a day in the summer and twice a day during the winter when it is cooler. Under warm conditions, the beans are ready for use in about 5 days. Under cooler winter temperatures, they may require 10 days or more. The bean sprouts are usually used when 3 in. long. They are stored in a cool, humid place until use.

BEANS, SOY OR EDIBLE SOYBEANS

Soybeans *(Glycine max)* are known as edible soybeans and vegetable soybeans. They appeared as a cultivated crop in northern China in 3000 B.C. (Metcalf and Burnham 1977). The cultivated type is not known in the wild, but it is believed to come from a viny North Asia type. Soybeans were brought to North America during colonial times but did not become a major crop until after World War II. They are used as an important food source in China, Korea, Japan, and Manchuria.

Plant Characteristics. Soybeans are a warm season crop with an erect bush-type growth. They grow slower than most types of beans and require from 75 to 170 days for maturity. They produce flowers that are self-fertile and usually self-pollinated. Most soybean seed is yellow, but green, black, black and yellow, and black and green seed are known.

Culture. The soil and cultural requirements are similar to snap beans. Field soybeans may be grown and harvested when the seeds are immature. The edible soybeans have larger seeds and are thus often preferred. Small early cultivars may be planted in rows 2 ft apart but later-maturing, larger cultivars are planted 3 ft apart. The seeds are planted 1 in. deep about 8 seeds per ft of row when the soil has warmed up (the time when tomatoes are transplanted).

If soybeans have never been grown on the soil, the seeds can be inoculated with bacteria which fix nitrogen from the air into a form the plants can use. This inoculum is available from various garden catalogues and at garden centers. Nitrogen fertilizer is seldom needed for soybeans in a garden.

Soybeans, as immature beans, should be ready to harvest from early cultivars about two months after planting. The time to maturity depends upon the day length of the area and the cultivar grown. As one goes from North to South, the summer day lengths become shorter; and soybean cultivars are classified into 10 maturity ranges. Cultivars in groups O and I are the earliest in maturity and adapted to the northern United States and southern Canada. Groups I and II are used in the northern areas of the United States. Groups II and III are adapted to the southern corn belt.

FIG. 8.1. Field soybeans may be harvested before the seed matures, and used as a vegetable.

Group III cultivars can be grown further south, but the plants will mature about a month earlier. Group VIII are late maturing cultivars grown near the Gulf Coast. The edible soybeans most frequently listed in garden catalogues are Giant Green and Kanrich.

Group O and I cultivars include Disoy, Early Hakucho, Okuhara Early, and Giant Green.

Group II and III cultivars are Kanrich, Sodefuri, and Verde.

Field soybeans adapted to the area may also be used. These cultivars may be purchased from a local elevator or farm supply store.

Harvest. Soybeans are usually harvested as the immature seed. The pods are harvested when the seeds are fully grown but before the pods turn yellow. Pods are harvested over a 10-day span. Soybean seeds are difficult to remove from the immature pods. The pods are usually placed in boiling water and cooked for 2 min. The pods can be removed and the seeds squeezed from the pods easily. These seeds are eaten similar to green shell beans.

If dry mature soybean seed is desired, the pods are allowed to remain on the plant. When the pods have turned brown, the entire plant is cut and placed on a rack or hung up to dry. This prevents loss of beans in the garden. The beans are shelled when the pods are dry. The beans should be thoroughly dried before storage.

Mature dried beans may be used to produce bean sprouts (Adjei-Twum *et al.* 1976) similar to mung beans or lightly roasted as a snack food.

Common Problems. Soybeans are seldom bothered by Mexican bean beetles and can be grown where this pest prevents the growing of snap beans. In some parts of the United States, soybeans may be attacked by various worms, caterpillars, grasshoppers, and mites.

BITTER MELON AND BALSAM APPLE

Bitter melon (*Momordica charantia*) is also called balsam pear, alligator pear, and bitter cucumber. Balsam apple (*M. balsamina*) is closely related. They are indigenous to the old world tropics and are an important food crop in India, China, and Southeast Asia.

Plant Characteristics. These cucurbits are similar in leaf shape and growth habit to the West Indian Gherkin (*see* Cucumber). They are vining crops with deep-lobed leaves and yellow flowers. Bitter melon fruit is 4–6 inches long, green turning to yellow and orange when ripe, oblong, and warted. Balsam apple fruit is 3 inches long, egg shaped, and orange colored.

Culture. These cucurbits are grown similar to cucumber. They can easily be grown on a trellis.

Harvest. The immature fruit is harvested and they are the least bitter at this stage. Fruits are often peeled and then soaked in salt water or parboiled to remove the bitter flavor. When the bitter melon is ripe it splits open. Bitter melon and balsam apple are used as a boiled vegetable or in soups.

Common Problems. These vegetables may have problems similar to cucumber. The bitter substances in the fruit and leaves often act as feeding stimulants for cucumber beetles.

BURDOCK, DOMESTIC

Domestic burdock (*Arcticum lappa*) or gobo is a native of Siberia and northeastern China. It was introduced by the early settlers and used by the American

Indians. Wild burdock (*A. minus*) occurs in much of the U.S.A., but the roots are small, fibrous, and strongly flavored. Domestic burdock is cultivated in Japan for its long tap root.

Plant Characteristics. Domestic burdock looks much like the wild plant. It is a biennial that will flower the second year, and may grow to 8 feet tall. It produces burrs as seed pods, which cause it to become a pest weed. The edible root somewhat resembles a carrot root, but is usually 2 feet long and is brown with white flesh. If not harvested, the root becomes tough.

Culture. Seeds require light to germinate and are planted ¼ inch deep in the soil when the ground has warmed up (50°F). A deep, well-drained sandy soil is recommended. Plants should be thinned to 6 inches apart in the row. As the root is long, it is often grown in a raised bed, which may be dismantled to remove the tap root. The tip of the main root or a feeder root may be left in the soil to produce next years crop. Burdock tops are killed when temperatures drop to just above freezing (35–40°F), but it is high-temperature tolerant. Roots will survive in the soil at temperatures well below freezing. As a result, burdock is sometimes planted in the fall. However, if the roots become larger than ¼ inch, they have enough energy to respond to a winter cold treatment and flower the next spring when daylengths become greater than 12.5 hours. In this case, the plants do not produce a satisfactory edible root.

Harvest. The young leaves may be harvested in the spring and eaten as greens, and the shoots may serve as a substitute for asparagus; or the shoots, with small roots attached, may be cooked like baby beets and beet greens.

The mature roots are best harvested after 2–4 months, when they are still young and tender. They may be washed, wrapped in damp cloth, placed in a plastic bag, and stored for several weeks in the refrigerator. To eat, the roots are scraped and cut into strips and soaked or cooked. The cooking water should be changed 2–3 times to improve the root's flavor. In Japan, burdock roots are often peeled, sliced, and placed in tubs of water until purchased. When cooked, the root is crunchier than carrots. It is used in stir-fry dishes, used as a side dish, or added to soups. It is also chopped fine and used as a condiment with rice or fish, or deep-fried.

Common Problems. Burdock has few insect or disease problems. It cannot withstand swampy conditions and should be irrigated infrequently to ensure that it produces a long rap root.

CACTUS, PRICKLY PEAR

Prickly pear cactus (*Opuntia* spp. and *Nopalea* spp.) are native from Canada and the U.S.A. to the Straits of Magellan. They are frequently grown as an ornamental plant.

Plant Characterisitics. The plants produce flat, padlike stems with many spines. The flowers are white, orange, or yellow. The fruits are whitish to yellow, but more commonly red or deep purple. The fruit shape and prickly skin have given them the name "prickly pear."

Culture. Where the cactus is native, seeds may be obtained and scattered on the garden area where the plants can grow permanently. They may be started indoors in peat and transplanted. However, several years must elapse before the plants are large enough to bear fruit. Most gardeners propagate them vegetatively from a pad or branch. About half of the pad can be inserted in a well-drained soil and left to grow. Prickly pear are often grown in rock gardens, and too much moisture causes them to mildew and rot. They are easy to grow and do not need much care. They do respond to a general garden fertilizer, but seldom should be irrigated.

Harvest. Both the pads and the fruit are eaten. Small pads can be harvested (with gloves) and the skin scraped with a blunt knife to remove the spines. They can then be peeled, cut into strips or diced. Pads may be boiled with onions until tender and used as a side dish with shellfish, pork or eggs. They may be deep-fried and used as a substitute for french-fried potatoes, although they taste somewhat like okra.

The fruits are harvested after they have turned red and begun to shrivel slightly. The fruits are rubbed between the hands, which are covered with a pair of heavy gloves, to remove the spines. The spines may also be removed by singeing them over an open flame. The fruits are eaten like a melon, although they contain many seeds. The fruit may also be cooked and made into a sugar syrup for cactus candy or fermented into a drink.

Common Problems. Prickly pear cactus has few insect or disease problems.

CALABAZA (*See* PUMPKIN, Chapter 7)

CELERIAC

Celeriac (*Apium graveolens* var. rapaceum) is also known as knob celery, celery root, and turnip-rooted celery. It is a strain of celery native to most of Europe and parts of Asia. It originated as a form of celery that grows well in marsh areas.

Plant Characteristics. Celeraic is a vegetable that somewhat resembles a beet. It is grown for its enlarged root, which develops at the ground level. This root has a brown skin with white interior. It tastes similar to celery but is not stringy.

FIG. 8.2. Celeriac tastes like celery and attains its full flavor after a frost. *Courtesy of W. Atlee Burpee Co.*

Culture. Celeriac grows best on heavy, moist, well-drained soils. It is less sensitive to heat and drought than celery but more sensitive than most garden vegetables. Celeriac requires a 120-day growing season and in most parts of the United States is grown from a transplant.

For transplants, the seed is planted indoors and grown for 2 months. It may require 2 or 3 weeks for the seed to germinate. When the plants are 2 to 3 in. high, they are transplanted into the garden. It has good frost tolerance and may be planted early in the spring. Plants are spaced 6 in. apart in 18 in. rows. A starter fertilizer should be added to establish the plants.

In mild climates, celeriac may be direct seeded. It is planted one-fourth inch deep and the seedbed kept moist until the seedlings emerge. The plant and row spacing is similar to transplanted celeriac.

Celeriac requires adequate fertility and moisture. It should receive 1 to 1½ in. of water each week during dry periods. In low fertility areas, additional nitrogen should be added when the plant is half grown. If its growth is interrupted, small, poorly shaped roots are produced.

Few cultivars are available, but catalogues frequently list Alabaster, Giant Prague, Large Smooth Prague, or Marble Ball.

Harvest. Celeriac attains its full flavor after it has received a frost, but it may be harvested when the root has attained a diameter of 2 in. When fully mature, they will be 2 to 5 in. in diameter, depending on the growing region. In areas with mild winters, roots can be mulched with leaves or straw, left in the ground, and harvested as needed. Roots may also be removed and stored in moist sand in a cool place.

Common Problems. Celeriac has few insect and disease problems in the garden.

CHAYOTE

Chayote *(Sechium edule)* is also known as mirliton and vegetable pear. It is indigenous to Central America and the West Indies. It was grown by the Aztecs before the Spanish Conquest.

Plant Characterstics. Chayote is a perennial vine that bears large fruit. The common cultivar produces light green, pear-shaped fruit. The fruit is often used as a substitute for summer squash and is related to pumpkins, squash, and gourds. Similar to these plants, chayote produces separate male and female flowers on the same plant.

Culture. Chayote can be grown in the mild winter regions of the south Atlantic and Gulf Coast states and in parts of California. In the southern United States, it is grown usually as an annual as freezing kills the plant. Heavy mulching will protect the plant roots so the plant will sprout and grow from the roots the following spring. It cannot be grown in the North because of the long day lengths that delay flowering until fall.

Chayote produces one seed per fruit, and the entire fruit is planted when the soil is warm and all danger of frost is past. The fruit is placed, stem end up, on a slant in the soil. The plant grows rapidly in well-drained soil rich in

FIG. 8.3. Chayote fruit is about 4 inches long (left), and the seed (right) may be saved to start new plants.

organic matter. Plants should be about 10 ft apart, but one plant is usually sufficient as it produces 30–35 fruits. Chayote grows best with little nitrogen but requires potassium. Excess nitrogen stimulates vine growth, but produces few fruits.

A large vigorous vine is produced that requires some type of trellis for support. The vine requires large amounts of water during dry periods.

Chayote seed is seldom sold as the whole fruit is planted. It is usually available only in areas where it is grown. To save the fruit for planting, the fruit should mature on the vine but be harvested before the seed sprouts. Individual fruits are wrapped in newspaper and stored in a cool ventilated area.

Harvest. Fruits are harvested about 30 days after the flowers are pollinated. Fruits are from 4 to 6 in. long and may weigh 2 lb.

Common Problems. Insect and disease problems are similar to pumpkin and squash. Common insect pests are cucumber beetle, squash bug, and squash vine borer. Mildew is the common disease. Pests are controlled with various chemical and botanical pesticides.

CHICORY AND FRENCH ENDIVE

Chicory *(Cichorium intybus)* is also known as Belgian endive, French endive, and witloof chicory. It is a close relative of endive and is native to Europe and Asia. The leaves from plants growing in the wild have been used as a salad vegetable since time immemorial. It was first listed as a cultivated vegetable in the thirteenth century.

Chicory has several uses. The leaves can be used for greens or a salad vegetable. The root can be used as a coffee adulterant. The root can also be harvested and forced to produce new shoots during the winter; it is then usually known as French endive.

Culture. Chicory grows best in rich, deep loam soil without too much organic matter. The seeds are sown one-fourth to one-half inch deep with 18 in. between rows.

For Greens. For greens, seeds are planted in the spring as early as the soil can be worked. Successive plantings can be made every several weeks for continuous salad greens. The plants are thinned to stand 8 in. apart in the row. Leaves will be curled and somewhat bitter and resemble dandelion leaves. They are ready for harvest 60 days after planting when they are 6 to 8 in. long. To reduce the bitter taste and the green color somewhat, the leaves are sometimes gathered and tied loosely together at the top when 10 in. long. This process is called blanching. Leaves should be dry before tying to prevent leaf rots. The blanching process requires 2 or 3 weeks.

Most cultivars withstand the summer heat and light frost. If seedstalks are produced, they are tender and can be eaten also.

For Coffee Adulterant. To grow the plants for roots, the seeds are planted at the time of the average date of the last frost in the spring. If seedstalks are produced, it reduces the size of the root. Plants are otherwise grown similarly to greens: planted one-half inch deep and thinned to 8 in. apart in rows 18 in. apart. Plants will be larger than those used for greens and will develop a large root 10 in. long. Some leaves may be harvested for greens during the summer. About 120 days after seeding, the roots are mature. They should be dug before frost. Roots are then cleaned, peeled, cubed, and roasted. After roasting, they are ground for a coffee substitute.

For French Endive. Most gardeners who grow chicory will use it for French endive. Seeds are planted in warm soil one-fourth to one-half inch deep. If seeds are planted too early, the plants will be exposed to cold temperatures and produce a flower stalk. Such plants cannot be used to produce French endive. Plants are thinned to stand 4 or 5 in. apart in rows 18 in. apart.

Roots are harvested before freezing weather in the fall about 110 days after seeding. Roots are washed and all the leaves are removed except the single center crown bud on top. If this bud is not easily identified, all the leaves may be cut off 2 in. from the top of the root. Roots are covered and stored in a cool, moist, frost-free area.

Roots vary considerably in size. Large roots produce large heads, but these are usually composed of several small divisions and are not of high quality. Medium-sized roots, 1 to 2 in. in diameter, produce the best French endive.

During the winter and early spring, the roots are removed from storage and the root tip cut off so that all the roots are 6 to 8 in. long. Roots are then placed in a box. Sand, fine soil, or sawdust is added around the roots and the roots are covered to a depth of 6 to 8 in. This covering excludes light and forces the heads to be compact. The box is watered thoroughly and placed at a temperature of 60°F. One additional watering may be required. In 3 to 4 weeks, the tips of the heads will begin to break through the surface. The heads are harvested by cutting them from the root. Heads are creamy yellow with tightly folded leaves which have not yet uncurled. Additional boxes of chicory roots can be made up every three weeks for successive harvests.

CHINESE CABBAGE

Chinese cabbage is also known as celery cabbage. Two types are available: a leafy type (*Brassica rapa*, Chinensis group) and a heading type (*Brassica rapa*, Pekinensis group). Both types probably originated in China, where they have been used since the fifth century. Chinese cabbage is more closely related to turnip than to cabbage or celery.

Plant Characteristics. Chinese cabbage is an annual that grows 10–20 in. tall. The leafy type has thin, long, dark green leaves that resemble Swiss chard in growth habit. The heading type resembles Cos lettuce but produces a larger, more elongated, and compact head.

Culture. Chinese cabbage is a hardy, cool weather crop. It is grown in the South as a winter crop and as a summer-planted fall crop in the North. If the plants are grown under long days, they produce seedstalks, and spring plantings of Chinese cabbage are frequently unsuccessful.

Chinese cabbage requires a rich, well-drained, moist soil. Seeds are planted one-fourth to one-half inch deep in rows 18–30 in. apart. Plants should be thinned to 8–15 in. apart. If planted during dry, summer conditions, additional water is required to establish the plants. It should be seeded 45 days before the first frost for leafy types and 75–85 days before frost for heading types. Chinese cabbage responds to nitrogen fertilizer, and its growth will be stimulated by adding nitrogen when it is half grown.

Harvest. The heads are harvested by cutting the entire head off at the ground line. They should be harvested before hard freezing weather occurs.

FIG. 8.4. Pak Choi Chinese cabbage. *Courtesy of W. Atlee Burpee Co.*

The leaves are crisp and the inner leaves are faintly green. It can be used as a substitute for lettuce and celery.

Common Problems. Chinese cabbage has few pest problems in the garden. Aphids, cabbageworms, cutworms, and flea beetles are a problem occasionally. These can be controlled with chemical, botanical, or biological control agents.

CHINESE AND LEMON CUCUMBER

Chinese cucumber (*Cucumis melo,* Flexuosus group) is also known as the Japanese, Oriental, Armenian, or snake cucumber. Lemon cucumber (*Cucumis melo,* Chito group) is also known as the vine peach and melon apple. They originated in southern Asia, and have been grown in India for centuries. The Greeks and Romans spread them into Europe and western Asia. Chinese and lemon cucumbers are more related to melons than to standard cucumbers.

Plant Characteristics. They are a vining plant with large leaves and resemble cucumber plants. Chinese cucumber fruit is smooth with deep furrows. They grow up to 3 feet long and turn yellowish when ripe. Lemon cucumber fruit is the size and shape of a lemon or orange and yellow to orange in color.

Culture. They are grown similar to cucumbers. The Chinese cucumber fruit is longer and thinner than a standard cucumber and tends to curl when grown on the ground. Plants should be grown on a trellis to encourage pollination by bees and ensure that long, straight fruit will be formed. They require ample water.

Harvest. Chinese cucumber fruit can be harvested when small for pickles or relish. The cucumbers become sweeter as they become larger and yellowish-green. They are commonly harvested when the entire fruit has filled out, but the ridges are still somewhat prominent. They have few seeds and do not store well. They soon become limp and lose their crispness. They should be harvested and eaten as soon as practical. They are burpless and seldom produce gas pains. As they are more related to winter melons, the fruit is often stir-fried or served in soups.

The lemon cucumber fruit is harvested when it is yellow or orange. It is used mostly in preserves and pickles.

Common Problems. The insect and disease problems are similar to cucumber, although the Chinese cucumber is more resistant to several diseases.

CHINESE RADISH (*See* RADISH, Winter, Chapter 7)

CHRYSANTHEMUM

The vegetable chrysanthemum (*Chrysanthemum coronarium*) is known as garland chrysanthemum, chop suey green, or crown daisy. It is native to the Mediterranean area and is eaten primarily in the Orient.

Plant Characteristics. The plant looks very much like the leafy portion of the ornamental chrysanthemum. The vegetable chrysanthemum is an annual plant that may grown to 3 feet tall. There are three types; plants with finely parted, narrow dark green leaves; plants with medium-sized leaves; and plants with large, slightly parted, pale green leaves. The smaller leafed plants are said to be more cold tolerant than the large-leafed types. Flowers are small, yellow, and daisy-like in appearance.

Culture. Seeds are sown in the garden or started as a transplant. Most grow best under cool conditions in early spring or fall. In the south, they can be grown as a winter vegetable. Plants should be 4 inches apart in the garden. Normal fertilizer, mulching, and soil preparation is suggested.

Harvest. Leaves or the entire plant may be cut at ground level 1 month after sowing. If the plant was not uprooted, it will regrow. Unless the plants are kept closely trimmed, they will require staking and will flower in midsummer. Removing the flower buds before they develop will help maintain the leaves in a tender and mild-flavored condition. Flowers may be eaten, but they have a strong aromatic flavor. Leaves are often eaten raw in salads, stir-fried, and used in soups and in various Chinese and Japanese dishes.

Common Problems. Vegetable chrysanthemums have few insect and disease problems.

CITRON (*See* WATERMELON, Chapter 7)

CRESS

Cress is a native to Ethiopia, where is has been grown for over 2000 years. Cress is now extensively naturalized in many parts of the world, and it was one of the first vegetables brought to America. Three major types can be grown as a vegetable.

Garden Cress or Pepper Grass. Garden cress *(Lepidium sativum)* or pepper grass is a hardy, cool season annual. It can be grown from as soon as the ground can be worked in spring until fall. In the South and Pacific Southwest, it can also be grown during the winter. It is easily grown indoors during the winter.

Seeds are sown one-fourth inch deep, 1 in. apart in rows 1 ft apart. Plants are not thinned. At 65–70°F, they germinate in 4 to 7 days and the leaves can be harvested in 10–14 days after planting. Successive plantings every few weeks must be made for a continuous supply. Leaves are harvested when they are 3 to 5 in. long. During hot weather, the plants tend to produce seedstalks quickly, and the leaves are reduced in quality.

Upland or Winter Cress. Upland cress *(Barbarea verna)* is also called winter cress, spring cress, and Belle Isle cress. It is a hardy biennial but is grown as an annual or winter perennial.

Seeds are planted one-fourth inch deep in rows 1 ft apart. After the plants are established, they are thinned to stand 4 in. apart. They can be planted as soon as the soil can be worked in the spring (hence the name spring cress), or in mild climates they can be planted in the late summer and grown through the winter (hence winter cress). The plants grow about 5 in. high and a foot across and can be harvested in 50 days. Plants require moist soil and can withstand fall frosts. It is frequently only listed in garden catalogues as Upland Cress or Winter Cress.

Watercress. Watercress *(Nasturtium officinale)* is a cool season semi-aquatic perennial widely distributed in Europe and western Asia. It grows naturally in clear, cold, shallow, slow-moving streams, either as a floating plant or rooted to the bottom.

To grow high-quality watercress requires considerable effort. It can be grown in a container (with drainage holes) filled with potting mix. Seeds are covered with one-sixteenth of an inch of fine sand and misted. Cuttings of watercress purchased for a salad may also be used. Cuttings are placed in the container and rooted. The container must be kept moist and should be placed in a tub of water and grown in partial shade. One plant is grown per 8 in. pot.

When the plants have grown about 6 in. high, the leading shoot is pinched off to encourage branching. Flowers should not be permitted to form. When buds are observed, the plant should be cut back.

It requires about 60–70 days for the plants to reach harvest maturity from seeds. Cuttings require a little less time. Plants are harvested by cutting about 6 in. of the leading shoots and side shoots.

Garden catalogues usually only list Watercress instead of cultivars.

DANDELION

Dandelion *(Taraxacum officinale)* is a native to Europe and Asia and was brought to America by early settlers for use as greens and in salads. The plant is a perennial with smooth or jagged, irregular dark green leaves.

Culture. Dandelion is a hardy cool season plant that withstands freezing temperatures. It grows in any well-drained soil in all parts of North

FIG. 8.5. Dandelion leaves are used in salads and may have smooth leaves (left) or saw-toothed leaves (right).

America. Dandelion is a perennial that should be planted to one side of the garden with the other perennial vegetables, such as rhubarb.

Seeds are planted one-fourth inch deep in spring in rows a foot apart. Plants are thinned to stand 6 to 10 in. apart in the row. A good supply of fertilizer in the spring and moisture during dry periods is necessary for a quality product.

Wild plants may be used but the named cultivars grow better and have a better flavor. Two such cultivars are Improved Thick Leaf and Thick Leaf.

Harvest. Leaves can be harvested in the fall by cutting a small piece of the root off with the leaves. The root will produce new leaves the next spring for the next year's harvest. Unharvested plants will produce a large amount of growth the next spring. These plants should be harvested in the spring before they produce flowers and the leaves become bitter.

The bitter taste may be reduced somewhat and the dandelion greens made to resemble endive. When the leaves are 8 to 10 in. long, the outer leaves are gathered and tied together at the top, or a tar paper collar is placed over the plant to exclude light. The inner rosette of leaves becomes light green in color in about 2 weeks.

DASHEEN

Dasheen *(Colocasia esculenta)* is also known as taro and has been used as a food crop in Oriental countries for 2000 years. It was brought into America from two separate continents. Some cultivars were brought into the southern states from Africa with slaves who used it as food. Later, better cultivars were introduced that originally came from China.

The dasheens look similar to and are easily confused with tanniers (Splittstoesser 1977). In southern Florida, tanniers are called cocoyam (eddo) or malanga *(Xanthosoma caracu)* but are a different species from dasheen. They are all grown similarly, however.

Plant Characteristics. Dasheen is related to ornamental elephant's ear, jack-in-the-pulpit, calla, and caladium (family Araceae). The plant resembles elephant's ear but produces edible corms and tubers underground. Some tropical cultivars grow well in wet areas and plants grow 5 ft tall.

Culture. Dasheen requires a growing period of 7 months under full sunlight. It is grown in the lowland coastal plains from South Carolina to Texas. In Hawaii, dasheen is grown to make poi.

Dasheen grows best in rich, loamy, well-drained soil with ample moisture. Small tubers weighing 2 to 5 oz are planted 3 in. deep 2 ft apart in rows 3 or 4 ft apart. The tubers are planted about 2 weeks before the last frost occurs in the spring. Tubers may also be planted indoors and then transplanted into the garden after danger of frost is past.

Fertilizer should be applied before planting and again when the plants are 2 ft tall. Adequate moisture is required, and mulching heavily will reduce water loss. Dasheen is shallow-rooted and produces poor quality corms during drought followed by regrowth.

Cultivars are seldom sold through garden catalogues. They are often available only in localities where the plant is grown.

Harvest. The corms and tubers are harvested after 7 months when the plant tops have died in the fall. They should be dug in dry weather to prevent injury to the corms. If the soil is well drained, the crop may be left in the ground and dug when needed. Each plant will normally have a large central corm surrounded by smaller tubers. Yields range from 2 to 8 lb, and both corms and tubers are edible. The tubers should be stored at room temperature for several days to cure bruises and then at 45°–50°F with good ventilation for long storage.

Young leaves just beginning to unroll can also be eaten. They are boiled 15 min with a teaspoon of baking soda to remove the calcium oxylate found in the leaves, which is harmful to humans. This water is then discarded and the leaves boiled until tender in fresh water.

Common Problems. Root-knot nematode can be a major problem. Tubers free of nematodes should be planted.

The tubers rot if they are harvested when immature. Storage rots also occur if tubers are not stored at the correct temperature with good ventilation.

DAYLILY

There are many species and cultivars of daylily (*Hemerocallis* spp.). They are native from Central Europe to China and Japan. They were introduced into the U.S.A. and Canada from the Orient. The daylily has escaped from cultivation and now grows wild throughout North America.

Plant Characteristics. Plants are perennials with long, rather grasslike leaves. The flowers are orange, yellow, pink, or red with six overlapping petals that curve backwards as the flower opens. Roots are fiberous or more or less tuberous. There are early, midseason, and late blooming cultivars.

Culture. Daylily plants should be planted in an area which will not be disturbed. Normal fertilizer or compost should be applied yearly. Every several years, in the spring or fall, the plants should be lifted and divided to keep them blooming well. Plants require little care.

Harvest. Leaves can be harvested in the spring when they are 3–5 inches long. They are used in stir-fry dishes and taste somewhat like onions. If the leaves are eaten in large amounts, they are said to have an hallucinatory effect.

The flower buds and blossoms are the most desirable edible part. Buds are often used in salads or stir-fried with edible podded peas and pork. Blossoms can be dipped in a light batter of flour and water and fried. The buds and flowers may be strung on a heavy thread and dried a week or two for later use. They are then stored in an airtight container. Before using, they are soaked in water for several minutes.

To prevent the plant from spreading, the plant roots or tubers should be harvested at the edge of the clump each fall, or when the plants are divided. These tubers are prepared and used similar to potatoes.

FLORENCE FENNEL

The fennel commonly grown as a vegetable is called (Anon. 1975) Florence fennel (*Foeniculum vulgare* var. azoricum), finocchio, or sweet anise. Fennel was used by the early Romans for food and medicinal properties. Florence fennel is related to celery and celeriac.

Plant Characteristics. Florence fennel somewhat resembles celery. The plants are 24–30 in. tall with fine feathery leaves. The stalks overlap at the base to form a solid bulb with a licorice flavor. The plant is not to be confused with the common fennel, which is used as an herb and does not form the large bulb at the base.

Culture. Fennel is a cool season crop grown in early spring, planted in the summer for a fall crop, or grown in the fall and winter in the South.

Seeds are planted one-fourth inch deep 2 weeks before the last frost in the spring. Plants are thinned to 6 to 10 in. apart in rows about 18 in. apart. Repeated plantings may be made. Plants withstand light frosts.

When the bases of the leafstalks have reached 2 in. in diameter, soil may be mounded up around these forming bulbs. This blanching process is not necessary, but, if done, the bulbs will become white.

Fennel grows best in well-fertilized soil with plenty of water. It has few insect and disease problems.

The only cultivar usually listed is Florence Fennel or Finocchio. It is frequently listed in the herb section of garden catalogues.

FIG. 8.6. Florence fennel (background) is grown for both the licorice-flavored bulb (left) and the feathery leaves (right). *Courtesy of W. Atlee Burpee Co.*

Harvest. Fennel should be eaten before it becomes tough and stringy, usually 80 days after planting. The compact bulb and petioles can be eaten like celery or cooked for their licorice flavor. The leaves are used in salads.

The plant will produce a seedstalk in midsummer if not harvested. Yellow flowers with a seed head resembling dill are produced. The seeds can be used similar to common fennel, the herb.

GARLIC AND ELEPHANT GARLIC

Garlic *(Allium sativum)* is a member of the onion family. It is native to Central Asia and northwest India and has been in use for over 2000 years. The Romans disliked the strong flavor of garlic, but it was used by the laborers and soldiers, who brought it to England in the sixteenth century.

Elephant garlic (*Allium ampeloprasum,* Ampeloprasum group) or great headed garlic is closely related to leeks. It is native to southern Europe, western Asia, and northern Africa. Elephant garlic plants that do not flower form a single large clove. Plants that do flower form a cluster of several large cloves around the central flower stalk. Plants produce few viable seeds and plants are grown from cloves, similar to garlic. Elephant garlic cloves are milder than regular garlic and they are sometimes eaten raw.

Culture. Garlic can be grown in any area where gardeners grow bulb or dry onions successfully. Garlic plants respond to day length and form a bulb under long days, regardless of the plant's size. In addition, if the soil

FIG. 8.7. Garlic produces a number of small cloves in a large bulb. *Courtesy of J.S. Vandemark II.*

temperature is above 68°F, the plants produce poor bulbs. Thus, in many areas, garlic is poorly adapted to home gardens.

Garlic is planted in the South and Southwest from fall till January. In the rest of the country, it is planted as soon as the soil can be worked in the spring, often 6 weeks before the frost-free date.

Garlic is propagated by planting the small cloves, which are divisions of the entire large bulb. The cloves are separated and planted separately. The larger the clove, the larger will be the size of the bulb produced for harvest. Cloves should not be separated until planting time. Separating them early and then storing them, reduces yield.

The cloves are planted in an upright position one-half to 1 in. deep, 4 in. apart in rows 12–24 in. apart. They should be planted in deep fertile soils with a high organic matter content. In order to produce large bulbs in areas outside the South and Southwest, high fertility is required to produce large plants before they bulb. The equivalent of 30 lb of 10-10-10 fertilizer per 1000 sq ft is often recommended.

Elephant garlic is a popular home garden vegetable that may produce a bulb six times larger than regular garlic. In the South and Southwest, it is fall planted and grown similarly to regular garlic. In the North, it is planted in the fall, allowed to mature, mulched, and then left to overwinter. The plants grow the next spring, and when they mature during the summer, they may be harvested.

Garlic needs ample water. If the soil becomes excessively dry, the bulb will be small. If the soil becomes compacted, the bulb will be irregular in shape.

Insect and disease problems are similar to those of onions.

There are many cultivars of garlic available. Many gardeners plant the cultivar available from their local seed supplier or from a food store's produce section. The seed cloves should be large, smooth, and disease free. Elephant garlic cloves usually must be ordered from a seed catalogue.

Harvest. Garlic is harvested when the top dries down. The bulbs are dug and cured for storage by placing them under cool, dry conditions for several days. The bulbs are then stored under dry conditions between 40° and 60°F.

GHERKIN (*See* CUCUMBER, Chapter 7)

HORSERADISH

Horseradish *(Armoracia rusticana)* is native to southeastern Europe, and both the leaves and roots were eaten in Germany during the Middle Ages. The word horseradish first appeared in 1597 in an English herbal on medicinal plants. It was grown in gardens of the early American settlers, and it is now well established as a wild plant.

FIG. 8.8. Horseradish is grown for its roots. *Courtesy of J.W. Courter.*

Culture. Horseradish is a hardy, cool season perennial that produces a whorl of large, rather coarse-textured leaves. Horseradish grows best in the northern regions of the United States and grows poorly in the South.

The plants grow on deep, rich, moist soils. The yields are reduced and the roots are malformed when grown on shallow, stony, or hard soils. Organic matter or manure should be worked into the soil at least 10 in. deep the fall before planting. The manure should not be added in the spring.

As horseradish is a perennial, it is frequently planted at one edge of the garden with the other perennial vegetables. Either crown divisions or root cuttings are planted. Crown divisions are a piece of root and the crown buds. These are removed from the old plant and planted. Root cuttings are small side roots from the main root, the size of a lead pencil and 10–12 in. long. The part of the cutting that was attached to the main root is considered the "top" and is planted near the soil surface.

Root cuttings are planted as soon as the soil can be worked in the spring. Cuttings are planted on a slant, 18–24 in. apart in rows 30 in. apart. The top of the root cutting should be about 3 in. below ground level.

To produce a large main root, the plants are "lifted" when the largest leaves are 10 in. long and again about 6 weeks later. The soil is carefully removed from around the main root, but the roots at the bottom end of the set are undisturbed. The small roots at the top or sides are rubbed off, and only those at the bottom are left. The crown is raised and all but the best sprout or crown of leaves is removed. The plant is returned to its original position and the soil replaced.

If horseradish is growing in a soil high in organic matter, it should receive enough moisture from rainfall. However, if the plant wilts during hot weather, it should be irrigated. Horseradish makes its largest amount of growth during the cooler weather of late summer and fall.

Horseradish cultivars are few. The gardener usually purchases the cultivar available but Bohemian and Maliner Kren are sometimes listed. Once the plants are established, new root cuttings can be obtained from the harvested plants. Usually the roots remaining in the soil after harvest are sufficient to reestablish the plants.

Occasionally horseradish is attacked by root rot. Use disease-free root cuttings, and do not plant in the same area for 3 years.

Harvest. Horseradish may be harvested anytime after a hard frost in the fall until the plants begin to grow again the next spring. Roots are dug with a shovel or spading fork, using the tops as a handle to help pull the roots from the soil. Roots may be stored in an airtight, black plastic bag in the refrigerator.

To prepare horseradish, the root is washed, peeled, and cut into small cubes. A blender is filled half full with the cubes and a small amount of cold water and crushed ice are added. The roots are ground and 2 or 3 tbsp of white vinegar (cider vinegar causes the root to turn brown), and one-half teaspoon of salt per cup of horseradish is added. For extra hot horseradish, wait 3 min before adding the vinegar. The mixture is stored in tightly capped jars in the refrigerator between uses. In about 6 weeks, the ground horseradish darkens and loses some flavor, when new horseradish should be prepared.

JAPANESE EGGPLANT (*See* EGGPLANT, Chapter 7)

JERUSALEM ARTICHOKE

Jerusalem artichoke *(Helianthus tuberosus)* or sunchoke is a relative of sunflower. It has no relationship either to Jerusalem or to (globe) artichoke. It is a native of North America and was being grown for food by the Indians when the early settlers arrived. It was taken to Europe in the 1700s where it is now well established.

Plant Characteristics. The plant resembles the sunflower. It is a perennial that grows over 6 ft high and produces tubers similar to potatoes. These tubers may become 4 in. long and 2 or 3 in. in diameter. The principal storage carbohydrate in the tuber is inulin, rather than starch. When eaten, fructose is released from inulin rather than glucose as occurs with common starch. Inulin starch is a type of carbohydrate usually tolerated by diabetics (Gibbons 1973). The inulin content is greatest just after frost and decreases during the winter.

Culture. Jerusalem artichoke can be grown in nearly all parts of the United States. It grows on soil too dry or too infertile to grow beets or potatoes. However, it grows best on good soils, and the tubers are easier to dig in sandy or loamy soils.

FIG. 8.9. Jerusalem artichokes are grown for their tubers. *Courtesy of J.S. Vandemark II.*

Jerusalem artichokes are planted where they will not shade other plants. The tubers are planted in the spring as soon as the soil can be worked, or in the fall at harvest time. Tubers or 2-oz pieces of tubers are planted 3 in. deep, 24 in. apart, in rows 36–40 in. apart.

Plants require irrigation and fertility similar to potatoes for best growth. They are seldom bothered by insects and diseases.

Several garden catalogues list Jerusalem artichoke or sunchoke; once they have been grown, the gardener can save tubers for seed. They are also found in the produce market and some health food stores.

Harvest. Tubers are dug in the fall after the tops have been killed by frost. They may be harvested at any time until growth begins in the spring. Spring-dug tubers have less inulin and taste sweeter.

Tops are removed before digging. The tubers are then dug with a fork or shovel. All the tubers should be harvested, and tubers are located some distance from the plant. Unharvested tubers will produce new Jerusalem artichoke plants the next spring, creating a weedy condition.

Tubers may be used to start a new planting of Jerusalem artichoke or stored. The tubers do not store as well as potato tubers and should be harvested as needed until freezing conditions are approaching. Tubers are then dug, cleaned, washed, and stored in airtight black plastic bags in a refrigerator. Tubers may also be placed in damp sand and stored in a root cellar or vegetable pit.

JICAMA

Jicama (pronounced hee-*kah*-mah) (*Pachyrhizus erosus*), or Yam bean, is native to Mexico and Central America. It is related to peas and beans.

Plant Characteristics. Jicama is a large vining plant that grows 15–25 feet in length, and resembles the sweet potato in its growth habit. Flowers are deep violet to white. The edible part is the very large tuberous root, which may reach 6 feet in length and weigh 50 pounds. Jicama sold in stores usually were harvested early and weigh 2–3 pounds. Roots are black-skinned with white flesh. The leaves, stems, mature pods, and seeds should not be eaten, as they are considered poisonous. They are said to contain rotenone, a botanical insecticide.

Culture. Seeds are soaked in warm water overnight before planting. They may also be started from small whole roots. They are planted about the same time as cucumbers, after the soil has warmed up. They are planted 6 inches apart in rows 1 foot apart. They should be trained to grow on a trellis. The soil should receive a fertilizer that is high in potassium (or wood ashes). Plants should be mulched after they have become established. Most gardeners cut off most of the flowers (the pods and seeds are not eaten) and allow only enough to set seed for next year's crop. This will direct the plant's energy into the root system. Jicama grows best in a warm climate with moderate rainfall, and is sensitive to frost. They can be grown throughout most of the U.S.A.

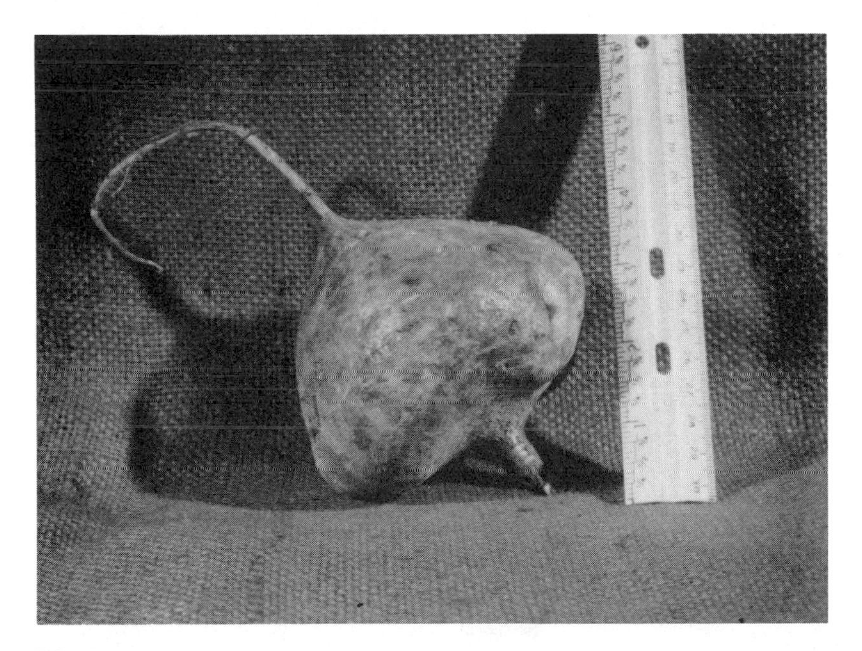

FIG. 8.10. Jicama roots are somewhat sweet and often eaten raw.

Harvest. About 8 months of warm weather is required for formation of large roots. However, many gardeners harvest the roots after 2–3 months, when the roots are under 5 pounds and are easy to dig and store. They can be stored and used similar to potatoes. They taste sweet and are used in salads or like a melon. Some cooks use them as a water chestnut substitute. The root is peeled and stored in a plastic container in the refrigerator. Pieces can be cut off as needed, as the root remains crisp and does not discolor.

LUFFA (*See* GOURDS, Chapter 7)

MUSTAKD AND MUSTARD SPINACH

Mustard *(Brassica juncea)* originated in China and Asia and is used for its tender leaves. Black mustard is a different species grown to produce seed for table mustard. The Tendergreen cultivar is a mustard spinach *(Brassica perviridis)*.

Culture. Mustard is grown similarly to collards and spinach and may be grown in all regions of the United States. It is grown in very early spring or fall in the North and in late fall and winter in the South. It grows at an optimum temperature of 60°–65°F and forms a seedstalk under long, warm summer days.

FIG. 8.11. Green Wave mustard has curled leaves.

It can be planted in the spring 3 to 4 weeks before the frost-free date, or 6 to 8 weeks before the first frost in the fall. Seeds are planted one-half inch deep in rows 15–24 in. apart. Seedlings are thinned to stand about 4 in. apart. Mustard responds to irrigation during dry weather.

If mustard is planted during hot weather, the leaves will become hot and peppery as their pungency increases with the daily temperature. It can be grown during the winter months in the south or in a 55°F greenhouse. It is also easily grown indoors in a pot during the winter. If mustard is allowed to reseed itself, it will revert to a tall-growing, very pungent plant, similar to its ancestors.

Mustard spinach withstands both hot and cold weather very well. It does not produce a seed stalk as rapidly as mustard, and its leaves become pungent at a much slower rate than mustard. It is best grown as a early spring, fall or winter crop.

There are few pest problems, but cabbageworms may be a problem, as with cabbage.

Harvest. Mustard is ready to harvest in about 40 days, when the leaves are 6 to 8 in. long and before they become tough and woody. The entire plant may be harvested or young, lower leaves harvested continuously, similar to collards and kale.

NASTURTIUM

Garden nasturtium *(Tropaedum majus)* or Indian, Mexican, or Peruvian Cress is either a large, tall plant, or a dwarf type *(T. minus)*. They originated in South America, probably Peru. It was brought to the U.S.A. in the mid-eighteenth century.

Plant Characteristics. The tall nasturtium will grow to a height of 8 feet. Both types are annuals which bear red, pink, orange, or yellow flowers. They are related to New Zealand spinach and grow well in hot climates.

Culture. Seeds are planted in the spring, about 3 inches apart. In the south they can be grown at all times of the year. The tall type should be trained to grow on a trellis; the dwarf type, however, can be used as a flower border. The plants respond to the general cultural practices used for most garden vegetables. In the fall, a plant tip may be removed, placed in sand in the house, and rooted. It can be transplanted into a pot and grown throughout the winter months.

Harvest. Young, tender leaves and tips, which taste like watercress, can be used in salads or as a garnish. Flowers, which taste somewhat like a radish, can also be used in salads, or used with herbs to make herb vinegar. The seeds are harvested in the fall and made into pickles.

PURSLANE

Purslane *(Portulaca oleracea)* is also called pusley, and is related to the moss rose. There are many different types worldwide and it is believed to have originated in India, southern Russia, Greece, and South America. Purslane was brought to the U.S.A. as a salad vegetable by the early settlers, but other types of purslane were already growing here. It is now often considered a weed in vegetable gardens, as it germinates late and responds well to the high potassium levels found in most gardens.

Plant Characteristics. Common purslane is a summar annual with red to purple, fleshy, succulent prostrate stems arising from a single taproot. It has small leaves and flowers.

Culture. Purslane is commonly collected as the wild form and is found in most gardens as a weed. It is difficult to eliminate, as the water in the succulent stems or stem pieces allow it to regenerate its root system and re-establish itself. Purslane, thus, needs little care. Cultivars are available in Europe, but they are seldom found in seed catalogues in America.

Harvest. Young, tender stems and leaves are harvested at any time. They are most commonly used in salads, but may also be used as a pot herb.

RADISH, WINTER (*See* RADISH, Chapter 7)

SHALLOT

Shallot (*Allium cepa,* Aggregatum group) is an onion-like plant that originated in western Asia. It was brought to Europe by French knights returning from the Crusades. De Soto brought it to Louisiana in 1532.

Culture. Shallots can be grown in nearly all parts of the country. They are planted as soon as the ground can be worked in the North and in the fall and winter in the South.

Shallots produce a cluster of bulbs somewhat like garlic. The individual bulbs are planted 2 in. deep, 4 in. apart in rows 12 in. apart. They are grown similarly to onions.

Usually no listed cultivars are available. Catalogues simply list "shallot," and bulbs purchased from the produce market or gourmet store may be planted. Once they are grown, the gardener can save bulbs for the next year.

Harvest. Shallots may be harvested and used similar to green onions. For dry bulbs, plants are harvested when the tops die down. Shallots are

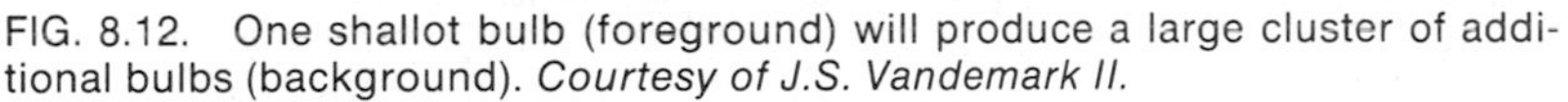

FIG. 8.12. One shallot bulb (foreground) will produce a large cluster of additional bulbs (background). *Courtesy of J.S. Vandemark II.*

hardy and can be left in the ground over the winter. Most types do not produce seed. For best results, the cluster of bulbs should be dug and the small ones planted at the desired time for another crop.

Dry bulbs are stored in a well-ventilated, cool, dry area, similarly to onions. Shallots are easily stored through the winter.

SQUASH FLOWERS (*See* SQUASH, Chapter 7)

TOMATILLO AND GROUND CHERRY

Ground cherry *(Physalis peruvianan)* is often called the husk tomato or poho berry. Tomatillo *(Physalis ixocarpa)* produces larger fruit. They both occur naturally in Latin American and Hawaiian gardens. They are naturalized in Mexico, but appear to have originated in Peru. These plants are related to the ornamental Chinese lantern, whose fruits are not eaten.

Culture. Plants are grown similar to tomatoes, but require less space. They have tomato-like leaves and grow to 2 feet tall. Seeds are planted ¼ inch deep in rows 2 feet apart after the soil has warmed up in the spring. Plants are thinned to stand 18 inches apart. As the seeds are small, transplants are easier to establish in the garden. Six-week old transplants are set out after the last frost.

FIG. 8.13. Tomatillo resembles the ground cherry, but is larger and harvested while it is still green. *Courtesy of W. Atlee Burpee Co.*

Both plants respond to fertility and irrigation similarly to tomato. Once the plant is grown, the unharvested fruits left in the garden are sufficient to produce volunteer plants the next year. Tomato fruit worms sometimes are a problem.

Harvest. The edible fruit is a berry enclosed in a thin husk. With the ground cherry, the husk turns from green to yellow to brown when the fruit is ripe, while the fruit turns from green to yellow. Ground cherries are eaten fresh or made into jam or sauce.

Tomatillo produces a fruit about the size of a walnut. The husk is usually split by the growing fruit, which then protrudes through the husk. The flavor is less sweet than the ground cherry, and the fruit is seedy, but solid and does not contain the juicy cavities found in tomatoes. Fruits are harvested when the husk has turned from green to tan and the fruit is still green. When ripe, they turn yellow or purple, but develop a bland taste, making them undesirable for use in chili sauce or salsa with tacos and other Mexican dishes. Fruits may be stored over the winter with their husks on, one layer deep in a cool, well-ventilated area.

VEGETABLE SPAGHETTI SQUASH

Vegetable spaghetti squash *(Cucurbita pepo)* is a type of winter squash that originated along with other squash. Like other squash, it has separate male and female flowers on the same plant, and it is cross-pollinated. It will cross-pollinate with any of the other *Cucurbita pepo* listed in Table 2.8.

Culture. Vegetable spaghetti will grow wherever winter squash is grown, and it is grown in a similar manner. Plants may be grown as transplants or seeded directly in the garden after all danger of frost is past. Four to 6 seeds are planted 1 in. deep, spaced 2 in. apart in groups (hills). The hills are 3 ft apart with rows 6 to 8 ft apart. The hills are thinned to 2 or 3 seedlings when the plants begin to crowd each other.

Vegetable spaghetti has disease and insect problems similar to those of squash. Cucumber beetles should be controlled from planting onward.

Vegetable spaghetti is the only cultivar listed. It may be listed under novelty items or with the fall and winter squash.

FIG. 8.14. Vegetable spaghetti squash can replace noodles or spaghetti. *Courtesy of Harris Seed Co.*

Harvest. Fruits are ready for harvest 100 days after planting, when they have turned from green to yellow. Fruits are 8–10 in. long and weigh 3 to 6 lbs. Production ends at the first frost, which does not harm the fruit. For storage, fruits are harvested with an inch of stem on the fruit and cured in a sunny place outdoors for a week. They are then stored in a cool, dry location.

Vegetable spaghetti may be boiled (30 min) or baked (1 hr) until the fruit surface yields to pressure. There is less water in the baked spaghetti. The fruit is cut lengthwise, the seeds scooped out and the spaghetti strands removed with a fork. It has few calories and can be used similar to noodles in soups or as spaghetti. Cooked vegetable spaghetti strands can be placed in plastic bags and stored frozen.

WAX GOURD OR WINTER MELON

Wax gourd *(Benineasa hispida)* is also called winter melon, white pumpkin, Chinese watermelon, and Chinese squash. It originated in Southeast Asia and is grown throughout Southeast Asia, China, and India. It is related to pumpkins and squash.

Plant Characteristics. The plant is a vining annual with a plant habit similar to pumpkin. It bears separate yellow male and female flowers on the same plant.

Culture. Culture practices for growing the crop are similar to winter squashes and pumpkins. They cannot withstand frost, but can tolerate drought. They grow best in well-drained soils high in organic matter.

Harvest. Chinese squash is harvested immature, when it is about 6 inches long, and used like summer squash.

Wax gourds are harvested when mature, and may weigh 40 pounds. The fruit produces a white waxy skin which increases with time, even after harvest. Cuts or bruises on mature fruit heal rapidly as the fruit produces a suberin layer, similar to potatoes. Fruits should be stored one layer deep at 55°F and 70% relative humidity. Fruits may be stored for over 6 months. They are used raw, sliced like cucumbers, or cut into pieces and cooked. Wax gourds are especially used in winter soups, hence the name winter melon.

WELSH AND EGYPTIAN ONION

Welsh onion *(Allium fistulosum)* is also known as Spanish onion, Japanese bunching onion, Japanese leek, and Evergreen Bunching onion. The plant originated in Siberia or eastern Asia. The term Welsh onion comes from the German word "Welshe," meaning foreign. Welsh onions have been grown in China and Japan for centuries.

Egyptian onions (*Allium cepa,* Proliferum Group) is also called top onion or walking onions. They were used by the people who built the pyramids in Egypt to increase their strength and stamina. They are presently found worldwide. The plants at the University of Illinois were brought from Denmark by the author's relatives in 1840.

Plant Characteristics. Welsh onions are related to leeks and never form a true bulb. During the second year of growth, the plant sends up a 20-inch seed stalk that produces yellowish-white flowers. The plant produces new plants at the base of the original plant.

Egyptian onions do not produce true bulbs, but produce new plants from the base of existing plants. The plant produces a large, 3-foot tall stalk, which then produces a number of small bulbs at the top, thus the name top onions. These bulbs may also grow and produce an additional stalk with more small bulbs. The stalk will fall over and these small bulbs will then grow, producing new plants; hence the name walking onion. The leaves are large and hollow.

Culture. Cultural practices used for growing these plants are similar to leeks. The Welsh onion is valued for a long, tender white stem. They are planted from seed and grown for a year. Some new plants will be produced from the original plants. The second year, the plants should be dug up, separated, and singly transplanted into a 6–10 inch deep trench containing compost. The trench is filled in to achieve the long, blanched white stem. Plants can be grown from extreme cold conditions to warm climates. In climates where the ground freezes, the young onion plants should be removed in the fall and stored in a cold frame or greenhouse, and planted into the trench the next spring.

Egyptian onions are propagated vegetatively into a permanent area. They should be mulched to control weeds and can be left in the same area for 20 years or more. They need little care except fertilizer and can withstand 30°F below zero without a mulch covering. Their spread needs to be controlled by yearly harvesting the top bulbs in the fall, and young plants in the spring.

Harvest. Welsh onion flower stalks are cut as they are produced and used in stir-fry dishes. Clumps of onions, for the long, tender white stem, are harvested as needed. The flavor is milder and sweeter than a regular onion.

Egyptian onion top bulbs are harvested when produced and can be eaten or used to grow scallions. In the fall, the bulbs can be stored under cool conditions and used to grow green onions in a greenhouse all winter. In the spring, young tender onion plants should be harvested and used like scallions. Once the temperatures becomes warm, the flavor of the onion plants become very strong, but can still be used in soups and stews. The large, hollow leaves can be harvested in the spring, slit and filled with cottage cheese or meat and used as an hors d'oeuvre.

YARDLONG BEAN (*See* SOUTHERN PEA, Chapter 7)

SELECTED REFERENCES

Adjei-Twum, D.C., Splittstoesser, W.E., and Vandemark, J.S. 1976. Use of soybeans as sprouts. HortScience *11:* 235–236.

Anon. 1975. Seed for Today. Asgrow Seed Co. Descriptive catalogue of vegetable varieties. *21*.

Metcalf, H.N. and Burnham, M. 1977. Miscellany, including celeriac, horseradish, artichoke, peanuts, vegetable soybeans. *In* Growing Your Own Vegetables, U.S. Dept. Agric. Bull. *409*.

Splittstoesser, W.E. 1977. Protein quality and quantity of tropical roots and tubers. HortScience *12:* 294–298.

9

Growing and Preserving Herbs

The botanical definition of an herb is a plant without a permanent woody stem. This would include nearly all plants except trees and shrubs. The popular definition of an herb is a group of plants grown for their flavors, essential oils, and scents. This definition would include sage and rosemary, which develop woody stems.

GROWING CONDITIONS

Herbs may be grown in formal or informal gardens, grown with vegetables, or grown as an individual plant. The herbs are frequently grown from seed planted indoors. The seeds are planted in individual containers or in a sterile soil mix about one-quarter in. deep. The seeds are sown thinly and may remain exposed to the air or covered with a thin layer of sand. The seeds are watered with a fine mist and the container covered with newspaper, glass, or damp burlap to prevent the seeds from rapidly drying out. Most herbs germinate best at 70°F (21°C) or warmer, and many seeds are slow and erratic to germinate. Rosemary, for example, may take 3 weeks to germinate. Many herbs such as mint, rosemary, thyme, sage, and tarragon are grown from rooted cuttings, and mints and chives are grown from divisions of existing plants.

The majority of herbs grow well in any well-drained soil. The soil should be moderately fertile and well supplied with organic matter. If soil drainage is poor, it is preferable to grow herbs in containers or raised beds. If grown in containers, the cold-sensitive herbs may be brought inside during the winter.

The herbs should be mulched to keep leaves clean during heavy rains. The leaves of parsley, for example, can be pushed into the soil during rains, and a mulch keeps the leaves and soil separated.

Most herbs grow best with a large amount of sunlight, low humidity, seasonal changes, and an average amount of rainfall, evenly distributed. Herbs are grown, therefore, in all locations in the United States. Herbs generally have few insect and disease problems but parsley is frequented by the caterpillar of the swallowtail butterfly and dill may also have an insect problem.

Most herbs should be grown in full sunlight, although some can be grown in partial shade. In the fall, an herb cutting may be rooted or seeds sown into pots and the plants grown indoors during the winter (Table 9.1). Herbs should be placed where they receive the maximum amount of sunlight. Fresh leaves may then be harvested all year long. If the biennials, such as parsley, are grown under cool conditions (55°F), they will produce flowers and seeds the next spring.

HARVEST

Most herbs are ready to harvest just before the flowers appear on the plant. At this time they contain the greatest amounts of essential oils and scents. Successive cuttings can be made over the growing season, but the lower leaves should be left on annuals to allow them to continue growth. At the end of the season, the entire annual plant can be harvested. Perennials should also be harvested over the growing season. Late harvests should be avoided as the plant needs to regrow and store food in order to survive the winter.

Herbs are best harvested on a clear day in early morning as the essential oils are greatest at this time. The oils diminish as sunlight and temperatures increase during the day. Herbs grown for seed such as dill and caraway may be harvested just before the seeds fall naturally.

DRYING HERBS

The foliage should be washed, the excess water shaken off, and then be dried rapidly. A dark, well-ventilated area where temperatures never exceed 100°F (37.8°C) is best for drying. When exposed to open-air, light, and high temperatures, the essential oils decline, flavor changes, and herb quality rapidly declines. Under proper conditions, the total shelf-life of most herbs is 1 or 2 years.

After the herbs are washed, they can be gathered into bunches and placed in a brown paper bag (to prevent them from getting dusty during drying). About 1 or 2 in. of stem should be left exposed, with the bag tied loosely and

TABLE 9.1. Some Herbs for Growing Indoors During the Winter.

Aloe vera	Chive and garlic chive	Marjoram
Angelica	Coriander	Mints
Basil	Dill	Parsley
Bee balm	Geranium	Pennyroyal
Borage	Lemon balm	Rosemary
Catnip	Lemon grass	Saffron
Chamomile	Lemon verbena	Savory, winter
Chervil	Licorice mint	Thyme
		Yarrow (Yerba buena)

hung in a warm, dry location. Seed heads from dill, anise, and caraway may also be dried in this manner. The seed heads can be placed in the bag as the seeds are becoming ripe (turning grey or brown), and the stems cut. Seeds and leaves should be dry in 2 or 3 weeks. The leaves can be pulverized by rubbing them between your hands, and the stems excluded.

Herb leaves may be dried on trays. The stems are removed and the washed leaves spread on a tray or window screen. The leaves should be turned to ensure uniform drying. If these screens are placed in the oven for fast drying, the temperature should not exceed 100°F (37.8°C), particularly when drying basil. The oven door is left slightly open to allow moisture to escape, and the herbs should be dry in 3 to 6 hr.

Herbs should be thoroughly dry and stored in airtight containers. If the leaves are stored intact and not pulverized, they usually retain their flavor longer. The leaves can be crushed just before use, but storing whole leaves requires more storage space. Dried herbs stored in airtight containers kept in the dark will retain their flavor for 1 to 2 years.

FREEZING HERBS

Herbs may be washed, shaken dry, and sealed in airtight containers for freezing. They may be chopped or left whole, but they should not be blanched before freezing. Frozen herbs should be used without defrosting. Chives, dill, mint, oregano, parsley, sweet marjoram, and tarragon freeze well. Caraway, anise, and dill seeds may also be stored frozen.

FRESH STORAGE

For fresh storage, the herbs should be harvested and placed in airtight containers. They will keep longer if the foliage is not washed until the herb is ready to be used. Properly stored fresh leaves will retain their quality for 2 to 4 months and may be harvested in the fall and stored fresh for use at Thanksgiving.

COMMON AND NOT-SO-COMMON HERBS

Interest in herbs has increased as people seek to change their dietary habits. Herbs are used as seasonings to replace seasonings such as monosodium glutamate and excess amounts of salt. The desire to drink caffeine-free beverages has awakened a renewed interest in herbal teas. The reduction in red meat consumption to reduce chloesterol has resulted in an increase in vegetable consumption, and with it a desire to change the taste of the common vegetable dishes into a new taste experience—and herb seasonings will certainly do that.

People are interested in the past, to help preserve the future; and with this interest comes a desire to use herb plants used by the original American settlers.

Herbs can be grown indoors, in hanging baskets, as a border around the garden or as a herb garden. Fresh herbs retain the subtleties of fragrance and flavor frequently lacking in dried herbs. However, about 2 to 3 times more fresh herbs are needed than dried herbs, as drying concentrates many flavors.

The separation of some specialty-use vegetables from herbs is somewhat arbitrary. Thus, plants such as cress, fennel, garlic, horseradish, Jerusalem artichokes, mustard spinach, tomatillo, and others are listed in Chapter 8. Almost all herbs will grow in an average soil under sunny conditions. Table 9.2 lists a large number of common and not-so-common herbs. Those more commonly grown herbs are also given additional attention in the sections that follow.

ALOE VERA

Aloe vera is a drought-resistant tropical plant, native to Africa, and is best grown indoors during the winter. It requires little care. If grown in a large gallon container indoors, it will produce a large flower stalk with orange flowers in late winter. It is propagated from offshoots of existing plants.

The leaves contain a resinous, yellow juice that has been used in cosmetic lotions and creams for over 3000 years. It also eliminates pain and swelling, if applied to burns. Every cook should have a plant of Aloe vera. It is a drug plant commercially and sold in suntan lotions, shampoos, creams, and is used to treat radiation burns.

ANGELICA

Angelica is a biennial, native to eastern Europe and Asia. It grows well in moist soil in semishade. If the flower stalks are removed, it will grow for several years.

Angelica is propagated from root cuttings or seeds. The seeds must be sown a few weeks after they ripen or they will become dormant. The seeds can be left on the stem to ripen and the plant will re-seed itself easily.

Angelica is known as the “Holy ghost” herb. Legend claims it was presented to a woman by an angel as a cure for the plague in the fifteenth century. The roots and stems are harvested the second year and can be eaten as a cooked vegetable, like celery. The seeds can be used in teas or soups. Stems and petioles are used in candy.

TABLE 9.2. Classification, Propagation, and Use of Some Herbs.

Common Name	Genus, Species	How Propagated	Life Cycle[1] Plant Size	Use
Aloe vera	*Aloe perryi*	suckers from existing plants	*P*, 3 feet tall. Grown in pots. Drought hardy.	Burn ointment, Creams, from leaf juice.
Angelica	*Angelica Archangelica*	seeds	*B*, 6 feet tall. Semishade, moist soil.	Stems and petioles for candy. Leaves as cooked vegetable. Seeds in soups.
Anise	*Pimpinella anisium*	seeds	*A*, 2 feet tall.	Seed in candy, cookies. Leaves in salads. Stalks as a cooked vegetable.
Basil, Opal	*Ocimum basilicum* var. Purpurascens	seeds	*A*, 2 feet tall. Purple leaves	Leaves in soups, salads, seasoning. Basil vinegar.
Basil, Sweet	*Ocimum basilicum*	seeds	*A*, 2 feet tall. green flowers.	Leaves for seasoning.
Bee balm	*Monarda didyma*	Cuttings, plant divisions	*P*, 2 feet tall. Grown like mint (*see* Mints).	Leaves for teas, salads.
Borage	*Borago officinalis*	seeds	*A*, 3 feet tall. Poor soil. Sun or shade.	Leaves in salads, as greens. Flowers in summer drinks and candy.
Burnet	*Poterium Sanguisorba*	seeds, plant divisions	*P*, 2 feet tall. Very hardy.	Leaves in soups, salads, and salad dressings.
Caraway	*Carum carvi*	seeds	*A*, 2 feet tall. Very hardy.	Seeds in breads or as a side dish. Leaves as a seasoning. Roots as carrots.
Catnip	*Nepeta cataria*	seeds, cuttings	*P*, 2 feet tall. Grown like mint.	Leaves used for tea. Many cats are addicted to it.
Chamomile, English or Roman	*Chamaemelum nobile*	seeds, plant divisions	*P*, 1 foot tall. Very hardy.	Flowers used in teas, creams, and hair rinses.

TABLE 9.2. *(Continued)*.

Common Name	Genus, Species	How Propagated	Life Cycle[1] Plant Size	Use
Chamomile, False, German, or Hungarian	*Matricaria recutita*	seeds, plant divisions	*P*, 1 foot tall.	Same as English types.
Chervil	*Anthriscus cerefolium*	seeds	*A*, 1 foot tall. Partial shade.	Used like parsley in salads, soups, or herb mixtures.
Chive	*Allium Schoenoprasum*	seeds, plants divisions	*P*, 1 foot tall.	Leaves used in soups, salads, and with cottage cheese.
Chive, Garlic	*Allium tuberosum*	seeds, plant divisions	*P*, 2 feet tall.	Leaves as chive. Young flower heads have garlic flavor.
Comfrey	*Symphytum officinale*	seeds, roots, plant divisions	*P*, 4 feet tall. Very hardy.	Leaves and roots for tea. Young leaves for salad. Said to be only vegetable source of Vitamin B_{12}.
Coriander	*Coriandrum sativum*	seeds	*A*, 3 feet tall.	Seeds for baking, curry, salsa.
Cumin	*Cuminum cyminum*	seeds	*A*, 3 feet tall. Plant in late spring, full sun.	Seeds used in chili and curry powder.
Dill	*Anethum graveolus*	seeds	*A*, 3 feet tall.	Leaves in salads, soups. Flowers in pickles. Seeds in soups and baking.
Dittany of Crete	*Origanum dictammus*	seed cuttings	*P*, 1 foot tall.	Used similar to oregano and marjoram, or in tea.
Geraniums, Scented	*Pelargonium* spp.	cuttings	*P*, 3–4 feet tall. Well-drained soil, sun or partial shade.	Leaves in teas, punch, salads, custards.
Almond	*P. quercifolium*			Almond fragrance.
Apple	*P. odoratissimum*			Apple fragrance.
Apricot	*P. scabrum*			Apricot or strawberry fragrance.
Coconut	*P. grossularioides*			Coconut or gooseberry fragrance

Lemon Lime Orange Peppermint Rose	*P. crispum* *P.* x *nervosum* *P.* x *citrosum* *P. tomentosum* *P. graveolens*			Lemon fragrance. Lime fragrance. Orange fragrance. Peppermint fragrance. Rose fragrance.
Ginseng	*Panax quinquefolium*	seed or roots	*P*, 1 foot tall. High organic, well-drained soil. 80% shade. Cold-hardy.	Roots harvested after 4–9 years. Widely used in Orient for tea.
Horehound	*Marrubium vulgare*	seeds, plant divisions	*P*, 3 feet tall. Well-drained soil, full sun.	Leaves used for candy or for tea to relieve coughs or colds.
Hyssop	*Hyssopus officinalis*	seed; stem or root cuttings	*P*, 2 feet tall. Grown like mint (*see* Mints).	Used as perfume base. Very strong fragrance. Leaves used sparingly for tea or stews.
Lavender	*Lavendula* spp.	seeds, cuttings	*P*, 3 feet tall. Sun, well-drained soil	Many species available. Flowers used for their fragrance.
Lemon balm	*Melissa officinalis*	cuttings, plant divisions	*P*, 2 feet tall. Grown like mint (*see* Mints).	Leaves used in tea. Lemon flavor.
Lemon grass or Fever grass	*Cymbopogon citratus*	plant divisions	*P*, 6 feet tall. Tropical, can be grown indoors.	Leaves used in tea. Lemon flavor.
Lemon verbena	*Aloysia triphylla*	cuttings	*P*, 6 feet tall. Tropical, can be grown indoors.	Lemon flavor, leaves used in tea.
Licorice	*Glycyrrhiza glabra*	seeds, rootstocks	*P*, 3 feet tall. Grows in rich, moist soil.	Roots used for commercial licorice; teas.
Licorice mint, Giant or Anise hyssop	*Agastahe foeniculum*	seeds, plant divisions	*P*, 3 feet tall. Grown like mint (*see* Mints).	Leaves used in fruit salads, teas.
Lovage	*Levisticum officinale*	seeds, plant divisions	*P*, 2 feet tall. Shade, moist area.	Steamed stalks used like celery. Sweet seeds in candy. Root sliced, preserved in honey, used as candy. Leaves in soups, salads.

TABLE 9.2. *(Continued)*.

Common Name	Genus, Species	How Propagated	Life Cycle[1] Plant Size	Use
Marjoram, Sweet	*Origanum majorana*	seeds, cuttings	*P*, 2 feet tall. Usually not hardy.	Leaves as seasoning.
Mints	*Mentha* spp.	seeds, cuttings	*P*, 2–3 feet tall. Hardy (*see* Mints).	Leaves used in teas, flavoring, jellies, juleps.
Apple	*M. suaveolens*			Fruity fragrance.
Creme de menthe	*M. requienii*			Used in liquers.
Lemon	*M.* x *piperita varcitrata*			Lemon fragrance.
Peppermint	*M.* x *piperita*			Peppermint fragrance.
Spearmint	*M. spicata*			Spearmint fragrance.
Oregano	*Origanum vulgare*	seeds, cuttings	*P*, 2 feet tall. Very hardy (*see* Marjoram).	Leaves used as seasoning.
Parsley, Italian	*Petroselinum crispum* var. neapolitanum	seeds	*B*, 1 foot tall.	Leaves as seasoning. Seeds, roots used in cooking.
Parsley, Triple Curled	*Petroselinum crispum*	seeds	*B*, 1 foot tall	Common parsley. Used in seasonings.
Pennyroyal	*Mentha pulegium*	seeds, cuttings	*P*, 1 foot tall. Grown like mint (*see* Mints).	Leaves used in teas.
Rosemary	*Rosmarinus officinalis*	seeds, cuttings	*P*, 3 feet tall. Not hardy.	Several varieties available. Leaves in soups, stews, sauces, and hair rinses.
Rue	*Ruta graveolens*	seeds, plant divisions	*P*, 3 feet tall. Hardy.	Leaves as condiments. Bitter taste. Causes dermatitis in some persons.
Saffron	*Crocus sativus*	bulbs or divisions	*P*, 1 foot tall	Stigmas of flowers dried to color cheese, flavor creams, sauces, fruit.
Sage	*Salvia* spp.	seeds, cuttings	*P*, 2–6 feet tall. Hardy, well-drained soils.	Leaves used in seasonings.
Garden	*S. officinalis*			Lavender flowers.
Dwarf	*S. officinalis* var. Minim			Dwarf plant. Lavender flowers.

Golden	*S. officinalis* var. Icterina			Golden variegated leaves.
Pineapple	*S. elegans*			Red flowers, pineapple fragrance. Very tall.
Purple	*S. officinalis* var. Purpurascens			Red or purple leaves.
Variegated	*S. tricolor*			White, purple, pink leaves.
Savory, Summer	*Satureia hortensis*	seeds, cuttings	*P*, 2 feet tall.	Leaves, flowers in salads, teas, soups, rice. Pepper-taste.
Savory, Winter	*Satureia montana*	seeds, cuttings	*P*, 2 feet tall.	Same as Summer Savory.
Sesame	*Sesamum indicum*	seeds	*A*, 3 feet tall. Tropical, drought resistant.	Seeds used on rolls, bread.
Tansy	*Tanacetum vulgare*	seed, plant divisions	*P*, 3–5 feet tall. Well-drained soil.	Leaves for seasoning. Plant for yellow or green dye for wool. Var. Crispum has fernlike leaves.
Tarragon, French	*Artemisia dracunculus*	root	*P*, 3 feet tall. Well-drained soil.	Fresh leaves in salad dressings. French cooking.
Thyme	*Thymus* spp.	seed, cuttings	*P*, 1 foot tall.	Very small leaves, used for seasoning.
Caraway	*T. herba-barone*			Caraway-flavored leaf.
Common, English	*T. vulgaris*			Variegated leaves, most common.
French	*T. vulgaris*			Unnamed variety with narrow leaves.
Lemon	*T. x citriodorus*			Lemon-flavored leaf.
Woodruff, Sweet	*Galium odoratum*	divisions	*P*, 1 foot tall. Hardy.	Leaves used in punch, tea, wine.
Wormwood	*Artemisia absinthium*	divisions	*P*, 3 feet tall. Dry, well-drained soil.	A type of sagebrush used to make absinthe (a narcotic). Bitter taste. Used in Vermouth and Liqueurs. Leaves used sparingly as a seasoning.
Yarrow, Yerba Buena	*Satureja douglasii*	cuttings, divisions	*P*, 1 foot tall. Grown similar to Savory (*see* Savory).	Not a true yarrow. Related to savory and used similarly.

[1] *A* = annual; *B* = biennial; *P* = perennial plant.

ANISE

Anise is found growing wild from Greece to Egypt. It is an annual 18–30 in. high. The plant produces an umbrella-like seed head resembling wild carrot. Both seeds and leaves are used. Anise may be started indoors or planted directly in the garden, 1 seed per in. about one-half inch deep. It should be thinned to 3 or 4 plants per ft. Seeds are harvested when they turn brown in late fall and have a sweet, spicy taste. Aniseed has a licorice-like flavor used in cakes, cookies, and candies. The leaves can be used in cooking or in fruit salads.

BASIL

Basil is native to tropical regions of Africa. It is an annual that grows 18–24 in. in height. The plants may be either green or purple (opal basil) and are started indoors from seed. They should be transplated about 10 in. apart. The flowers should be continually removed before the seeds mature to stimulate continual foliage development. Leaves can be harvested until the first frost, which kills the plant. The leaves are sensitive to temperature and should be air-dried in a shady area. If the leaves are not dry within 3 days, they should be dried in the oven at 90°F (32.2°C) or the leaves will turn brown. Basil may also be grown indoors in a container for fresh use during the winter. Basil is used in soups, stews, omelets, egg dishes, and salads.

BEE BALM

Bee Balm is a hardy perennial, also called Oswego tea. It is a native to North America, but the botanical name comes from Monardes, a sixteenth century

FIG. 9.1. Basil is a popular home garden herb.

Spanish horticulturist. It grows 2 to 3 feet tall with a very thick, shallow root system. It is propagated from cuttings or plant divisions when the plants are divided, every 3 to 4 years. There are a number of cultivars available with different colored flowers. It is grown like mint (*see* Mints).

The leaves have a strong mint flavor and are used in tea, salad, or in candy. The flowers are used in flower arrangements.

BORAGE

Borage is native to Europe and northern Africa. It is an annual propagated from seeds planted directly into the garden. It is a large plant with rough, coarse leaves, which requires considerable space in the garden. Medieval literature called borage "the herb of courage," and was placed under the zodiac sign of Leo the Lion.

Borage grows under direct or filtered sunlight, in a well-drained soil, and is drought resistant. It easily re-seeds itself, and may become a weed.

Leaves have a cucumber-like taste and are used in salads, in pickling, or as greens. The blue to purple flowers are candied or used in summer drinks for a cooling effect; for this reason, in Europe the plant is known as "cool-tankard." The flowers are also attractive in floral arrangements.

BURNET

Burnet is a native to Europe and western Asia. It is a perennial, is very hardy, and is a semishrub related to roses. It is propagated from seeds and re-seeds itself easily and could become a weed. It grows best in full sun, on well-drained soil, with routine irrigation. Young leaves are harvested as an herb.

Burnet leaves can be used in salads, soups and home-made French dressing. The leaves have a fresh cucumber-like taste, and can be made into teas. It is reported that soldiers of the American Revolutionary War made this herb into a tea and drank it before battles to prevent bleeding, if they were wounded.

CARAWAY

Caraway is native to Europe and has now been naturalized in North America. It is a biennial plant usually grown for its seeds. It grows about 18–24 in. high and is planted from seed indoors or directly in the garden. Seeds are planted one-half inch deep in rows 2 ft apart and plants are thinned to 6 plants per ft of row. The plants should be grown in an area that will not be spaded up the following season, as the plants require 2 years to produce seed. Plants grow the first season, die down when winter comes, and regrow the following season. They produce seedstalks the second year. Seeds should be harvested when they are brown and then dried in the sun or shade.

Seeds are used in breads and cakes and with cabbage, coleslaw, carrots, cheese, and potatoes. The roots may be prepared and eaten like carrots or turnips. Leaves can be used in soups and stews.

CHAMOMILE

There are two major species available: Garden, English, or Roman chamomile and False, Hungarian, or German chamomile. The latter reportedly makes the better tea. The plants originated in western Europe and northern Africa.

It is a very hardy, frost-tolerant perennial that grows best in well-drained soil, in full sun or partial shade. The plant produces runners underground, which causes the plant to spread. It should be grown outdoors in a container, similar to mints, to prevent its spread.

The flowers appear from late spring onwards and are daisy-like, with white or yellow centers. The entire young flower or the petals may be harvested and used in teas or hair rinses.

CHERVIL

Chervil is a native of southeastern Europe and western Asia. It has finely cut and divided leaves similar to parsley. The plant has been used as a salad herb since the Middle Ages.

Chervil is a hardy annual that is grown as a spring or fall crop, and in mild climates during the winter. It does not withstand hot summer weather. It easily re-seeds itself in late fall. It is started from seed and is best grown in partial shade.

The roots may be used as a vegetable, similar to carrots. The leaves are used in salads, soups, and as a garnish, similar to parsley.

CHIVE AND GARLIC CHIVE

Chive, or schnitlach, and garlic chive, Chinese chive, or oriental chive are native to Siberia and Southeast Asia. It is a perennial that grows about 1 ft high. It grows in clumps of small, bulbous plants and is usually propagated by dividing the clumps into 5 bulblets for transplanting. The clumps can be planted in the spring or fall and should be divided every 3 years to prevent overcrowding. If chives are started from seed, they should be planted one-fourth inch deep as early in the spring as possible.

Leaves may be harvested throughout the growing season. The plants should be cut close to the ground regularly to encourage new bulblets to develop, prevent the leaves from becoming tough, and prevent flower for-

FIG. 9.2. Chive leaves are cut close to the ground regularly to encourage new growth. Cut flowers are in the foreground.

mation. Flowers are often desirable, however, as an attractive purple flower is produced in early spring.

The harvested leaves can be used immediately, chopped and frozen for later use, or dried. They should be dried completely and placed in airtight containers. Any moisture will be absorbed, causing the chives to lose color and flavor.

Chives may be grown inside during the winter and used fresh, but chives need a rest period to rejuvenate. A clump may be dug in late January and planted and grown indoors; or a clump may be planted in a pot in late summer, the pot sunk into the ground and brought into the house 90 days after the first killing frost.

Chives have an onion-like flavor and are used in soups, salads, sauces, and with cottage cheese.

Garlic chive is grown similar to chive, with some notable exceptions. The hard, black seeds produced readily form new plants, and the plant tends to become a weed. It is frequently grown in a container and can be grown indoors year-round. The flat leaves are used like chive, but are much milder. The young, white flower heads are harvested for stir-fry dishes, and have a strong garlic flavor. Once the seeds begin to mature, the flavor disappears.

CORIANDER

Coriander is also known as Chinese parsley, and is native to southern Europe. Its seeds have been found in Egyptian tombs. It is a relative of parsley, and has a tap root with a single flower stalk with many flowers branching from it.

FIG. 9.3. Both garlic chive leaves and flowers can be used. *Courtesy of W. Atlee Burpee Co.*

Coriander grows best in full sunlight in well-drained soil. It is grown similar to parsley. It is planted in the spring for its seeds, and fall and winter, in pots indoors, for its leaves. Seeds are harvested in midsummer, as soon as they are ripe. The seed stalks are tall and the seeds heavy, causing the stalk to bend to the ground and lose its seeds.

Seeds are used in baking and curry. The young leaves are used in Mexican salsa, Far Eastern dishes, and Oriental foods.

DILL

Dill is native to Southwest Asia and is now naturalized in Europe and North America. It is a hardy annual that grows 2 or 3 ft in height. The plant has

FIG. 9.4. The seed head of dill. *Courtesy of Ferry Morse Seeds.*

feathery leaves with an open umbrella-shaped seed head that resembles wild carrot in appearance. The umbels produce yellow flowers and eventually produce seed. Young dill plants are difficult to transplant and should be planted directly in the garden. They are planted one-fourth inch deep and thinned to 1 plant every 5–10 in.

Dill leaves, seed heads, and seeds are used. Leaves may be harvested 8 weeks after seeding by cutting the outer leaves close to the stem. They may be dried or frozen fresh. For pickling, the flowering umbels are used and should be harvested with a few leaves when in full bloom. These may be bagged and dried or used fresh. Seeds should be harvested when they are light brown. The umbels are cut in the early morning when there is less chance of the seeds being shaken loose and lost. The umbles are placed in a bag, and, when dried, the seeds are shaken loose.

Dill easily reseeds itself for the next year, and a few plants may be left for this purpose. Frequently, new plants will develop from these seeds in late summer or fall. The leaves of these plants are of excellent quality and are harvested before the flower heads appear.

The leaves are used in salads, soups, and omelets. The flowering umbels are used in making pickles. Seeds are used in soups and baking.

MARJORAM AND OREGANO

There are many species of these herbs, but sweet marjoram and oregano are the most widely grown. They are native to Northern Africa and Asia and can now be found in the wild in many parts of America. The major herb difference

FIG. 9.5. Oregano with flowers. *Courtesy of W. Atlee Burpee Co.*

between these perennial plants is that oregano is more hardy, but both may not overwinter in cold northern climates. Sweet marjoram is frequently grown as an annual. The plants are grown in a similar manner, and both should be heavily mulched during the winter and uncovered in the spring to ensure their survival in cold climates.

These herbs are grown from seed, root cuttings, or crown divisions. They should be started indoors if grown from seed and transplanted about 6 in. apart when the soil has warmed. The plants may be dug up in the fall and grown indoors during the winter and planted outdoors the following spring.

Leaves may be harvested throughout the season. When flowers appear, the plants should be cut back to about 4 in. above ground level to stimulate new growth. This second growth is the main crop. The herbs may be cut back two or three times a season.

The harvested plant should be dried rapidly. When dry, the leaves will powder and can be sifted through a screen to remove the woody stems. It may then be stored for later use.

Marjoram and oregano are used with green vegetables, salads, soups, in herb butter, and with various meat and egg dishes.

MINTS

There are many herbs that are members of the mint family, and they are all grown in a similar manner. They include Bee balm, Hyssop, Lemon balm, Licorice mint, Pennyroyal, and the common mints. In addition to Peppermint and Spearmint, several different flavored mints are also available (*see* Table 9.2—Mints).

FIG. 9.6. Mints are frequently grown in a tile partially sunken in the ground to prevent the plants from spreading.

Mints are propagated from roots, rooted cuttings, or entire plants. Seed often germinates well, but the plants produced are often not true to the parent plants. The underground stems spread rapidly and may easily occupy too much area. To contain the mints, a large tile or plastic container with holes in the bottom is buried in the ground with about 3 in. above ground. The mints are grown in this container, which limits their spread.

Mint leaves may be harvested throughout the year, with the best-quality leaves being produced in late summer. Mints are best harvested when they begin to flower, or the lower leaves become yellow. The plants can be cut back to 1 in. above ground level, and all stems and leaves removed to reduce the possibility of diseases. They may be completely harvested twice a season. The leaves are removed and (preferably) frozen; or dried in warm shade and then stored in a sealed container. After the final harvest, the plants should be covered with mulch or compost to protect the plants from frost damage and provide nutrients for next season. Peppermint has the most potent flavor of the mints.

PARSLEY

Parsley is a biennial that grows about 1 ft high. The second year it produces seeds and thus is frequently grown as an annual. There are two distinct types of

FIG. 9.7. Curly-leaf parsley produces the most leaves for seasoning. *Courtesy of Harris Seeds.*

parsley. The flat leaf or Italian parsley is used for its leaves and roots. The curly-leaved or triple-curled parsley is used for its leaves only. Both plants are native to Eastern Europe and Western Asia.

Parsley was said to be sacred to Pluto, the god of the underworld, and the ultimate victor. In the ancient days of Greece and Rome, a wreath of parsley was presented to the winners of many of the ancient games.

Parsley is best started from seed indoors, but it may be sown directly in the garden in early spring, about 1 seed per in., and later thinned to stand 5 in. apart.

Leaves may be harvested throughout the growing season, and the plants remain green until early winter. The entire plant is usually harvested and dried in late fall. The leaves should be dried in a short time in the shade and may be finally oven-dried. Once dried, parsley leaves are stored in a dry, tight, dark container. When parsley leaves lose color, they also lose flavor.

Parsley is the most widely grown herb used for garnishing and flavoring. It can be blended with other herbs such as basil, marjoram, oregano, rosemary, summer savory, and thyme. The combination of these herbs imparts flavor as a unit, rather than as a single herb flavor.

ROSEMARY

Rosemary is a perennial evergreen shrub that grows 2 ft in height. It probably originated in the Mediterranean regions of Spain and Portugal. Rosemary cannot stand temperature below 27°F (−2.8°C), and may be grown as an annual in cold climates or heavily mulched and covered to protect it during the winter. Plants may be potted and taken indoors for winter use and to produce cuttings.

FIG. 9.8. Rosemary is an attractive shrubby evergreen plant.

Rosemary can be propagated from seeds, cuttings, or layering. Seeds should be started indoors in early spring, but frequently only 10–20% of the seeds germinate, and it may take 2 or 3 years to produce a cuttable bush.

Cuttings are produced by removing a 6 in. tip of new growth and placing the lower 4 in. in sand or vermiculite. To produce a new plant by layering, a lower branch of the bush is buried in the soil. When the roots have formed on the branch, the new plant is cut from the mother plant.

Rosemary should be planted 1 ft apart and may be trimmed several times each season to harvest the leaves. When the removed stems are thoroughly dry, the leaves are stripped off and stored in a closed container.

Rosemary leaves are used as an accent in soups, stews, meats, sauces, and with leafy greens. It is frequently picked fresh and cuttings laid directly on roasts and poultry. Rosemary may also be added as an ingredient of mixed herbs.

SAGE

Sage is a shrubby perennial plant that grows 15–38 in. tall. It is found growing wild from Central Spain to the Balkans and Asia Minor. In the wild, it is found growing in dry and stony areas. It grows poorly on clay soils. Sage will withstand most American winters and grows best on well-drained soil. How well sage overwinters depends upon when and how many leaves were harvested in the fall. The plant needs some leaves to provide energy for winter survival. The last harvest should occur no later than late summer or early fall, and then only leaves and stems high up on the plant should be harvested. The first year of growth, only one light harvest is possible if the plant is to survive the winter.

Sage may be propagated from seeds, stem cuttings, or divisions of the plant itself (crown divisions). Seeds may be planted indoors a month earlier than planted outside in the garden. The seedlings may be transplanted when they are 3 in. high and spaced 12–18 in. apart.

Each spring the woody growth should be trimmed back severely to eliminate flowering that prevents vegetative growth. However, some gardeners use little sage and grow the plant as an ornamental. It produces fragrant purple flowers the second and following years.

The top 6 to 8 in. of growth should be harvested at least twice during the growing season. Frequently, the leaves and stems are washed, placed in a bag, and hung up to dry. The sage is not removed until ready to use. Sage leaves may also be dried in the shade until they are crisp. If the leaves are to be used as a tea, the leaves can be broken up by hand. For seasoning, the leaves should be rubbed through a fine screen.

Sage leaves are used as a tea and for stuffing pork, duck, and geese. The powdered leaves are used to flavor cheese, gravies, sauces, and sausages.

FIG. 9.9. One summer's growth of a sage plant started from a cutting.

SAVORY AND YARROW

Savory was also known as "Poor Man's Pepper" due to its taste. Summer savory is an annual that grows 18 in. tall, while winter savory is a perennial that grows 2 feet tall. They are native to the Mediterranean region. Yerba buena yarrow is also a type of savory grown as a perennial. It is native to the Pacific Coast and was the herb after which San Francisco was originally named. They grow rapidly from seeds or cuttings and may be started from seed either indoors or planted directly in the garden. Seeds should be planted in rows about one-fourth inch deep and 1 in. apart. If the topsoil dries out quickly, the soil should be watered lightly to keep the seeds moist. The plants should be thinned to 6 or 8 in. apart.

Summery savory is best grown in light soil, high in organic matter. It is easily grown in containers. Winter savory grows best in well-drained soil with average moisture. Yerba buena yarrow grows best in well-drained soil with little moisture. Except for the Pacific Coast, it should be grown in partial shade.

Plants are harvested when they are 6 in. high, and harvesting may continue throughout the growing season. This will delay flowering and promote vegetative growth. When the plant finally does flower, the entire plant may be harvested when the flowers open. If the plant is not harvested at this time, the leaves will turn yellow, curl, and drop. Only one-third of the winter savory or yerba buena yarrow plants should be harvested, to allow it to over-winter.

The harvested leaves and stems dry rapidly in warm shade. They may be tied into small bundles and dried on paper or fine screens. When dry, the leaves are removed and stored in closed containers.

Savory is used in egg, rice, and bean dishes; as a tea; in soups, sauces, and stuffings; on salads; and with nearly all kinds of meat and fowl. It is also used in various herb blends and mixtures.

TARRAGON, FRENCH

French tarragon is a hardy perennial plant that may grow to 3 ft in height. The desirable species is French or German tarragon and seldom has flowers that produce viable seed. The less desirable species is Russian tarragon, which produces large amounts of seed and lacks the essential oils. Tarragon is related to sagebrush and wormwood and is native to Southern Europe, Asia, and west of the Mississippi River in the United States. French tarragon is propagated by cuttings or by splitting the root cluster of a mature plant into 2 or 3 new plants (crown divisions) in early spring. These root clusters are planted 12–20 in. apart and should be split into new plants every 4 years to rejuvenate them.

Tarragon should be mulched to protect the plant over the winter. A root cluster may be placed in a 10 in. or larger pot and grown indoors for winter use.

Mature plants can be harvested throughout the growing season and are preferably used fresh. The top growth should be harvested two or three times a season to encourage branching of the stems. Tarragon leaves are sensitive to temperature and should not be dried above 90°F (32.2°C) Leaves should be removed from the stem and dried rapidly in the dark. Tarragon leaves are stored in tight, dry, dark containers.

Tarragon has the sweet aroma of fresh-mown hay. It is often used in herb blends, but tarragon flavor can easily overpower other herbs. It is used in sauces and dressings, with green vegetables, and in tartar sauce for fish dishes.

Tarragon vinegar is a popular use of this herb. A wide-mouth bottle is filled with fresh leaves and stems, and the bottle is filled with apple or wine vinegar and closed. After a few weeks, the tarragon vinegar is ready for use in various salads, salad dressings, and sauces.

THYME

Thyme is a shrublike perennial that grows 6–12 in. in height. Thyme is native to the Mediterranean region of southern Italy. There are many closely related species of thyme which are used as ornamentals and as ground covers. Several types of thyme are available: English (variegated leaves), French (narrow leaves), Caraway (caraway-flavored leaves), and Lemon (lemon-flavored leaves).

Thyme is propagated from seed or cuttings or by dividing the mature plant. Seeds should be planted indoors in early spring and will take about 2 weeks to germinate at 70°F (21°C). When the plants are 3 in. high, they can be transplanted in the garden about 1 ft apart. When plants are 3 or 4 years old, they should be divided into new plants to prevent the mature plants

FIG. 9.10. Thyme is a small plant with small leaves.

from becoming excessively woody, with a resulting reduction in leaf production. Many gardeners begin new plants from cuttings or seed rather than divide the mature plants.

Thyme should be harvested just before the pink or violet flowers appear. Mature plants can be harvested by cutting the entire plant 2 in. above the ground. A second crop can be harvested in late summer, but only the upper third of the plant should be removed. Removing too much foliage may prevent thyme from overwintering, particularly if temperatures fluctuate widely and the plant is not mulched. Thyme is hardy but needs a mulch to overwinter better in cold climates.

Thyme may be blended in with other herbs, and many recommend using thyme in practically everything. It is used with meat, poultry, fish, soups, gravies, cheese, chowders, egg dishes, and almost all vegetables.

Index